Centenary of the Borel Conjecture

CONTEMPORARY MATHEMATICS

755

Centenary of the Borel Conjecture

Marion Scheepers
Ondřej Zindulka
Editors

AMS AMERICAN MATHEMATICAL SOCIETY

2010 *Mathematics Subject Classification.* Primary 03E17, 03E35,03E65, 22A10, 22B05, 22C05, 54C50, 54D20, 54H11, 91A44.

Library of Congress Cataloging-in-Publication Data

Names: Scheepers, Marion, 1957– editor. | Zindulka, Ondřej, 1961– editor.
Title: Centenary of the Borel conjecture / Marion Scheepers, Ondřej Zindulka, editors.
Description: Providence, Rhode Island : American Mathematical Society, [2020] | Series: Contemporary mathematics, 0271-4132 ; volume 755 | Includes bibliographical references.
Identifiers: LCCN 2019054325 | ISBN 9781470450991 (paperback) | ISBN 9781470456382 (ebook)
Subjects: LCSH: Borel, Emile, 1871-1956. | Set theory. | Numbers, Real. | Topological groups. | AMS: Mathematical logic and foundations – Set theory – Cardinal characteristics of the continuum. | Mathematical logic and foundations – Set theory – Consistency and independence results. | Mathematical logic and foundations – Set theory – Other hypotheses and axioms. | Topological groups, Lie groups {For transformation groups, see 54H15, 57Sxx, 58-XX. For abstract harmonic analysis, see 43-XX} – Topological and differentiable algebraic systems {For topological ring | Topological groups, Lie groups {For transformation groups, see 54H15, 57Sxx, 58-XX. For abstract harmonic analysis, see 43-XX} – Locally compact abelian groups (LCA groups) – General properties and | Topological groups, Lie groups {For transformation groups, see 54H15, 57Sxx, 58-XX. For abstract harmonic analysis, see 43-XX} – Compact groups – Compact groups. | General topology {For the topology of manifolds of all dimensions, see 57Nxx} – Maps and general types of spaces defined by maps – Special sets defined by functions [See also 26A21]. | General topology {For the topology of manifolds of all dimensions, see 57Nxx} – Fairly general properties – Noncompact covering properties (paracompact, etc.) | General topology {For the topology of manifolds of all dimensions, see 57Nxx} – Connections with other structures, applications – Topological groups [See also 22A05]. | Game theory, economics, social and behavioral sciences – Game theory – Games involving topology or set theory.
Classification: LCC QA248 .C357 2020 | DDC 516–dc23
LC record available at https://lccn.loc.gov/2019054325

Contemporary Mathematics ISSN: 0271-4132 (print); ISSN: 1098-3627 (online)

DOI: https://doi.org/10.1090/conm/755

∞ The paper used in this book is acid-free and falls within the guidelines
established to ensure permanence and durability.
Visit the AMS home page at `https://www.ams.org/`

10 9 8 7 6 5 4 3 2 1 25 24 23 22 21 20

Contents

Preface

It is a common phenomenon in mathematics that an innocuous observation made in the process of a wider research project stimulates long term and broad impact on the development of Mathematics. What has been called the Borel Conjecture is a prime example of this phenomenon. In a 1919 paper the mathematician Émile Borel, exploring the then recently established notion of measurability in terms of the Lebesgue measure, noted that countable sets of real numbers have a certain covering property, and he speculated that only countable sets of real numbers have that property.

In 1919 set theory was still a young field, with axioms recently formulated, and driven by several very basic questions, including the question whether the Continuum Hypothesis holds. Less than ten years after Borel's paper appeared, information about Borel's Conjecture emerged from the exploration of consequences of the Continuum Hypothesis: Sierpiński proved that the Continuum Hypothesis implies the negation of Borel's Conjecture. When Gödel established a few years later that the Continuum Hypothesis is consistent, one of the consequences was of course that Borel's Conjecture cannot be proven from the adopted axiom system. Almost half a century later, in 1976, using the breakthrough technique of forcing, Laver proved that also the negation of Borel's Conjecture cannot be proven from the adopted axiom system: Borel's Conjecture is independent of the axioms adopted as foundation for Mathematics. This independence proof also introduced the new technique of countable support iterated forcing onto the mathematical arena.

The independence of Borel's Conjecture from the foundational axioms opens a large field of exploration of alternative versions of the mathematical universe, depending on whether Borel's Conjecture is adopted as an axiom, or whether its negation is adopted as an axiom. One might ask whether Borel's Conjecture or its negation would have any significant or interesting impact on the corresponding mathematical universe. And to a large extent the unabated research on the mathematical objects Borel's Conjecture is about sheds light on this question. As the reader will see from the papers in this volume, the impact of Borel's Conjecture on mathematics is broad and interesting: There are equivalents of Borel's Conjecture in terms of infinite two-person games, there are equivalents of Borel's Conjecture in terms of properties of certain topological groups, in terms of Ramsey theory, and numerous other mathematical notions that at first glance appear unrelated. In addition, the basic idea underlying Borel's Conjecture inspired, in light of the many mathematical statements found to be equivalent to Borel's Conjecture, several innovative mathematical concepts, and related questions of consistency and independence, as the reader will see from the papers in this volume.

It was not possible to represent all the developments related to the Borel Conjecture in the time-frame envisioned for putting this volume together. For example, the reader will notice that there is no paper that reports on BMax (for those familiar with this specialized terminology), or on the several generalizations of the covering property at the center of Borel's Conjecture beyond the world of metrizable topological spaces and groups. On the time-scale of mathematics as a field, exploration of the Borel Conjecture and its relatives is still in its infancy. With this volume we hope to initiate a tradition of updating, at appropriate times, the developments in mathematics inspired by the Borel Conjecture.

The papers in this volume are listed, as is the standard in mathematics, alphabetically by last name of the authors.

We thank the American Mathematical Society for publishing this volume in the Contemporary Mathematics series. In particular, we thank Christine Thivierge of the American Mathematical Society for her excellent guidance throughout the whole process of preparing this volume and we thank Mike Saitas for his very efficient and accommodating management of the production of this volume. Last, but not least, we thank the referees of the papers appearing in this volume for their selfless work in helping to maintain standards towards a continuing credible record of mathematical research.

Marion Scheepers
Ondřej Zindulka

Contemporary Mathematics
Volume **755**, 2020
https://doi.org/10.1090/conm/755/15178

Game-theoretical aspects of the Borel conjecture

Leandro F. Aurichi and Rodrigo R. Dias

ABSTRACT. Having the strong measure zero game of Mycielski and Solovay as a starting point, we discuss results from the bibliography of topological games that establish game-theoretical characterizations of strong measure zero sets and game-theoretical versions of the Borel Conjecture, as well as related games and notions that emerge from the previous ones. This includes generalizations of strong measure zero to other contexts, such as Rothberger-boundedness and topological measure zero. We finish with a brief discussion on some higher-cardinal versions of the Borel Conjecture appearing in the literature.

1. Introduction

In 1919, Émile Borel [**6**] introduced a new property for sets of real numbers, which he called "asymptotic measure less than every series given in advance."[1] Then, after remarking that every countable set had that property, he conjectured that the converse must also hold. In current terminology, this has become:

DEFINITION 1.1 (Borel [**6**]). A set $X \subseteq \mathbb{R}$ has **strong measure zero** if, for every sequence $\langle \varepsilon_n : n \in \omega \rangle$ of positive real numbers, there is a sequence $\langle I_n : n \in \omega \rangle$ of open intervals satisfying $\ell(I_n) \leq \varepsilon_n$ for all $n \in \omega$ and $X \subseteq \bigcup_{n\in\omega} I_n$.

The **Borel Conjecture** (BC) is the statement "Every strong measure zero subset of $\mathbb{R}$ is countable."

In 1928, Sierpiński [**40**] noted that every Luzin set [**26**] — whose existence was already known to follow from the Continuum Hypothesis (CH) — is an uncountable set of reals that has strong measure zero. A decade later, when Gödel [**15**] proved (by means of the Axiom of Constructibility) the Continuum Hypothesis to be consistent with the Zermelo–Fraenkel axioms for set theory (ZFC), this yielded the consistency of $\neg$BC. Only in 1976 was the independence of the Borel Conjecture settled, when Laver [**25**] used the technique of forcing to prove that BC is also consistent with ZFC.

It is easy to see that every strong measure zero subset of $\mathbb{R}$ has Lebesgue measure zero; a simple example showing that the converse does not hold is the ternary Cantor set, where the failure of strong measure zero can be witnessed by taking $\varepsilon_n = \frac{1}{3^{n+1}}$ for each $n \in \omega$. The big leap from Lebesgue measure zero to

2010 *Mathematics Subject Classification.* Primary 91A44; Secondary 03E65, 22C05, 54B10, 54D20, 54E45.

The first author was supported by FAPESP (2017/09252-3).

[1]In the original, "*mesure asymptotique inférieure à toute série donné à l'avance.*"

strong measure zero is that, in the latter, the lengths of the intervals go to zero as fast as we want: the sequence $\langle\varepsilon_n : n \in \omega\rangle$ can be viewed as the "speed" of the approximation by intervals of small length. One question that may rise is: What if we do not know the speed in advance? This is the idea that motivates an infinite game introduced by Mycielski and Solovay, which became a starting point for the study of topological games associated with strong measure zero. Our aim in this text is to provide an account of the relationships between strong measure zero and some games that have already been explored in the literature; in particular, several game-theoretical equivalences of BC emerge from this study.

Although strong measure zero is a notion originally introduced for subsets of the real line, a number of generalizations of this definition to broader contexts have been made. Let us briefly mention the ones we will be dealing with in this text — each of them has a topological game naturally associated to it, and that will be our main focus in the coming sections.

Perhaps the most immediate generalization is the extension of the definition to metric spaces, which is the version the Mycielski–Solovay game is based upon: $\langle X, d\rangle$ is said to have strong measure zero if, for every sequence $\langle\varepsilon_n : n \in \omega\rangle$ of positive reals, we can cover X with sets $\langle A_n : n \in \omega\rangle$ such that $\operatorname{diam}(A_n) \leq \varepsilon_n$ for all $n \in \omega$. Surprisingly enough, this generalization has no effect on the Borel Conjecture. Indeed, Carlson proved in [**7**, Theorem 3.2]:

THEOREM 1.2 (Carlson [**7**]). BC *is equivalent to the statement "Every strong measure zero metric space is countable."*

Another context in which there is a natural extension of the notion of strong measure zero is the realm of topological groups. For a topological group $\langle G, *\rangle$ with identity element e, call a subset $X \subseteq G$ **Rothberger-bounded** [**2**, **30**][2] if, for every sequence $\langle U_n : n \in \omega\rangle$ of open neighbourhoods of e, one can cover X with one left translate of each neighbourhood in the sequence — i.e. one can find a sequence $\langle x_n : n \in \omega\rangle$ of elements of G with $X \subseteq \bigcup_{n\in\omega} x_n * U_n$. Clearly, a subset of $\mathbb{R}$ has strong measure zero if and only if it is a Rothberger-bounded subset of $\mathbb{R}$ seen as an additive topological group. As it turns out, unlike the metric space scenario, the Borel Conjecture fails to be equivalent to its more straightforward version for topological groups — that is, the statement obtained when "strong measure zero subset of the real line" is replaced with "Rothberger-bounded subset of a topological group". In fact, such statement is provably false:

EXAMPLE 1.3 (Scheepers [**38**]). For every infinite cardinal κ, the σ-product

$$\sigma(\mathbb{Z}^\kappa) = \{f \in \mathbb{Z}^\kappa : |\{\alpha \in \kappa : f(\alpha) \neq 0\}| < \aleph_0\},$$

where $\mathbb{Z}$ is given the discrete topology and the product $\mathbb{Z}^\kappa$ is given the Tychonoff topology, is a Rothberger-bounded additive group of cardinality κ.

PROOF. It is clear that $|\sigma(\mathbb{Z}^\kappa)| = \kappa^{<\aleph_0} \cdot \aleph_0 = \kappa$. To see that $\sigma(\mathbb{Z}^\kappa)$ is Rothberger-bounded, fix a sequence $\langle U_n : n \in \omega\rangle$ of basic open neighbourhoods of $\bar{0} \in \sigma(\mathbb{Z}^\kappa)$. For each $n \in \omega$, let F_n be the finite subset of κ such that $U_n = \{f \in \sigma(\mathbb{Z}^\kappa) : \forall\alpha \in F_n\ (f(\alpha) = 0)\}$. Let then $C = \bigcup_{n\in\omega} F_n \subseteq \kappa$. As C is countable, it

[2]It should be noted that this property was also considered independently both by Galvin and Kočinac in unpublished work.

follows that $\sigma(\mathbb{Z}^C)$ is countable — say $\sigma(\mathbb{Z}^C) = \{g_n : n \in \omega\}$. Now, for each $n \in \omega$, define $f_n \in \sigma(\mathbb{Z}^\kappa)$ by

$$f_n(\alpha) = \begin{cases} g_n(\alpha) & \text{if } \alpha \in C; \\ 0 & \text{otherwise.} \end{cases}$$

We claim that $\sigma(\mathbb{Z}^\kappa) = \bigcup_{n\in\omega}(f_n + U_n)$. Indeed, let $h \in \sigma(\mathbb{Z}^\kappa)$. Let $n \in \omega$ be such that $h \restriction C = g_n$, and define $h' \in \sigma(\mathbb{Z}^\kappa)$ by

$$h'(\alpha) = \begin{cases} 0 & \text{if } \alpha \in C; \\ h(\alpha) & \text{otherwise.} \end{cases}$$

Note that $h' \in U_n$, whence $h = f_n + h' \in f_n + U_n$. □

Yet, as we shall see in Theorem 5.4, by adding "second-countable" to the statement, we do obtain an equivalent form of BC in terms of topological groups.

Now, a question that naturally suggests itself is: Can we define a similar property in the realm of general topological spaces? The first idea that comes to mind is possibly the Rothberger property, introduced by Rothberger in [**30**] under the name $\mathbf{C}''$:

DEFINITION 1.4 (Rothberger [**30**]). A topological space X is **Rothberger** if, for every sequence $\langle \mathcal{C}_n : n \in \omega \rangle$ of open covers of X, there is a sequence $\langle U_n : n \in \omega \rangle$ such that each $U_n \in \mathcal{C}_n$ and $X = \bigcup_{n\in\omega} U_n$.

It is clear that every Rothberger metric space has strong measure zero — just note that, for each $\varepsilon \in \mathbb{R}_{>0}$, the collection of all open sets with diameter less than ε forms an open cover. As in the case of topological groups, there are ZFC examples of uncountable Rothberger topological spaces — e.g. the topology on the set $\kappa + 1$ (for κ an arbitrary uncountable cardinal) in which every $\alpha \in \kappa$ is isolated and open neighbourhoods of κ are cocountable. We shall see equivalent formulations of BC in terms of the Rothberger property in Theorems 3.9 and 9.7.

As it turns out, the Rothberger property is (consistently) strictly stronger than strong measure zero, even for sets of reals; this was proven by Rothberger in [**31**, Theorem 3] — the fact that consistency is the best one could attain will be a consequence of Theorem 3.9.

THEOREM 1.5 (Rothberger [**31**]). CH *implies that there is a strong measure zero set of reals that does not satisfy the Rothberger property.*

PROOF. We will construct a strong measure zero set $X \subseteq \mathbb{R}$ that can be continuously mapped onto the Cantor space 2^ω; the fact that X is not a Rothberger space will then follow from two straightforward observations: that the Rothberger property is preserved by continuous images, and that 2^ω is not a Rothberger space.

Using CH, fix an enumeration $2^\omega = \{h_\alpha : \alpha < \omega_1\}$, and let $\langle g_\alpha : \alpha < \omega_1 \rangle$ be an unbounded sequence in ω^ω such that $\alpha < \beta < \omega_1 \to g_\alpha \leq^* g_\beta$ — this can easily be accomplished by transfinite recursion. For each $\alpha < \omega_1$, let $f_\alpha \in \omega^\omega$ be defined by $f_\alpha(n) = 2 \cdot g_\alpha(n) + h_\alpha(n)$. Now set $Z = \{f_\alpha : \alpha < \omega_1\}$. It is easy to see that the function $\psi : \omega^\omega \to 2^\omega$ defined by $\psi(f)(n) = f(n) \mod 2$ for every $n \in \omega$ is continuous; therefore, $\psi \restriction Z : Z \to 2^\omega$ is also continuous — and it is clearly onto. Finally, fix a homeomorphism $\varphi : \omega^\omega \to [0, 1] \setminus \mathbb{Q}$, and let $X = \varphi[Z]$. As Z maps continuously onto 2^ω, so does X. It remains to check that X has strong measure zero.

Let $\langle \varepsilon_n : n \in \omega \rangle$ be a sequence of positive reals. Fix an enumeration $[0,1] \cap \mathbb{Q} = \{q_k : k \in \omega\}$, and for each $k \in \omega$ let $I_{2k} = \left(q_k - \frac{\varepsilon_{2k}}{2}, q_k + \frac{\varepsilon_{2k}}{2}\right)$. Now, as $[0,1] \setminus \bigcup_{k\in\omega} I_{2k}$ is a compact subset of $[0,1] \setminus \mathbb{Q}$, it follows that $K = \varphi^{-1}\left[[0,1] \setminus \bigcup_{k\in\omega} I_{2k}\right]$ is a compact subset of ω^ω, hence there is $F \in \omega^\omega$ such that $K \subseteq \{f \in \omega^\omega : \forall n \in \omega\ (f(n) \leq F(n))\}$. By choosing $\eta < \omega_1$ such that $g_\eta \not\leq^* F$, we obtain that every $\alpha \in \omega_1 \setminus \eta$ is such that $g_\alpha \not\leq^* F$, whence $f_\alpha \not\leq^* F$; therefore, $K \cap \{f_\alpha : \alpha < \omega_1\} \subseteq \{f_\alpha : \alpha < \eta\}$, so the set $X \setminus \bigcup_{k\in\omega} I_{2k} \subseteq \varphi[K \cap \{f_\alpha : \alpha < \omega_1\}] \subseteq \varphi[\{f_\alpha : \alpha < \eta\}]$ is countable. Now we can enumerate this set as $X \setminus \bigcup_{k\in\omega} I_{2k} = \{x_m : m \in \omega\}$ and define $I_{2m+1} = \left(x_m - \frac{\varepsilon_{2m+1}}{2}, x_m + \frac{\varepsilon_{2m+1}}{2}\right)$ for each $m \in \omega$, and it will follow that $X \subseteq \bigcup_{n\in\omega} I_n$. □

An alternative candidate for a purely topological property that captures the notion of strong measure zero was proposed by Scheepers in [**37**]. Let us say that a subset X of a topological space Y is of **topological measure zero** if, for every sequence $\langle \mathcal{V}_n : n \in \omega \rangle$ of open covers of Y, one can select $V_n \in \mathcal{V}_n$ for $n \in \omega$ so that $X \subseteq \bigcup_{n\in\omega} V_n$. Note that, if X is a Rothberger space, then X has topological measure zero in every superspace — and we have already seen that there are very simple examples of uncountable Rothberger spaces. Yet, as we shall see in Theorem 4.1 and Corollary 4.2, topological measure zero is equivalent to strong measure zero for a wide range of metric spaces (namely, σ–totally-bounded metric spaces — or, as will often be the case of interest, σ-compact metric spaces[3]).

A word on (σ-)compactness is now in order. As the original context in the study of strong measure zero sets is the real line, adding σ-compactness to the hypotheses of results on this property is generally deemed reasonable. In fact, taking into consideration that the Borel Conjecture is the underlying context motivating the study of this property, in a number of results it is assumed that the space under consideration is compact — indeed, BC is easily seen to be equivalent to its corresponding versions in which the ambient space is the interval $[0,1]$ or the circle $\mathbb{S}^1$ or even the Cantor space[4] 2^ω instead of $\mathbb{R}$. In many cases, the proof of a result about σ-compact spaces can be understood more clearly in the simpler setting of compact spaces, and going from the compact case to the σ-compact case is a matter of adapting the argument with the addition of (rather cumbersome) technicalities; this is illustrated in the proof of Theorem 4.1, which employs essentially the same reasoning used for proving Lemma 4.3. Having this in mind, some results that could be done for σ-compact spaces are often stated and proven for compact spaces instead, for the sake of sparing the reader the intricacies needed in the proof of the more general case. A key feature of compact metric spaces for the results requiring (σ-)compactness will be the classical Lebesgue Covering Lemma:

[3]Recall that a metric space is **totally bounded** if, for every $\varepsilon \in \mathbb{R}_{>0}$, it can be covered by finitely many sets of diameter at most ε; it is σ–**totally-bounded** if it is a union of countably many totally bounded subsets. When the property we are dealing with refers to subsets of metric spaces, assuming that the ambient space is compact is not more restrictive than assuming total boundedness, as the completion of a totally bounded metric space is compact; the same holds for σ–total-boundedness vs. σ-compactness.

[4]On the set 2^ω, we consider the metric d defined by $d(f,g) = \max\{\frac{1}{n+1} : n \in \omega \text{ and } f(n) \neq g(n)\}$ for all distinct $f, g \in 2^\omega$.

LEBESGUE COVERING LEMMA. Let X be a compact metric space and $\mathcal{C}$ be an open cover of X. Then there is $\varepsilon \in \mathbb{R}_{>0}$ such that every $A \subseteq X$ with $\operatorname{diam}(A) \leq \varepsilon$ is a subset of some element of $\mathcal{C}$. (Any such ε is said to be a **Lebesgue number** for the cover $\mathcal{C}$.)

2. The strong measure zero game of Mycielski and Solovay

Probably the most natural game translation of the strong measure zero property is the following:

DEFINITION 2.1 (Galvin–Mycielski–Solovay [**12**]). The **Mycielski–Solovay game** on a metric space X ($MS(X)$) is played between players ALICE and BOB as follows. In each inning $n \in \omega$, ALICE chooses $\varepsilon_n \in \mathbb{R}_{>0}$, and BOB answers with $A_n \subseteq X$ such that $\operatorname{diam}(A_n) \leq \varepsilon_n$. BOB is declared the winner if $\bigcup_{n\in\omega} A_n = X$, otherwise ALICE wins.

The following theorem will be a first step in establishing a direct relation between the Mycielski–Solovay game and the Borel Conjecture:

THEOREM 2.2 (Galvin–Mycielski–Solovay [**12**]). *Let X be a metric space. Then X is countable if and only if* BOB *has a winning strategy in $MS(X)$.*

PROOF. If X is countable, this is clear. On the other hand, suppose θ is a winning strategy for BOB. For each $x \in X$, let $s_x \in (\mathbb{Q}_{>0})^{<\omega}$ be such that $x \in \theta(s_x{}^\frown q)$ for every $q \in \mathbb{Q}_{>0}$ — there is such an s_x, since otherwise we could construct a play of the game in which BOB would make use of θ and fail to cover x, thus contradicting the fact that θ is a winning strategy. Note that $x \mapsto s_x$ is injective, as the condition that $\operatorname{diam}(\theta(s_x{}^\frown q)) \leq q$ for each $q \in \mathbb{Q}_{>0}$ implies $\bigcap_{q\in\mathbb{Q}_{>0}} \theta(s_x{}^\frown q) = \{x\}$. □

Surprisingly enough, under the additional hypothesis of σ–total-boundedness, strong measure zero is equivalent to ALICE not having a winning strategy in $MS(X)$:

THEOREM 2.3 (Galvin–Mycielski–Solovay [**12**]). *Let X be a σ–totally-bounded metric space. Then X has strong measure zero if and only if* ALICE *does not have a winning strategy in $MS(X)$.*

PROOF. It is clear that, if X does not have strong measure zero, then ALICE has a winning strategy in $MS(X)$: if $\langle \varepsilon_n : n \in \omega \rangle$ is such that there is no sequence $\langle A_n : n \in \omega \rangle$ of subsets of X satisfying $\operatorname{diam}(A_n) \leq \varepsilon_n$ for each $n \in \omega$ and $\bigcup_{n\in\omega} A_n = X$, then playing ε_n in the n-th inning constitutes a winning strategy for ALICE in $MS(X)$.

Now suppose that X has strong measure zero, and let $\theta : \wp(X)^{<\omega} \to \mathbb{R}_{>0}$ be a strategy for ALICE in $MS(X)$. Fix an increasing sequence $\langle X_n : n \in \omega \rangle$ of totally bounded subsets of X with $X = \bigcup_{n\in\omega} X_n$. We will define $\varepsilon_n \in \mathbb{R}_{>0}$ and $\mathcal{C}_n \in [\wp(X)]^{<\aleph_0}$ for $n \in \omega$ recursively as follows.

Set $\varepsilon_0 = \theta(\langle\rangle)$. Now, suppose that $n \in \omega$ is such that $\langle \varepsilon_k : k \leq n \rangle$ and $\langle \mathcal{C}_k : k < n \rangle$ have already been defined. Since X_n is totally bounded, there exists a finite set $\mathcal{F}_n \subseteq \wp(X_n)$ such that $\operatorname{diam}(Y) \leq \frac{\varepsilon_n}{3}$ for all $Y \in \mathcal{F}_n$ and $\bigcup \mathcal{F}_n = X_n$. Let

$$\mathcal{C}_n = \left\{ \bigcup_{y\in Y} B\left[y, \frac{\varepsilon_n}{3}\right] : Y \in \mathcal{F}_n \right\}$$

and

$$\varepsilon_{n+1} = \min\Big\{\theta(\langle C_k : k \leq n\rangle) : \langle C_k : k \leq n\rangle \in \prod_{k\leq n} \mathcal{C}_k\Big\}.$$

This defines ε_n and $\mathcal{C}_n$ for all $n \in \omega$. Note that, for each $n \in \omega$,

(†) every nonempty $A \subseteq X_n$ with $\operatorname{diam}(A) \leq \frac{\varepsilon_n}{3}$ is included in some element of $\mathcal{C}_n$.

(In order to see this, just pick $Y \in \mathcal{F}_n$ with $A \cap Y \neq \emptyset$, and it will follow that $A \subseteq \bigcup_{y\in Y} B\Big[y, \frac{\varepsilon_n}{3}\Big] \in \mathcal{C}_n$.)

Let ω be partitioned into infinite subsets as $\omega = \dot{\bigcup}_{n\in\omega} M_n$ in such a way that $\min(M_n) \geq n$ for each n. As every X_n has strong measure zero, there is for each n a sequence $\langle A_m : m \in M_n\rangle$ of subsets of X_n such that $\operatorname{diam}(A_m) \leq \frac{\varepsilon_m}{3}$ for each $m \in M_n$ and $\bigcup_{m\in M_n} A_m = X_n$. Note that each $m \in M_n$ satisfies $A_m \subseteq X_n \subseteq X_m$; hence, by (†), there is $C_m \in \mathcal{C}_m$ with $A_m \subseteq C_m$ — and here note that $C_m \in \mathcal{C}_m$ implies $\operatorname{diam}(C_m) \leq \varepsilon_m \leq \theta(\langle C_k : k < m\rangle)$ by construction. Therefore,

$$\langle \theta(\langle\rangle), C_0, \theta(\langle C_0\rangle), C_1, \theta(\langle C_0, C_1\rangle), C_2, \ldots, \theta(\langle C_k : k < m\rangle), C_m, \ldots\rangle$$

is a play of $MS(X)$ in which ALICE plays according to θ and loses, since $X = \bigcup_{n\in\omega} X_n = \bigcup_{n\in\omega}\bigcup_{m\in M_n} A_m \subseteq \bigcup_{m\in\omega} C_m$. □

Therefore, we have:

COROLLARY 2.4. *The following statements are equivalent:*[5]

(1) BC;
(2) *the game $MS(X)$ is determined on every $X \subseteq \mathbb{R}$;*
(3) *the game $MS(X)$ is determined on every σ–totally-bounded metric space X.*

PROOF. The implication (3) → (2) is trivial.

To see that (2) → (1), let X be a strong measure zero subset of $\mathbb{R}$. Then Theorem 2.3 tells us that ALICE does not have a winning strategy in the game $MS(X)$; therefore, by (2), BOB has a winning strategy in $MS(X)$, so X must be countable by Theorem 2.2.

Finally, assume BC, and let X be a σ–totally-bounded metric space. Suppose that ALICE does not have a winning strategy in $MS(X)$. Then X has strong measure zero; thus, by Theorem 1.2, X must be countable — hence BOB has a winning strategy in $MS(X)$. Therefore, (1) → (3). □

3. The Rothberger game

The Rothberger property, mentioned in the Introduction, is a classical example of what is known today as a selective property. These are properties defined by the so-called selection principles, such as the one we will now define:

DEFINITION 3.1 (Scheepers [**32**]). Let $\mathcal{A}$ and $\mathcal{B}$ be nonempty families of nonempty sets. We denote by $\mathsf{S}_1(\mathcal{A}, \mathcal{B})$ the following statement:

For every sequence $\langle A_n : n \in \omega\rangle$ of elements of $\mathcal{A}$, there is a sequence $\langle x_n : n \in \omega\rangle$ such that

- $x_n \in A_n$ for all $n \in \omega$

and

[5] A game is **determined** if one of the players has a winning strategy.

- $\{x_n : n \in \omega\} \in \mathcal{B}$.

Thus, the statement that a topological space X is a Rothberger space can be expressed by $\mathsf{S}_1(\mathcal{O}(X), \mathcal{O}(X))$, where $\mathcal{O}(X)$ is the set of all the open covers of X. When there is no chance of misinterpretation, we shall just write $\mathcal{O}$ instead of $\mathcal{O}(X)$.

To each property defined by the selection principle S_1, there is a naturally associated infinite game:

DEFINITION 3.2 (Scheepers [**34**]). Let $\mathcal{A}$ and $\mathcal{B}$ be nonempty families of nonempty sets. The game $\mathsf{G}_1(\mathcal{A}, \mathcal{B})$ is defined as follows. In each inning $n \in \omega$, ALICE plays $A_n \in \mathcal{A}$, and then BOB picks $x_n \in A_n$. BOB is declared the winner if $\{x_n : n \in \omega\} \in \mathcal{B}$, otherwise ALICE wins.

Thus, the game naturally associated to the Rothberger property on a topological space X — often referred to as the **Rothberger game** on X — is $\mathsf{G}_1(\mathcal{O}, \mathcal{O})$, defined by: in each inning $n \in \omega$, ALICE chooses an open cover $\mathcal{C}_n$ of X, and BOB answers with $U_n \in \mathcal{C}_n$; the winner is BOB if $\bigcup_{n\in\omega} U_n = X$, and ALICE otherwise.

Note that, if $\mathcal{A}$ and $\mathcal{B}$ are such that ALICE does not have a winning strategy in the game $\mathsf{G}_1(\mathcal{A}, \mathcal{B})$, then $\mathsf{S}_1(\mathcal{A}, \mathcal{B})$ holds: if, by way of contradiction, there were a sequence $\langle A_n : n \in \omega\rangle$ of elements of $\mathcal{A}$ witnessing the failure of $\mathsf{S}_1(\mathcal{A}, \mathcal{B})$, then playing the set A_n in the inning $n \in \omega$ would constitute an obvious winning strategy for ALICE in $\mathsf{G}_1(\mathcal{A}, \mathcal{B})$. When the converse implication holds (which of course depends on the families $\mathcal{A}$ and $\mathcal{B}$), we have a very powerful result, as will be seen later on.

We have previously mentioned the Rothberger property keeping in mind the line drawn between this property and strong measure zero. Let us now turn to the similar behaviour they have under some circumstances. We begin by noting some natural relationships between the games $MS(X)$ and $\mathsf{G}_1(\mathcal{O}, \mathcal{O})$.

PROPOSITION 3.3. *Let $\langle X, d\rangle$ be a metric space. The following hold:*

(1) *if* ALICE *has a winning strategy in $MS(X)$, then* ALICE *has a winning strategy in $\mathsf{G}_1(\mathcal{O}, \mathcal{O})$;*
(2) *if* BOB *has a winning strategy in $\mathsf{G}_1(\mathcal{O}, \mathcal{O})$, then* BOB *has a winning strategy in $MS(X)$.*

Moreover, if X is also compact, then the converses of the two above statements also hold.

PROOF. The direct implications follow directly from the fact that the collection of all open subsets of X with diameter less than a fixed ε forms an open cover of X.

Now suppose that X is compact.

(1) Let θ be a winning strategy for ALICE in $\mathsf{G}_1(\mathcal{O}, \mathcal{O})$. We will use this strategy to create a winning strategy ϱ for ALICE in $MS(X)$. Let $\varrho(\langle\rangle)$ be a Lebesgue number for $\theta(\langle\rangle)$. Let A_0 be BOB's choice in the first inning of $MS(X)$. Since $\operatorname{diam}(A_0) < \varrho(\langle\rangle)$, there is a $V_0 \in \theta(\langle\rangle)$ such that $A_0 \subseteq V_0$. So let $\varrho(\langle A_0\rangle)$ be a Lebesgue number for $\theta(\langle V_0\rangle)$. Note that, no matter which A_1 BOB chooses, there is a $V_1 \in \theta(\langle V_0\rangle)$ with $A_1 \subseteq V_1$, so we can define $\varrho(\langle A_0, A_1\rangle)$ to be a Lebesgue number for $\theta(\langle V_0, V_1\rangle)$; and so forth. In general, if $\langle A_0, \dots, A_n\rangle \in \wp(X)^{<\omega}$ is such that $\varrho(\langle A_0, \dots, A_n\rangle)$ has been defined to be a Lebesgue number for $\theta(\langle V_0, \dots, V_n\rangle)$ — where $\langle V_0, \dots, V_n\rangle$ is such that $A_k \subseteq V_k \in \theta(\langle V_0, \dots, V_{k-1}\rangle)$ for each $k \leq n$ —, then,

for each $A_{n+1} \subseteq X$ with $\operatorname{diam}(A_{n+1}) \leq \varrho(\langle A_0, \ldots, A_n\rangle)$, we may choose $V_{n+1} \in \theta(\langle V_0, \ldots, V_n\rangle)$ such that $A_{n+1} \subseteq V_{n+1}$, and then declare $\varrho(\langle A_0, \ldots, A_n, A_{n+1}\rangle)$ to be a Lebesgue number for $\theta(\langle V_0, \ldots, V_n, V_{n+1}\rangle)$. Once such a play is finished, $\bigcup_{n\in\omega} V_n$ does not cover X (since θ is winning), so neither does $\bigcup_{n\in\omega} A_n$.

(2) Let θ be a winning strategy for BOB in $MS(X)$. We will define a winning strategy ϱ for BOB in $\mathsf{G}_1(\mathcal{O}, \mathcal{O})$ as follows. First, for each $\mathcal{C} \in \mathcal{O}$, let a Lebesgue number $\varepsilon_{\mathcal{C}}$ for $\mathcal{C}$ be fixed. After ALICE plays an open cover $\mathcal{C}_0$ in $\mathsf{G}_1(\mathcal{O}, \mathcal{O})$, choose $\varrho(\langle \mathcal{C}_0\rangle) \in \mathcal{C}_0$ such that $\theta(\langle \varepsilon_{\mathcal{C}_0}\rangle) \subseteq \varrho(\langle \mathcal{C}_0\rangle)$. In general, if $\langle \mathcal{C}_0, \ldots, \mathcal{C}_n\rangle \in \mathcal{O}^{<\omega}$ is the sequence of ALICE's moves thus far, choose $\varrho(\langle \mathcal{C}_0, \ldots, \mathcal{C}_n\rangle) \in \mathcal{C}_n$ such that $\theta(\langle \varepsilon_{\mathcal{C}_0}, \ldots, \varepsilon_{\mathcal{C}_n}\rangle) \subseteq \varrho(\langle \mathcal{C}_0, \ldots, \mathcal{C}_n\rangle)$. Note that, after such a play is finished, the fact that θ is winning yields $X = \bigcup_{n\in\omega} \theta(\langle \varepsilon_{\mathcal{C}_0}, \ldots, \varepsilon_{\mathcal{C}_n}\rangle) \subseteq \bigcup_{n\in\omega} \varrho(\langle \mathcal{C}_0, \ldots, \mathcal{C}_n\rangle)$. □

With this we obtain:

COROLLARY 3.4. *Let X be a compact metric space. The following are equivalent:*

(1) *X has strong measure zero;*
(2) ALICE *does not have a winning strategy in $MS(X)$;*
(3) ALICE *does not have a winning strategy in $\mathsf{G}_1(\mathcal{O}, \mathcal{O})$;*
(4) *X is Rothberger.*

PROOF. It is clear that (3) → (4) → (1). The implications (1) → (2) and (2) → (3) follow from Theorem 2.3 and Proposition 3.3 respectively. □

As we have already observed, (3) → (4) in the above corollary holds for topological spaces in general. The quite nontrivial fact that the converse also holds for every topological space was proven in Lemma 2 of [**29**]:

THEOREM 3.5 (Pawlikowski [**29**]). *A topological space X is Rothberger if and only if* ALICE *does not have a winning strategy in $\mathsf{G}_1(\mathcal{O}, \mathcal{O})$.*

We will now direct our attention to the Rothberger property and the game $\mathsf{G}_1(\mathcal{O}, \mathcal{O})$, and explore further how they relate to the Borel Conjecture. First, let us mention another very closely related topological game. The **point-open game** on a topological space X was independently introduced by Telgársky [**45**] and Galvin [**11**], and is defined as follows: in each inning $n \in \omega$, ALICE chooses x_n, and then BOB chooses an open set V_n such that $x_n \in V_n$; ALICE wins if $\bigcup_{n\in\omega} V_n = X$, and BOB wins otherwise. Also in [**11**], Galvin proved the following relation between this game and the Rothberger game:

THEOREM 3.6 (Galvin [**11**]). *The game $\mathsf{G}_1(\mathcal{O}, \mathcal{O})$ and the point-open game are dual, i.e.* ALICE *has a winning strategy in one of the games if and only if* BOB *has a winning strategy in the other.*

This kind of duality is often very useful. To illustrate this, consider the following:

THEOREM 3.7 (Telgársky [**45**], Galvin [**11**]). *Let X be a topological space in which every point is a G_δ. Then* BOB *has a winning strategy in the game $\mathsf{G}_1(\mathcal{O}, \mathcal{O})$ played on X if and only if X is countable.*

PROOF. Note that, by the duality, this is equivalent to showing that ALICE has a winning strategy in the point-open game if and only if X is countable. If X is countable, this is immediate. Conversely, assume that X is uncountable, and let θ be a strategy for ALICE in the point-open game on X. For each $x \in X$, let $\{V_k^x : k \in \omega\}$ be a set of open neighbourhoods of x satisfying $\{x\} = \bigcap_{k\in\omega} V_k^x$. Define recursively:

- $x_{\langle\rangle} = \theta(\langle\rangle)$;
- $W_{\langle k\rangle} = V_k^{x_{\langle\rangle}}$ for every $k \in \omega$;
- $x_s = \theta(\langle W_{s\restriction 1}, \ldots, W_{s\restriction n}\rangle)$ for $s \in \omega^n$;
- $W_{s^\frown k} = V_k^{x_s}$ for each $k \in \omega$.

As $\{x_s : s \in \omega^{<\omega}\}$ is countable and X is uncountable, we may pick $y \in X \setminus \{x_s : s \in \omega^{<\omega}\}$. Consider the following play of the point-open game. ALICE starts with $x_{\langle\rangle} = \theta(\langle\rangle)$. Since $y \neq x_{\langle\rangle}$, there is $k_0 \in \omega$ be such that $y \notin W_{\langle k_0\rangle}$. Let $W_{\langle k_0\rangle}$ be BOB's answer. As a reply to this, ALICE will play $x_{\langle k_0\rangle} = \theta(\langle W_{\langle k_0\rangle}\rangle)$. Then, as $y \neq x_{\langle k_0\rangle}$, pick $k_1 \in \omega$ such that $y \notin W_{\langle k_0,k_1\rangle}$, and let $W_{\langle k_0,k_1\rangle}$ be BOB's move. Now ALICE plays $x_{\langle k_0,k_1\rangle} = \theta(\langle W_{\langle k_0\rangle}, W_{\langle k_0,k_1\rangle}\rangle)$. In general, for each $n \in \omega$, if $s \in \omega^n$ is such that ALICE plays x_s in the n-th inning, we may pick $k_n \in \omega$ such that $y \notin W_{s^\frown k_n}$, and let BOB respond with $W_{s^\frown k_n}$. Note that this describes a play where ALICE follows θ and loses (since the point y is never covered), which shows that θ is not a winning strategy. □

We are now getting closer to the equivalence of BC in terms of Rothberger spaces and the game $\mathsf{G}_1(\mathcal{O}, \mathcal{O})$.

DEFINITION 3.8. Let us call the **R-Borel Conjecture** the statement "Every Rothberger subspace of the real line is countable."

Although it is consistent with ZFC that the Rothberger property is strictly stronger than strong measure zero (as we have seen in Theorem 1.5), one surprisingly has the following theorem — which is a consequence of Theorem 4 of [**31**] (see also [**27**, Proposition 8]):

THEOREM 3.9 (Rothberger [**31**]). *The Borel Conjecture and the R-Borel Conjecture are equivalent.*

In view of previous results, an immediate consequence of this theorem is the following game-theoretical equivalence of the Borel Conjecture:

COROLLARY 3.10 (Pawlikowski [**29**]). *The Borel Conjecture holds if and only if the game $\mathsf{G}_1(\mathcal{O}, \mathcal{O})$ — equivalently, the point-open game — is determined on every subspace of the real line.*

PROOF. By Theorem 3.5, the R-Borel Conjecture translates as "If $X \subseteq \mathbb{R}$ is such that ALICE does not have a winning strategy in the game $\mathsf{G}_1(\mathcal{O}, \mathcal{O})$ played on X, then X is countable." By Theorem 3.7, this is equivalent to "If $X \subseteq \mathbb{R}$ is such that ALICE does not have a winning strategy in the game $\mathsf{G}_1(\mathcal{O}, \mathcal{O})$ played on X, then BOB has a winning strategy in the game $\mathsf{G}_1(\mathcal{O}, \mathcal{O})$ played on X" — i.e. the game $\mathsf{G}_1(\mathcal{O}, \mathcal{O})$ is determined on every $X \subseteq \mathbb{R}$. The result then follows from Theorem 3.9. □

For the proof of Theorem 3.9, we will need two results by Rothberger. The first one is a consequence of Theorem 2 of [**31**]:

THEOREM 3.11 (Rothberger [**31**]). *If $\mathfrak{b} = \aleph_1$, then the R-Borel Conjecture fails.*

PROOF. We will show that $\mathfrak{b} = \aleph_1$ implies the existence of a subset of $\mathbb{R}$ that has size $\aleph_1$ and is concentrated on a countable subset.[6]

Let $\langle f_\alpha : \alpha < \omega_1 \rangle$ be an unbounded sequence in ω^ω with $f_\alpha \leq^* f_\beta$ whenever $\alpha < \beta < \omega_1$. Writing $X = \{f_\alpha : \alpha \in \omega_1\} \subseteq \omega^\omega$ and fixing a homeomorphism $\varphi : \omega^\omega \to [0,1] \setminus \mathbb{Q}$, we now claim that $Y = \varphi[X] \cup ([0,1] \cap \mathbb{Q})$ is concentrated on $[0,1] \cap \mathbb{Q}$, which shall end the proof.

Let $U \subseteq \mathbb{R}$ be an open subset including $[0,1] \cap \mathbb{Q}$. Then $F = [0,1] \setminus U \subseteq [0,1] \setminus \mathbb{Q}$ is compact, whence there exists $g \in \omega^\omega$ such that $\varphi^{-1}[F] \subseteq \prod_{n\in\omega} g(n) \subseteq \omega^\omega$. As X is an unbounded subset of ω^ω, there is $\alpha \in \omega_1$ such that $f_\alpha \not\leq^* g$. Note that $X \cap \varphi^{-1}[F] \subseteq X \cap \prod_{n\in\omega} g(n) \subseteq \{f_\xi : \xi < \alpha\}$ is countable; therefore, so is $\varphi[X] \cap F = \varphi[X] \setminus U = Y \setminus U$. □

REMARK 3.12. The reader surely noticed that the argument in the previous proof is practically the same we used in Theorem 1.5 for showing that our set had strong measure zero. The difference is that, in Theorem 1.5, the set of reals we constructed was not exactly concentrated on a subset of itself: we have embedded Z in a superset Y in such a way that there existed a countable $C \subseteq Y \setminus Z$ such that $|Z \setminus U| \leq \aleph_0$ for every open subset U of Y satisfying $C \subseteq U$. Indeed, it could not be the case that Z is concentrated on a countable subset of itself, as Z is not Rothberger. In a sense, this parallels the idea of topological measure zero characterizing strong measure zero: we make use of open sets of a superspace in order to conclude that the set we are working on has strong measure zero. This phenomenon will also occur in Theorem 4.1.

The second result is Lemma 4 of [**31**]:

LEMMA 3.13 (Rothberger [**31**]). *Every strong measure zero subset of $\mathbb{R}$ of size less than $\mathfrak{b}$ is Rothberger.*

PROOF. Let X be a strong measure zero subset of $\mathbb{R}$ with $|X| < \mathfrak{b}$. Let $\langle \mathcal{C}_n : n \in \omega \rangle$ be a sequence of covers of X by open subsets of $\mathbb{R}$. For each $x \in X$ and $n \in \omega$, let $f_x(n) \in \omega \setminus \{0\}$ and $U_x(n) \in \mathcal{C}_n$ be such that $\left(x - \frac{1}{f_x(n)}, x + \frac{1}{f_x(n)}\right) \subseteq U_x(n)$. Since $\{f_x : x \in X\} \subseteq \omega^\omega$ has size less than $\mathfrak{b}$, there is $g \in \omega^\omega$ such that, for each $x \in X$, there is $m_x \in \omega$ with $\{n \in \omega : f_x(n) > g(n)\} \subseteq m_x$.

Now consider a partition $\omega = \dot{\bigcup}_{k\in\omega} A_k$ of ω into infinite sets such that every $k \in \omega$ satisfies $\min(A_k) \geq k$. For each $k \in \omega$, the set $X_k = \{x \in X : m_x = k\}$ has strong measure zero; hence, there is a sequence $\langle x_n : n \in A_k \rangle$ of points of X_k satisfying $X_k \subseteq \bigcup_{n\in A_k} \left(x_n - \frac{1}{g(n)}, x_n + \frac{1}{g(n)}\right)$. Note that, if $x \in X_k$ and $n \in A_k$, then $n \geq k = m_x$, whence $f_x(n) \leq g(n)$; thus, $X_k \subseteq \bigcup_{n\in A_k} \left(x_n - \frac{1}{g(n)}, x_n + \frac{1}{g(n)}\right) \subseteq$

[6]We say that Y is **concentrated** on $A \subseteq Y$ if $|Y \setminus U| \leq \aleph_0$ for every open subset $U \subseteq Y$ satisfying $A \subseteq U$. Every space Y that is concentrated on a countable subset $A = \{a_n : n \in \omega\}$ is Rothberger [**30**]: given a sequence $\langle \mathcal{C}_n : n \in \omega \rangle$ of open covers of Y, one can pick $C_{2n} \in \mathcal{C}_{2n}$ with $a_n \in C_{2n}$ for each $n \in \omega$; note that, as $Y \setminus \bigcup_{n\in\omega} C_{2n}$ is countable, one can now cover all of the points in this set by selecting one element out of each open cover $\mathcal{C}_{2n+1}$ with $n \in \omega$.

$\bigcup_{n\in A_k}\left(x_n - \frac{1}{f_{x_n}(n)}, x_n + \frac{1}{f_{x_n}(n)}\right)$. Therefore,

$$X = \bigcup_{k\in\omega} X_k \subseteq \bigcup_{k\in\omega}\bigcup_{n\in A_k}\left(x_n - \frac{1}{f_{x_n}(n)}, x_n + \frac{1}{f_{x_n}(n)}\right) =$$

$$= \bigcup_{n\in\omega}\left(x_n - \frac{1}{f_{x_n}(n)}, x_n + \frac{1}{f_{x_n}(n)}\right) \subseteq \bigcup_{n\in\omega} U_{x_n}(n),$$

thus the sequence $\langle U_{x_n}(n) : n \in \omega\rangle$ witnesses the fact that X is Rothberger. □

Now we only have to put these ingredients together:

PROOF OF THEOREM 3.9. Suppose that the R-Borel Conjecture holds. By Theorem 3.11, it follows that $\mathfrak{b} > \aleph_1$. Now Lemma 3.13 implies that a strong measure zero subset of $\mathbb{R}$ of size $\aleph_1$ would be Rothberger, thus contradicting the R-Borel Conjecture. Hence, the Borel Conjecture must hold. □

Another connection between the Rothberger property and strong measure zero was proven by Fremlin and Miller in [**10**, Theorem 1]:

THEOREM 3.14 (Fremlin–Miller [**10**]). *Let X be a metrizable topological space. The following are equivalent:*

(1) *X is Rothberger;*
(2) *X has strong measure zero with respect to each metric that generates the topology of X.*

We can present the proof of this theorem in a more general setting. Let us first mention that Rothberger T_3 spaces and strong measure zero metric spaces share another feature: both imply zero-dimensionality.

THEOREM 3.15 (Scheepers–Tall [**39**]). *Every Rothberger T_3 space is zero-dimensional.*

PROOF. Let X be a Rothberger T_3 space. In particular, X is a Lindelöf T_3 space, hence Tychonoff; therefore, X is homeomorphic to a subspace of $[0,1]^\kappa$ for some κ. We may then assume that $X \subseteq [0,1]^\kappa$.

For each $\alpha \in \kappa$, the projection of X in the α-th coordinate is a continuous function; hence, its image is a Rothberger — in particular, a strong measure zero — subset of $[0,1]$. By a classical theorem due to Szpilrajn-Marczewski [**42**, Theorem 2], every strong measure zero metric space is zero-dimensional; therefore, the image of X under the projection on each coordinate is a zero-dimensional subset of $[0,1]$. Thus X is a subspace of a product of zero-dimensional spaces, whence X is itself zero-dimensional. □

Let X be a topological space. We say that a pseudometric ρ on X is **admissible** if the open balls defined by ρ are open in the topology of X. For this next result only, let us relax the definition of strong measure zero from metric spaces to pseudometric spaces.

THEOREM 3.16 (Fremlin–Miller [**10**]). *Let X be a T_3 space. The following are equivalent:*

(1) *X is Rothberger;*
(2) *X is Lindelöf, zero-dimensional and $\langle X,\rho\rangle$ has strong measure zero for every admissible pseudometric ρ on X.*

PROOF. (1) $\to$ (2). Suppose that X is Rothberger. Then X is zero-dimensional by Theorem 3.15. Given a sequence $\langle \varepsilon_n : n \in \omega \rangle$ of positive real numbers, note that $\mathcal{C}_n = \{B_\rho(x, \frac{\varepsilon_n}{2}) : x \in X\}$ is an open cover of X — therefore, the Rothberger property implies that $\langle X, \rho \rangle$ has strong measure zero.

(2) $\to$ (1). Let $\langle \mathcal{C}_n : n \in \omega \rangle$ be a sequence of open covers of X. As X is zero-dimensional and Lindelöf, we may assume that each $\mathcal{C}_n$ is countable and consists of clopen sets; therefore, we may further assume that these sets are pairwise disjoint. We may also assume that each $\mathcal{C}_{n+1}$ is a refinement of $\mathcal{C}_n$. Then, note that the function $\rho : X \times X \to \mathbb{R}_{\geq 0}$ defined by $\rho(x, y) = \frac{1}{n+1}$ if $n = \min\{k \in \omega :$ there is a $C \in \mathcal{C}_k$ such that $x \in C$ and $y \notin C\}$ if there is such an n, and $\rho(x, y) = 0$ otherwise, is an admissible pseudometric on X. Finally, note that a set whose diameter is at most $\frac{1}{n+1}$ must be a subset of some element of $\mathcal{C}_n$. Therefore, the result follows from applying strong measure zero to the sequence $\left\langle \frac{1}{n+1} : n \in \omega \right\rangle$. □

The proof of Theorem 3.14 is analogous to the previous one, with the extra care of making the refinements in the final part having also diameter going to zero — so we obtain a metric, not only a pseudometric. It may be noted that the only thing we needed from "Lindelöf and zero-dimensional" in the proof of Theorem 3.16 is the fact that every open cover can be refined to a cover whose elements are pairwise disjoint clopen sets — so we could have replaced "Lindelöf and zero-dimensional" by "ultraparacompact" in the statement.

We also remark that Theorems 1.5 and 3.14 together imply that, consistently (which is the best one can hope for, in view of Theorem 1.2), strong measure zero is not a topological property: If X is a non-Rothberger set of reals having strong measure zero, then, by Theorem 3.14, there is a metric on X that preserves its topology but in which X does not have strong measure zero; thus, the identity function on X witnesses that strong measure zero is (consistently) not preserved by homeomorphisms.

4. The topological measure zero game (and the weak Rothberger game on subspaces of the Alexandroff double)

We now turn to topological measure zero and its game version. For a subset $X \subseteq Y$, we write $\mathcal{O}_X(Y) = \{\mathcal{V} \in \mathcal{O}(Y) : X \subseteq \bigcup \mathcal{V}\}$. We shall also write $\mathcal{O}_X$ in place of $\mathcal{O}_X(Y)$ when the ambient space being considered is clear from the context.

The next result and its corollary follow from Theorems 9 and 10 of [**37**].

THEOREM 4.1 (Scheepers [**37**]). *Let Y be a σ-compact metric space and $X \subseteq Y$. The following are equivalent:*

(1) *X has strong measure zero;*
(2) ALICE *does not have a winning strategy in the game $\mathsf{G}_1(\mathcal{O}, \mathcal{O}_X)$ on Y;*
(3) *Y satisfies $\mathsf{S}_1(\mathcal{O}, \mathcal{O}_X)$.*

COROLLARY 4.2 (Scheepers [**37**]). *Let X be a σ-totally-bounded metric space and Y be the completion of X. The following are equivalent:*

(1) *X has strong measure zero;*
(2) ALICE *does not have a winning strategy in the game $\mathsf{G}_1(\mathcal{O}, \mathcal{O}_X)$ on Y;*
(3) *Y satisfies $\mathsf{S}_1(\mathcal{O}, \mathcal{O}_X)$.*

PROOF. Write $X = \bigcup_{j\in\omega} F_j$, where each F_j is totally bounded and the sequence $\langle F_j : j \in \omega\rangle$ is increasing. For each $j \in \omega$, the completion G_j of F_j (that is, the closure of F_j in Y) is also totally bounded, hence compact; thus, $Y = \bigcup_{j\in\omega} G_j$ is σ-compact. Now apply Theorem 4.1. □

It is worth mentioning that Theorem 2.3 is also a direct consequence of Corollary 4.2: one only has to note that every strategy for ALICE in $MS(X)$ naturally produces a strategy for ALICE in $\mathsf{G}_1(\mathcal{O}, \mathcal{O}_X)$ in any superspace Y of X, since for every $\varepsilon \in \mathbb{R}_{>0}$ the set of all open balls of radius $\frac{\varepsilon}{2}$ is a cover of Y by open sets of diameter at most ε. (This is essentially the argument that will appear in the proof of (3) $\to$ (1) in Proposition 4.5.)

For the proof of Theorem 4.1, we will first tackle a simpler case:

LEMMA 4.3. *Let Y be a compact metric space and X be a strong measure zero subset of Y. Then* ALICE *does not have a winning strategy in the game $\mathsf{G}_1(\mathcal{O}, \mathcal{O}_X)$ on Y.*

PROOF. Let $\theta : \tau^{<\omega} \to \mathcal{O}$ be a strategy for ALICE in $\mathsf{G}_1(\mathcal{O}, \mathcal{O}_X)$ played on Y, where τ is the topology of Y. As Y is compact, we may assume that every open cover produced by θ is finite, since BOB may always choose a finite subcover to select from anyway. Let $\mathcal{U}_0 = \theta(\langle\rangle)$. Now let[7] $\mathcal{U}_1 = \bigwedge_{V\in\theta(\langle\rangle)} \theta(\langle V\rangle)$. In general, let

$$\mathcal{U}_n = \bigwedge_{s\in S_n} \theta(s)$$

where S_n is the collection of all $s \in \tau^n$ such that

$$\langle\theta(s \restriction 0), s(0), \theta(s \restriction 1), s(1), \theta(s \restriction 2), s(2), \ldots, \theta(s \restriction (n-1)), s(n-1)\rangle$$

is a partial play of this game — that is, we are looking at all the (finitely many) possible partial plays of length n in which ALICE is following θ, and then putting the corresponding sequences of BOB's moves in S_n. Since S_n is finite (as every open cover given by θ is finite), $\mathcal{U}_n$ is an open cover of Y. For each $n \in \omega$, let ε_n be a Lebesgue number for $\mathcal{U}_n$. As X has strong measure zero, there is a sequence $\langle A_n : n \in \omega\rangle$ of subsets of X such that $\bigcup_{n\in\omega} A_n = X$ and each A_n has diameter at most ε_n.

Now we will define a play of the game $\mathsf{G}_1(\mathcal{O}, \mathcal{O}_X)$. ALICE starts with $\theta(\langle\rangle) = \mathcal{U}_0$. By the choice of ε_0, there is $U_0 \in \theta(\langle\rangle)$ such that $A_0 \subseteq U_0$; let then $V_0 = U_0$ be BOB's response in the initial inning. Then ALICE's next move is $\theta(\langle V_0\rangle)$. By the choice of ε_1, there is $U_1 \in \mathcal{U}_1$ such that $A_1 \subseteq U_1$; now, as $\mathcal{U}_1$ is a refinement of $\theta(\langle V_0\rangle)$, there is $V_1 \in \theta(\langle V_0\rangle)$ such that $U_1 \subseteq V_1$. Let V_1 be BOB's answer, and so on. In general, in the n-th inning, ALICE's choice $\theta(\langle V_0, \ldots, V_{n-1}\rangle)$ is refined by $\mathcal{U}_n$ with Lebesgue number ε_n, so BOB may select $V_n \in \theta(\langle V_0, \ldots, V_{n-1}\rangle)$ with $A_n \subseteq V_n$. Thus we obtain a play

$$\langle\theta(\langle\rangle), V_0, \theta(\langle V_0\rangle), V_1, \theta(\langle V_0, V_1\rangle), V_2, \ldots\rangle$$

in which ALICE follows θ and loses — since $X = \bigcup_{n\in\omega} A_n \subseteq \bigcup_{n\in\omega} V_n$. Therefore, θ is not winning. □

Now we are going to use the idea of the *proof* of Lemma 4.3 to prove Theorem 4.1:

[7] For $I \neq \emptyset$ and $\langle \mathcal{C}_i : i \in I\rangle$, we define $\bigwedge_{i\in I} \mathcal{C}_i = \left\{\bigcap_{i\in I} A_i : \langle A_i : i \in I\rangle \in \prod_{i\in I} \mathcal{C}_i\right\}$.

PROOF OF THEOREM 4.1. It is clear that the implications (2) $\to$ (3) and (3) $\to$ (1) hold for every metric space Y and every subset $X \subseteq Y$. It remains to prove that, under the hypothesis that Y is σ-compact, the implication (1) $\to$ (2) holds as well.

Write $Y = \bigcup_{n\in\omega} K_n$ with each K_n compact. Now let $\theta : \tau^{<\omega} \to \mathcal{O}$ be a strategy for ALICE in $\mathsf{G}_1(\mathcal{O}, \mathcal{O}_X)$ played on Y, where τ is the topology of Y. We will show that this strategy can be defeated. Since Y is not compact, we cannot mimic the proof of Lemma 4.3 yet — there, it was important that each open cover given by ALICE was finite. So our first task will be to fix some options for BOB in advance, in order to make finite the relevant open covers provided by ALICE. To accomplish that, first let the set ω be partitioned into infinitely many infinite sets as $\omega = \dot{\bigcup}_{n\in\omega} M_n$. For each $j \in \omega$, let $n(j) \in \omega$ be such that $j \in M_{n(j)}$.

Let $\mathcal{F}(\langle\rangle) \subseteq \theta(\langle\rangle)$ be finite and such that $K_{n(0)} \subseteq \bigcup \mathcal{F}(\langle\rangle)$, and set $\mathcal{U}_0 = \mathcal{F}(\langle\rangle)$. Now, for each $U_0 \in \mathcal{F}(\langle\rangle)$, let $\mathcal{F}(\langle U_0\rangle) \subseteq \theta(\langle U_0\rangle)$ be finite and such that $K_{n(1)} \subseteq \bigcup \mathcal{F}(\langle U_0\rangle)$; then, define $\mathcal{U}_1 = \bigwedge_{U_0\in\mathcal{F}(\langle\rangle)} \mathcal{F}(\langle U_0\rangle)$ — which is a finite open cover of $K_{n(1)}$. In the next step, let $S_2 = \{\langle U_0, U_1\rangle \in \tau^2 : U_0 \in \mathcal{F}(\langle\rangle) \text{ and } U_1 \in \mathcal{F}(\langle U_0\rangle)\}$ and, for each $\langle U_0, U_1\rangle \in S_2$, let $\mathcal{F}(\langle U_0, U_1\rangle) \subseteq \theta(\langle U_0, U_1\rangle)$ be finite and such that $K_{n(2)} \subseteq \bigcup \mathcal{F}(\langle U_0, U_1\rangle)$. Define, then, $\mathcal{U}_2 = \bigwedge_{\langle U_0,U_1\rangle\in S_2} \mathcal{F}(\langle U_0, U_1\rangle)$ — which is a finite open cover of $K_{n(2)}$, as S_2 is finite and $\mathcal{F}(\langle U_0, U_1\rangle)$ is finite for every $\langle U_0, U_1\rangle \in S_2$. Next, define $S_3 = \{\langle U_0, U_1, U_2\rangle \in \tau^3 : U_0 \in \mathcal{F}(\langle\rangle) \text{ and } U_1 \in \mathcal{F}(\langle U_0\rangle) \text{ and } U_2 \in \mathcal{F}(\langle U_0, U_1\rangle)\}$ and, for each $\langle U_0, U_1, U_2\rangle \in S_3$, pick a finite $\mathcal{F}(\langle U_0, U_1, U_2\rangle) \subseteq \theta(\langle U_0, U_1, U_2\rangle)$ with $K_{n(3)} \subseteq \bigcup \mathcal{F}(\langle U_0, U_1, U_2\rangle)$; then, we set $\mathcal{U}_3 = \bigwedge_{\langle U_0,U_1,U_2\rangle\in S_3} \bigcup \mathcal{F}(\langle U_0, U_1, U_2\rangle)$. And so forth.

By proceeding in this fashion, we obtain a sequence $\langle \mathcal{U}_j : j \in \omega\rangle$ of finite families of open subsets of Y such that each $\mathcal{U}_j$ covers $K_{n(j)}$. For each $j \in \omega$, let then ε_j be a Lebesgue number for the open cover $\{U \cap K_{n(j)} : U \in \mathcal{U}_j\}$ of $K_{n(j)}$. For each $n \in \omega$, it follows from (1) that the set $X \cap K_n$ has strong measure zero; hence, there is a sequence $\langle A_j : j \in M_n\rangle$ of subsets of $X \cap K_n$ such that $\bigcup_{j\in M_n} A_j = X \cap K_n$ and each A_j has diameter at most ε_j.

Now define a play of $\mathsf{G}_1(\mathcal{O}, \mathcal{O}_X)$ on Y as follows. ALICE begins the play with $\theta(\langle\rangle)$; making use of the choice of ε_0, BOB picks $U_0 \in \mathcal{U}_0 \subseteq \theta(\langle\rangle)$ such that $A_0 \subseteq U_0 \cap K_{n(0)} \subseteq U_0$, and then plays the set $V_0 = U_0 \in \mathcal{F}(\langle\rangle)$. Now ALICE's next move is $\theta(\langle V_0\rangle)$; by the choice of ε_1, there is $U_1 \in \mathcal{U}_1$ such that $A_1 \subseteq U_1 \cap K_{n(1)} \subseteq U_1$; then, using the fact that $\mathcal{U}_1$ is a refinement of $\mathcal{F}(\langle V_0\rangle)$, BOB replies with $V_1 \in \mathcal{F}(\langle V_0\rangle) \subseteq \theta(\langle V_0\rangle)$ such that $V_1 \supseteq U_1$. In the next inning, ALICE's move is $\theta(\langle V_0, V_1\rangle)$; again, there is $U_2 \in \mathcal{U}_2$ such that $A_2 \subseteq U_2 \cap K_{n(2)} \subseteq U_2$, so — as $\mathcal{U}_2$ is a refinement of $\mathcal{F}(\langle V_0, V_1\rangle)$ — BOB can respond with $V_2 \in \mathcal{F}(\langle V_0, V_1\rangle) \subseteq \theta(\langle V_0, V_1\rangle)$ such that $V_2 \supseteq U_2$. And so on.

Thus, we obtain a play of $\mathsf{G}_1(\mathcal{O}, \mathcal{O}_X)$ on Y in which ALICE follows the strategy θ and loses, as the sequence $\langle V_j : j \in \omega\rangle$ of sets played by BOB satisfies $\bigcup_{j\in\omega} V_j \supseteq \bigcup_{j\in\omega} A_j = \bigcup_{n\in\omega}\bigcup_{j\in M_n} A_j = \bigcup_{n\in\omega} X \cap K_n = X \cap \bigcup_{n\in\omega} K_n = X \cap Y = X$. Therefore, θ is not a winning strategy. $\square$

We will now aim at obtaining further equivalences of BC in terms of topological games. In the remaining results of this section, we will refer to the hyperspace defined as follows.

For a topological space Y, let $A(Y)$ be the Alexandroff double of Y; this is the set $Y \times \{0, 1\}$ endowed with the topology in which the points in $Y \times \{1\}$ are isolated and the basic open sets meeting $Y \times \{0\}$ are of the form $(V \times \{0\}) \cup ((V \setminus \{y\}) \times \{1\})$ for V an open subset of Y and $y \in Y$ — for a set B of this form, the point $\langle y, 1\rangle$ will be called the *centre* of B, and denoted by $c(B)$. If $U \subseteq A(Y)$ is an open set, we will write U^0 to mean the (open) subset of Y satisfying $U \cap (Y \times \{0\}) = U^0 \times \{0\}$.

PROPOSITION 4.4. *Let Y be a compact metric space and $X \subseteq Y$. The following are equivalent:*

(1) BOB *has a winning strategy in* $\mathsf{G}_1(\mathcal{O}, \mathcal{O}_X)$ *played on* Y*;*
(2) BOB *has a winning strategy in* $\mathsf{G}_1(\mathcal{O}, \mathcal{O}_{X\times\{1\}})$ *played on* $A(Y)$*;*
(3) BOB *has a winning strategy in* $MS(X)$*.*

PROOF. (1) → (3). Let $\theta : \mathcal{O}^{<\omega} \setminus \{\langle\rangle\} \to \tau$ be a winning strategy for BOB in the game $\mathsf{G}_1(\mathcal{O}, \mathcal{O}_X)$ played on Y, where τ is the topology of Y. We can now define a winning strategy ϱ for BOB in $MS(X)$ as follows.

If ALICE starts the play of $MS(X)$ by playing $\varepsilon_0 \in \mathbb{R}_{>0}$, BOB responds with $\varrho(\langle\varepsilon_0\rangle) = \theta(\langle\mathcal{U}_0\rangle) \cap X$, where $\mathcal{U}_0 = \{B(y, \frac{\varepsilon_0}{2}) : y \in Y\}$. If ALICE replies with $\varepsilon_1 \in \mathbb{R}_{>0}$ in the next inning, then BOB plays $\varrho(\langle\varepsilon_0, \varepsilon_1\rangle) = \theta(\langle\mathcal{U}_0, \mathcal{U}_1\rangle) \cap X$, where $\mathcal{U}_1 = \{B(y, \frac{\varepsilon_1}{2}) : y \in Y\}$. In general, after the n-th inning, if $\varrho(\langle\varepsilon_0, \ldots, \varepsilon_n\rangle) = \theta(\langle\mathcal{U}_0, \ldots, \mathcal{U}_n\rangle) \cap X$ — where $\mathcal{U}_k = \{B(y, \frac{\varepsilon_k}{2}) : y \in Y\}$ for every $k \leq n$ —, then, for each $\varepsilon_{n+1} \in \mathbb{R}_{>0}$ that can be played by ALICE in the next inning of $MS(X)$, BOB replies with $\varrho(\langle\varepsilon_0, \ldots, \varepsilon_{n+1}\rangle) = \theta(\langle\mathcal{U}_0, \ldots, \mathcal{U}_{n+1}\rangle) \cap X$, where $\mathcal{U}_{n+1} = \{B(y, \frac{\varepsilon_{n+1}}{2}) : y \in Y\}$. It is clear that this is a valid strategy for BOB in $MS(X)$, and BOB wins by playing in this fashion since θ is a winning strategy in $\mathsf{G}_1(\mathcal{O}, \mathcal{O}_X)$.

(3) → (2). Let $\theta : (\mathbb{R}_{>0})^{<\omega} \setminus \{\langle\rangle\} \to \wp(X)$ be a winning strategy for BOB in $MS(X)$. Define a strategy ϱ for BOB in $\mathsf{G}_1(\mathcal{O}, \mathcal{O}_{X\times\{1\}})$ played on $A(Y)$ as follows.

First, note that we can assume that ALICE plays only basic open covers of $A(Y)$, and let us fix for each basic open cover $\mathcal{U}$ of $A(Y)$ a Lebesgue number $\varepsilon(\mathcal{U})$ for the cover $\{U^0 : U \in \mathcal{U}\}$ of Y.

If ALICE begins the play of $\mathsf{G}_1(\mathcal{O}, \mathcal{O}_{X\times\{1\}})$ with a basic open cover $\mathcal{U}_0$ of $A(Y)$, let BOB pick $\varrho(\langle\mathcal{U}_0\rangle) = U_0 \in \mathcal{U}_0$ such that $\theta(\langle\varepsilon(\mathcal{U}_0)\rangle) \subseteq U_0^0$. Now, if $\mathcal{U}_1$ is the cover played by ALICE in the next inning, BOB will cover the point $c(U_0)$ with $\varrho(\langle\mathcal{U}_0, \mathcal{U}_1\rangle) = U_1 \in \mathcal{U}_1$. The play proceeds with a basic open cover $\mathcal{U}_2$ of $A(Y)$ played by ALICE, to which BOB responds by selecting $\varrho(\langle\mathcal{U}_0, \mathcal{U}_1, \mathcal{U}_2\rangle) = U_2 \in \mathcal{U}_2$ satisfying $\theta(\langle\varepsilon(\mathcal{U}_0), \varepsilon(\mathcal{U}_1)\rangle) \subseteq U_2^0$. Now BOB covers $c(U_2)$ in the next inning with $\varrho(\langle\mathcal{U}_0, \mathcal{U}_1, \mathcal{U}_2, \mathcal{U}_3\rangle) = U_3 \in \mathcal{U}_3$, and so forth. In general, in the $2k$-th inning, if $\langle\mathcal{U}_i : i \leq 2k\rangle$ is the sequence of basic open covers played by ALICE thus far, BOB will play $\varrho(\langle\mathcal{U}_i : i \leq 2k\rangle) = U_{2k} \in \mathcal{U}_{2k}$ such that $\theta(\langle\varepsilon(\mathcal{U}_{2j}) : j \leq k\rangle) \subseteq U_{2k}^0$; next, in the $(2k+1)$-th inning, BOB's move $\varrho(\langle\mathcal{U}_i : i \leq 2k+1\rangle) = U_{2k+1} \in \mathcal{U}_{2k+1}$ will be chosen so that $c(U_{2k}) \in U_{2k+1}$.

As θ is a winning strategy for BOB in $MS(X)$, it follows that, at the end of a play of $\mathsf{G}_1(\mathcal{O}, \mathcal{O}_{X\times\{1\}})$ in which BOB follows ϱ,

$$\begin{aligned}\bigcup_{n\in\omega} \varrho(\langle\mathcal{U}_i : i \leq n\rangle) = \bigcup_{n\in\omega} U_n &= \bigcup_{k\in\omega}(U_{2k} \cup U_{2k+1}) \supseteq \\ &\supseteq \bigcup_{k\in\omega}(\theta(\langle\varepsilon(\mathcal{U}_{2j}) : j \leq k\rangle) \times \{1\}) \supseteq X \times \{1\};\end{aligned}$$

hence, ϱ is a winning strategy for BOB.

(2) → (1). Left to the reader. (It follows straight from the fact that, if $\mathcal{U}$ is an open cover of Y, then $\{U \times \{0, 1\} : U \in \mathcal{U}\}$ is an open cover of $A(Y)$.) □

PROPOSITION 4.5. *Let Y be a compact metric space and $X \subseteq Y$. The following are equivalent:*

(1) ALICE *has a winning strategy in* $\mathsf{G}_1(\mathcal{O}, \mathcal{O}_X)$ *played on* Y*;*
(2) ALICE *has a winning strategy in* $\mathsf{G}_1(\mathcal{O}, \mathcal{O}_{X\times\{1\}})$ *played on* $A(Y)$*;*
(3) ALICE *has a winning strategy in* $MS(X)$*.*

PROOF. (2) $\to$ (3). Let τ be the topology of $A(Y)$, and let $\mathcal{O}$ stand for $\mathcal{O}_{A(Y)}$. Fix a winning strategy $\theta : \tau^{<\omega} \to \mathcal{O}$ for ALICE in the game $\mathsf{G}_1(\mathcal{O}, \mathcal{O}_{X\times\{1\}})$ played on $A(Y)$; we may assume that θ provides only basic open covers of $A(Y)$. Now define a strategy ϱ for ALICE in $MS(X)$ as follows.

ALICE kicks off by playing a Lebesgue number $\varrho(\langle\rangle)$ for the open cover $\{U^0 : U \in \theta(\langle\rangle)\}$ of Y. If BOB's move is $A_0 \subseteq X$, pick $U_0 \in \theta(\langle\rangle)$ with $A_0 \subseteq U_0^0$, and let $U_1 \in \theta(\langle U_0\rangle)$ be such that $c(U_0) \in U_1$. In the next inning, let ALICE play a Lebesgue number $\varrho(\langle A_0\rangle)$ for the open cover $\{U^0 : U \in \theta(\langle U_0, U_1\rangle)\}$ of Y. If BOB responds with $A_1 \subseteq X$ and $U_2 \in \theta(\langle U_0, U_1\rangle)$ is such that $A_1 \subseteq U_2^0$, pick $U_3 \in \theta(\langle U_0, U_1, U_2\rangle)$ with $c(U_2) \in U_3$; and so on. In general, in the k-th inning of $MS(X)$, assume that $\langle A_i : i < k\rangle \in \wp(X)^k$ and $\langle U_j : j < 2k\rangle \in \tau^{2k}$ are such that $\varrho(\langle A_i : i < k\rangle)$ is a Lebesgue number for the open cover $\{U^0 : U \in \theta(\langle U_j : j < 2k\rangle)\}$ of Y. If BOB then plays $A_k \subseteq X$ in the k-th inning of $MS(X)$, choose $U_{2k} \in \theta(\langle U_j : j < 2k\rangle)$ with $A_k \subseteq U_{2k}^0$, and then choose $U_{2k+1} \in \theta(\langle U_j : j \leq 2k\rangle)$ with $c(U_{2k}) \in U_{2k+1}$; finally, let $\varrho(\langle A_i : i \leq k\rangle)$ be a Lebesgue number for $\{U^0 : U \in \theta(\langle U_j : j < 2k+2\rangle)\}$.

By proceeding in this fashion, we obtain a play

$$\langle \varrho(\langle\rangle), A_0, \varrho(\langle A_0\rangle), A_1, \varrho(\langle A_0, A_1\rangle), A_2, \ldots\rangle$$

of $MS(X)$ in which ALICE makes use of the strategy ϱ, and also a play

$$\langle \theta(\langle\rangle), U_0, \theta(\langle U_0\rangle), U_1, \theta(\langle U_0, U_1\rangle), U_2, \theta(\langle U_0, U_1, U_2\rangle), U_3, \theta(\langle U_0, U_1, U_2, U_3\rangle), U_4, \ldots\rangle$$

of $\mathsf{G}_1(\mathcal{O}, \mathcal{O}_{X\times\{1\}})$ on $A(Y)$ in which ALICE employs the strategy θ. As θ is a winning strategy, we have $X \times \{1\} \not\subseteq \bigcup_{n\in\omega} U_n$. Now, since $A_k \times \{0,1\} \subseteq U_{2k}^0 \times \{0,1\} \subseteq U_{2k} \cup U_{2k+1}$ for each $k \in \omega$, it follows that $X \times \{0,1\} \not\subseteq \bigcup_{k\in\omega} A_k \times \{0,1\}$, i.e. $X \not\subseteq \bigcup_{k\in\omega} A_k$. Hence, ϱ is a winning strategy for ALICE in $MS(X)$.

(3) $\to$ (1). Let $\theta : \wp(X)^{<\omega} \to \mathbb{R}_{>0}$ be a winning strategy for ALICE in $MS(X)$. For each $\varepsilon \in \mathbb{R}_{>0}$, let $\mathcal{U}_\varepsilon = \{B(y, \frac{\varepsilon}{2}) : y \in Y\}$. A winning strategy ϱ for ALICE in the game $\mathsf{G}_1(\mathcal{O}, \mathcal{O}_X)$ played on Y can then be defined as follows.

First, ALICE begins the play with $\varrho(\langle\rangle) = \mathcal{U}_{\theta(\langle\rangle)}$. If BOB responds with $U_0 \in \varrho(\langle\rangle)$, then ALICE plays $\varrho(\langle U_0\rangle) = \mathcal{U}_{\theta(\langle U_0 \cap X\rangle)}$ in the next inning. If BOB's reply is now $U_1 \in \varrho(\langle U_0\rangle)$, then ALICE's next move is $\varrho(\langle U_0, U_1\rangle) = \mathcal{U}_{\theta(\langle U_0\cap X, U_1\cap X\rangle)}$; and so forth. In general, ϱ can be defined recursively by declaring $\varrho(s^\frown U_n) = \mathcal{U}_{\theta(t^\frown (U_n\cap X))}$, where s and t are such that $\varrho(s) = \mathcal{U}_{\theta(t)}$.

Thus we obtain a play

$$\langle \theta(\langle\rangle), U_0 \cap X, \theta(\langle U_0 \cap X\rangle), U_1 \cap X, \theta(\langle U_0 \cap X, U_1 \cap X\rangle), \ldots\rangle$$

of $MS(X)$ in which ALICE plays according to θ. Since θ is a winning strategy for ALICE in $MS(X)$, it follows that $X \not\subseteq \bigcup_{n\in\omega}(U_n \cap X)$, whence $X \not\subseteq \bigcup_{n\in\omega} U_n$. Therefore, ϱ is a winning strategy for ALICE in the game $\mathsf{G}_1(\mathcal{O}, \mathcal{O}_X)$ played on Y.

(1) $\to$ (2). Left to the reader. □

For $X \subseteq Y$, we will denote by $T(X, Y)$ the subspace $(Y \times \{0\}) \cup (X \times \{1\})$ of $A(Y)$. Note that, if Y is compact T_2, then $T(X, Y)$ is compact T_2 for every $X \subseteq Y$.

For a topological space Y with topology τ, we write $\mathcal{D}(Y) = \{\mathcal{V} \subseteq \tau : \overline{\bigcup \mathcal{V}} = Y\}$ — and, as usual, we write $\mathcal{D}$ instead of $\mathcal{D}(Y)$ whenever possible. The game $\mathsf{G}_1(\mathcal{O}, \mathcal{D})$ is a game version of the weak Rothberger property introduced by Daniels in [**8**]; this game has been first introduced by Tkachuk in [**46**] (under a different yet equivalent formulation) and later by Babinkostova, Pansera and Scheepers in [**3**].

The following easy lemma is left to the reader.

LEMMA 4.6. *If Y is a topological space and $X \subseteq Y$ is dense in Y, then the game $\mathsf{G}_1(\mathcal{O}, \mathcal{O}_{X\times\{1\}})$ played on $A(Y)$ is equivalent*[8] *to the game $\mathsf{G}_1(\mathcal{O}, \mathcal{D})$ played on $T(X, Y)$.*

We remark that, in the same spirit in which Theorem 4.1 was proven by adapting the proof of Lemma 4.3, both of Propositions 4.4 and 4.5 could have been done for σ-compact metric spaces instead. Briefly:

- For (3) $\to$ (2) in Proposition 4.4 (which is the only implication where compactness was required), let θ be a winning strategy for BOB in $MS(X)$. Write $Y = \bigcup_{n\in\omega} K_n$ with K_n compact for every $n \in \omega$, and consider a partition $\omega = \dot{\bigcup}_{n\in\omega} M_n$ with M_n infinite for every $n \in \omega$; let each M_n be enumerated increasingly as $M_n = \{m(n, j) : j \in \omega\}$. For every $n \in \omega$ and every basic open cover $\mathcal{U}$ of $A(Y)$, fix a Lebesgue number $\varepsilon_n(\mathcal{U})$ for the cover $\{U^0 \cap K_n : U \in \mathcal{U}\}$ of K_n. Now define a strategy for BOB in the game $\mathsf{G}_1(\mathcal{O}, \mathcal{O}_{X\times\{1\}})$ played on $A(Y)$ as follows. For each $k \in \omega$, if $n_k, j_k \in \omega$ are such that $k = m(n_k, j_k)$, let BOB respond to a basic open cover $\mathcal{U}_{2k}$ played by ALICE in the $2k$-th inning by selecting $U_{2k} \in \mathcal{U}_{2k}$ with $\theta(\langle \varepsilon_{n_k}(\mathcal{U}_{2m(n_k,i)}) : i \leq j_k\rangle) \cap K_{n_k} \subseteq U^0_{2k}$, and then covering $c(U_{2k})$ in the next inning by selecting U_{2k+1} from the open cover $\mathcal{U}_{2k+1}$ played by ALICE. At the end of the play, we will have

$$\begin{aligned}
\bigcup_{n\in\omega} U_n \quad &= \bigcup_{k\in\omega}(U_{2k} \cup U_{2k+1}) \supseteq \\
&\supseteq \bigcup_{k\in\omega}(\theta(\langle \varepsilon_{n_k}(\mathcal{U}_{2m(n_k,i)}) : i \leq j_k\rangle) \cap K_{n_k}) \times \{1\} = \\
&= \bigcup_{n\in\omega} \bigcup_{k\in M_n}(\theta(\langle \varepsilon_n(\mathcal{U}_{2m(n,i)}) : i \leq j_k\rangle) \cap K_n) \times \{1\} \supseteq \\
&= \bigcup_{n\in\omega} \bigcup_{j\in\omega}(\theta(\langle \varepsilon_n(\mathcal{U}_{2m(n,i)}) : i \leq j\rangle) \cap K_n) \times \{1\} \supseteq \\
&= \bigcup_{n\in\omega}(K_n \cap \bigcup_{j\in\omega}(\theta(\langle \varepsilon_n(\mathcal{U}_{2m(n,i)}) : i \leq j\rangle)) \times \{1\} \supseteq \\
&\supseteq \bigcup_{n\in\omega}(K_n \cap X) \times \{1\} = \\
&= X \times \{1\};
\end{aligned}$$

  hence, the strategy just described is a winning one for BOB.
- For extending Proposition 4.5 to the case in which Y is σ-compact, we will make use of previous results. By Theorems 2.3 and 4.1, both of (1) and (3) are equivalent to "X does not have strong measure zero". Now

[8] We say that two games G and G' played between ALICE and BOB are **equivalent games** if either ALICE has a winning strategy in both games, or BOB has a winning strategy in both games, or both games are undetermined.

the observation that (1) and (2) are equivalent (regardless of whether or not Y is σ-compact) is left to the reader.[9]

Furthermore, as the completion of every σ–totally-bounded metric space is σ-compact, the above remark together with Lemma 4.6 yield:

COROLLARY 4.7. *Let X be a σ–totally-bounded metric space and Y be its completion. Then the following games are equivalent:*

- $MS(X)$;
- $\mathsf{G}_1(\mathcal{O}, \mathcal{O}_X)$ *on* Y;
- $\mathsf{G}_1(\mathcal{O}, \mathcal{D})$ *on* $T(X, Y)$.

Thus, we have more game-theoretical equivalences of BC. The equivalences of BC with statements (3) and (4) in the next result are essentially due to Scheepers ([**37**] and [**35**] respectively).

COROLLARY 4.8 (Galvin–Mycielski–Solovay [**12**], Scheepers [**35**, **37**]). *The following statements are equivalent:*

(1) BC;
(2) *the game $MS(X)$ is determined on every σ–totally-bounded metric space;*
(3) *for every σ–totally-bounded metric space X, the game $\mathsf{G}_1(\mathcal{O}, \mathcal{O}_X)$ played on the completion of X is determined;*
(4) *for every σ–totally-bounded metric space X with completion Y, the game $\mathsf{G}_1(\mathcal{O}, \mathcal{D})$ played on $T(X, Y)$ is determined.*

PROOF. It is immediate from Corollary 4.7 that (2) $\leftrightarrow$ (3) $\leftrightarrow$ (4), and the equivalence (1) $\leftrightarrow$ (2) was done in Corollary 2.4. □

We end this section by mentioning (without proof) a recent result from [**48**] that characterizes a metric space having strong measure zero without recurring to a superspace or a hyperspace of it. In order to state it, let us call an open cover $\mathcal{C}$ of a metric space X an **uniform cover** (resp. **uniform n-cover**, where $n \in \omega$) if there exists $\varepsilon \in \mathbb{R}_{>0}$ such that $\{B(x, \varepsilon) : x \in X\}$ (resp. $\{\bigcup_{x \in F} B(x, \varepsilon) : F \in [X]^n\}$) is a refinement of $\mathcal{C}$; we also call $\mathcal{C}$ a **uniform ω-cover** if $\mathcal{C}$ is a uniform n-cover for every $n \in \omega$. For a fixed metric space X, the sets of all uniform covers of X, all uniform n-covers of X and all uniform ω-covers of X are denoted, respectively, by $\mathcal{O}^{\mathrm{unif}}$, $\mathcal{O}_n^{\mathrm{unif}}$ and Ω^{unif}. Finally, for a sequence $\langle \mathcal{A}_n : n \in \omega \rangle$ of nonempty families of nonempty sets, we extend Definition 3.1 and write $\mathsf{S}_1(\langle \mathcal{A}_n \rangle, \mathcal{B})$ to abbreviate the following statement:

For every sequence $\langle A_n : n \in \omega \rangle$ such that $A_n \in \mathcal{A}_n$ for all $n \in \omega$, there is a sequence $\langle x_n : n \in \omega \rangle$ satisfying

- $x_n \in A_n$ *for all* $n \in \omega$
  and
- $\{x_n : n \in \omega\} \in \mathcal{B}$.

Likewise, we can extend Definition 3.2 and introduce the game $\mathsf{G}_1(\langle \mathcal{A}_n \rangle, \mathcal{B})$ similarly. We then have:

THEOREM 4.9 (Zindulka [**48**]). *Let X be a σ–totally-bounded metric space. The following conditions are equivalent:*

[9] The implication (2) $\to$ (1) can be obtained by adapting the same idea used in the argument presented for (2) $\to$ (3) — that is, to play two consecutive innings $2k$ and $2k+1$ of $\mathsf{G}_1(\mathcal{O}_{A(Y)}, \mathcal{O}_{X \times \{1\}})$ in order to define ALICE's move in the $(k+1)$-th inning of $\mathsf{G}_1(\mathcal{O}_Y, \mathcal{O}_X)$.

(1) Alice *does not have a winning strategy in the game* $\mathsf{G}_1(\mathcal{O}^{\mathrm{unif}}, \mathcal{O})$ *played on* X*;*
(2) X *satisfies* $\mathsf{S}_1(\mathcal{O}^{\mathrm{unif}}, \mathcal{O})$*;*
(3) Alice *does not have a winning strategy in the game* $\mathsf{G}_1(\langle \mathcal{O}_n^{\mathrm{unif}} \rangle, \mathcal{O})$ *played on* X*;*
(4) X *satisfies* $\mathsf{S}_1(\langle \mathcal{O}_n^{\mathrm{unif}} \rangle, \mathcal{O})$*;*
(5) Alice *does not have a winning strategy in the game* $\mathsf{G}_1(\Omega^{\mathrm{unif}}, \mathcal{O})$ *played on* X*;*
(6) X *satisfies* $\mathsf{S}_1(\Omega^{\mathrm{unif}}, \mathcal{O})$*;*
(7) X *has strong measure zero.*

5. The Rothberger-boundedness game

We will now turn to a game naturally associated with the property of Rothberger-boundedness. In this work, by "topological group" we will always mean "T_0 topological group" — equivalently (by a result due to Pontryagin [**47**]), "$T_{3\frac{1}{2}}$ topological group". For a group $\langle G, * \rangle$, $x \in G$ and $A, B \subseteq G$, we also write $A^{-1} = \{a^{-1} : a \in A\}$, $x * A = \{x * a : a \in A\}$ and $A * B = \bigcup_{a \in A} a * B$.

We start with a rewriting of Theorem 33 of [**21**] stated in the particular case of topological groups. It may be seen as an analogue of the Lebesgue Covering Lemma, and its role in the next theorem will be essentially the same as the role of the Lebesgue Covering Lemma in Theorem 4.1.

Lemma 5.1 (e.g. Kelley [**21**]). *Let* K *be a compact subset of a topological group* G *with identity element* e*, and let* $\mathcal{C}$ *be a cover of* K *by sets open in* G*. Then there is an open neighbourhood* U *of* e *in* G *such that, for every* $y \in K$*, the set* $y * U$ *is included in some element of* $\mathcal{C}$*. Furthermore,* U *may be taken symmetric — i.e. such that* $U^{-1} = U$*.*

Proof. For each $x \in K$, choose $C_x \in \mathcal{C}$ with $x \in C_x$, and then fix an open neighbourhood V_x of e in G such that $x * V_x * V_x \subseteq C_x$. As K is compact, there exists a finite $F \subseteq K$ satisfying $\bigcup_{x \in F} x * V_x \supseteq K$; define, then, $U = \bigcap_{x \in F} V_x$.

Now let $y \in K$ be arbitrary. By picking $x' \in F$ such that $y \in x' * V_{x'}$, we obtain $y * U \subseteq x' * V_{x'} * U \subseteq x' * V_{x'} * V_{x'} \subseteq C_{x'} \in \mathcal{C}$.

Finally, in order to obtain a symmetric neighbourhood of e, take $U^{-1} \cap U$ instead of U. □

Let G be a topological group with identity element e, and fix a subset $X \subseteq G$. The game $RB(X, G)$ we will consider is, as one might expect, defined as follows: in each inning $n \in \omega$, Alice plays an open neighbourhood V_n of e in G, and then Bob chooses $x_n \in G$; the winner is Bob if $\bigcup_{n \in \omega} x_n * V_n \supseteq X$, and Alice otherwise. Note that $RB(X, G)$ is equivalent to the game $\mathsf{G}_1(\mathcal{O}_{\mathrm{nbhd}}, \mathcal{O}_X)$ played on G, where $\mathcal{O}_{\mathrm{nbhd}}$ is the family of all of the covers of form $\mathcal{U}(V) = \{x * V : x \in G\}$ for $V \subseteq G$ an open neighbourhood of e. Note also that $\mathsf{S}_1(\mathcal{O}_{\mathrm{nbhd}}, \mathcal{O}_X)$ is equivalent to the statement that X is a Rothberger-bounded subset of G.

The next result provides a characterization of Rothberger-boundedness in terms of these games.

Theorem 5.2 (Scheepers [**38**]). *Let* G *be a* σ*-compact topological group and* $X \subseteq G$*. The following conditions are equivalent:*

(1) X *is Rothberger-bounded;*

(2) ALICE *does not have a winning strategy in the game* $RB(X, G)$*;*
(3) ALICE *does not have a winning strategy in the game* $\mathsf{G}_1(\mathcal{O}_{\mathrm{nbhd}}, \mathcal{O}_X)$ *on* G.

PROOF. We have already noted that (2) $\leftrightarrow$ (3), and (2) $\rightarrow$ (1) is clear. It remains to prove that (1) $\rightarrow$ (2).

Let $\theta : G^{<\omega} \to \tau_e$ be a strategy for ALICE in $RB(X, G)$, where $\tau_e = \{U \subseteq G : U \text{ is open and } e \in U\}$. Let also $\tau'_e = \{U \in \tau_e : U = U^{-1}\}$. Write $G = \bigcup_{n\in\omega} K_n$ with each K_n compact, and split the set of innings into infinitely many infinite sets as $\omega = \dot{\bigcup}_{n\in\omega} M_n$. For each $j \in \omega$, let $n(j) \in \omega$ be such that $j \in M_{n(j)}$.

We will now define recursively $\langle V_j : j \in \omega\rangle$, $\langle F_j : j \in \omega\rangle$ and $\langle W_j : j \in \omega\rangle$ as follows. Let $j \in \omega$ be such that $\langle V_i : i < j\rangle$, $\langle F_i : i < j\rangle$ and $\langle W_i : i < j\rangle$ have already been defined so that each F_i is a finite subset of $K_{n(i)}$. Define $V_j = \bigcap_{s\in\prod_{i<j} F_i} \theta(s)$. Then $\{x * V_j : x \in K_{n(j)}\}$ is an open cover of $K_{n(j)}$; thus, as $K_{n(j)}$ is compact, there is a finite $F_j \subseteq K_{n(j)}$ such that $K_{n(j)} \subseteq \bigcup_{x\in F_j}(x * V_j)$. Now, by Lemma 5.1, there is $W_j \in \tau'_e$ such that, for every $y \in K_{n(j)}$, there is $x \in F_j$ satisfying $y * W_j * W_j \subseteq x * V_j$.

For each $n \in \omega$, we have that $X \cap K_n$ is a Rothberger-bounded subset of G; thus, there is a sequence $\langle y_j : j \in M_n\rangle$ of elements of G with $\bigcup_{j\in M_n} y_j * W_j \supseteq X \cap K_n$.

<u>Claim.</u> For every $j \in \omega$ there is $z_j \in F_j$ such that $(y_j * W_j) \cap K_{n(j)} \subseteq z_j * V_j$.

PROOF OF CLAIM. If $y_j \in K_{n(j)}$, the claim is a direct consequence of the choice of W_j, since $W_j \subseteq W_j * W_j$. Suppose, then, that $y_j \notin K_{n(j)}$. Let $\hat{y} \in (y_j * W_j) \cap K_{n(j)}$ be arbitrary — if there is no such $\hat{y}$, then the claim is trivially true. We will first show that $y_j * W_j \subseteq \hat{y} * W_j * W_j$.

Indeed, let $w \in y_j * W_j$ be arbitrary. Then $y_j^{-1} * w \in W_j$. In addition, $y_j^{-1} * \hat{y} \in W_j$; hence, as W_j is symmetric, $\hat{y}^{-1} * y_j \in W_j$. Thus, $\hat{y}^{-1} * w = (\hat{y}^{-1} * y_j) * (y_j^{-1} * w) \in W_j * W_j$, so $w \in \hat{y} * W_j * W_j$, as required.

Now, as $\hat{y} \in K_{n(j)}$, from the choice of W_j we have that there is $z_j \in F_j$ such that $y_j * W_j \subseteq \hat{y} * W_j * W_j \subseteq z_j * V_j$, and the claim follows. $\square$

From the above claim we obtain a play

$$\langle \theta(\langle\rangle), z_0, \theta(\langle z_0\rangle), z_1, \theta(\langle z_0, z_1\rangle), z_2, \ldots\rangle$$

of $RB(X, G)$ in which ALICE follows θ and loses — since

$$\begin{aligned}
\textstyle\bigcup_{j\in\omega} z_j * \theta(\langle z_i : i < j\rangle) \supseteq\ & \textstyle\bigcup_{j\in\omega} z_j * V_j \supseteq \\
\supseteq\ & \textstyle\bigcup_{j\in\omega} (y_j * W_j) \cap K_{n(j)} = \\
=\ & \textstyle\bigcup_{n\in\omega} \bigcup_{j\in M_n} (y_j * W_j) \cap K_{n(j)} = \\
=\ & \textstyle\bigcup_{n\in\omega} (\bigcup_{j\in M_n} y_j * W_j) \cap K_n \supseteq \\
\supseteq\ & \textstyle\bigcup_{n\in\omega} X \cap K_n = \\
=\ & X.
\end{aligned}$$

Therefore, θ is not a winning strategy. $\square$

We can also characterize the existence of a winning strategy for BOB in these games by paralleling Theorem 2.2:

PROPOSITION 5.3. *Let G be a first-countable topological group and $X \subseteq G$. The following conditions are equivalent:*

(1) *X is countable;*
(2) BOB *has a winning strategy in the game $RB(X,G)$;*
(3) BOB *has a winning strategy in the game $\mathsf{G}_1(\mathcal{O}_{\text{nbhd}}, \mathcal{O}_X)$ on G.*

PROOF. It is clear that (1) $\to$ (2) $\leftrightarrow$ (3). We will prove that (2) $\to$ (1).

Fix a countable local base $\mathcal{N} = \{V_n : n \in \omega\}$ for e in G with $V_{n+1} \subseteq V_n$ for all $n \in \omega$. We may assume that ALICE only chooses elements of $\mathcal{N}$ when playing the game $RB(X,G)$. So let $\theta : \mathcal{N}^{<\omega} \to G$ be a winning strategy for BOB in $RB(X,G)$, where τ is the topology of G. As θ is a winning strategy, for each $x \in X$ there is $s_x \in \mathcal{N}^{<\omega}$ such that $x \in \bigcap_{n\in\omega} \theta(s_x{}^\frown V_n) * V_n$ — otherwise, if $\bar{x} \in X$ were a counterexample to this, we could build a play of $RB(X,G)$ in which BOB plays according to θ and $\bar{x}$ is never covered, thus contradicting the assumption that θ is a winning strategy. We will show that the map $x \mapsto s_x$ is one-to-one, which will then finish the proof.

Let $x, y \in G$ be such that $s_x = s_y$. For each $n \in \omega$, define $z_n = \theta(s_x{}^\frown V_n) \in G$. We then have $\{x,y\} \subseteq \bigcap_{n\in\omega} z_n * V_n$. Thus $z_n^{-1} * x \in V_n$ for all $n \in \omega$, whence $z_n^{-1} * x \xrightarrow{n} e$, that is, $z_n^{-1} \xrightarrow{n} x^{-1}$; similarly, $z_n^{-1} \xrightarrow{n} y^{-1}$. As we are assuming that our topological groups are Hausdorff, this yields $x^{-1} = y^{-1}$, i.e. $x = y$. □

Now, the next result yields a statement about topological groups that is equivalent to the Borel Conjecture — which we will put together with the previous results in this section to obtain yet another game-theoretical statement equivalent to BC.

THEOREM 5.4 (Galvin–Scheepers [**13**]). *The following statements are equivalent:*

(1) BC*;*
(2) *every Rothberger-bounded subset of a second-countable topological group is countable;*
(3) *every Rothberger-bounded subset of a first-countable topological group is countable.*

PROOF. It is clear that (3) $\to$ (2) $\to$ (1). Furthermore, since every Hausdorff first-countable group admits a left-invariant metric that generates its topology (this is the Birkhoff–Kakutani Theorem [**4**, **20**]), the implication (1) $\to$ (3) follows from Theorem 1.2. □

Therefore, we have:

COROLLARY 5.5. *The following statements are equivalent:*

(1) BC*;*
(2) *the game $RB(X,G)$ is determined for every first-countable σ-compact topological group G and every $X \subseteq G$;*
(3) *on every first-countable σ-compact topological group G, the game $\mathsf{S}_1(\mathcal{O}_{\text{nbhd}}, \mathcal{O}_X)$ is determined for every $X \subseteq G$.*

PROOF. As before, (2) $\leftrightarrow$ (3) is clear. Now, in view of Theorem 5.2 and Proposition 5.3, (2) is equivalent to the statement "for every first-countable topological group G and every $X \subseteq G$, either X is not Rothberger-bounded or X is countable", which in turn is equivalent to (1) by Theorem 5.4. $\square$

6. A summary of game-theoretical equivalences of BC

We are now ready to gather the game-theoretical equivalences of BC involving the games discussed in the previous results.

THEOREM 6.1. *The following statements are equivalent:*

(1) BC*;*
(2) *the game* $MS(X)$ *is determined on every* σ*–totally-bounded metric space;* (Galvin–Mycielski–Solovay [**12**])
(2′) *the game* $MS(X)$ *is determined on every* $X \subseteq \mathbb{R}$*;* (Galvin–Mycielski–Solovay [**12**])
(3) *the game* $\mathsf{G}_1(\mathcal{O}, \mathcal{O})$ *is determined on every metric space;* (Pawlikowski [**29**])
(3′) *the game* $\mathsf{G}_1(\mathcal{O}, \mathcal{O})$ *is determined on every* $X \subseteq \mathbb{R}$*;* (Pawlikowski [**29**])
(4) *the point-open game is determined on every metric space;* (Pawlikowski [**29**])
(4′) *the point-open game is determined on every* $X \subseteq \mathbb{R}$*;* (Pawlikowski [**29**])
(5) *for every* σ*–totally-bounded metric space* X*, the game* $\mathsf{G}_1(\mathcal{O}, \mathcal{O}_X)$ *played on the completion of* X *is determined;* (Scheepers [**37**])
(5′) *the game* $\mathsf{G}_1(\mathcal{O}, \mathcal{O}_X)$ *played on* $\overline{X}$ *is determined for every* $X \subseteq \mathbb{R}$*;* (Scheepers [**37**])
(5″) *the game* $\mathsf{G}_1(\mathcal{O}, \mathcal{O}_X)$ *played on* $\mathbb{R}$ *is determined for every* $X \subseteq \mathbb{R}$*;* (Scheepers [**37**])
(6) *for every* σ*–totally-bounded metric space* X *with completion* Y*, the game* $\mathsf{G}_1(\mathcal{O}, \mathcal{D})$ *played on* $T(X, Y)$ *is determined;* (Scheepers [**35**])
(6′) *the game* $\mathsf{G}_1(\mathcal{O}, \mathcal{D})$ *played on* $T(X, \overline{X})$ *is determined for every* $X \subseteq \mathbb{R}$*;* (Scheepers [**35**])
(6″) *the game* $\mathsf{G}_1(\mathcal{O}, \mathcal{D})$ *played on* $T(X, \mathbb{R})$ *is determined for every dense* $X \subseteq \mathbb{R}$*;* (Scheepers [**35**])
(7) *the game* $RB(X, G)$ *is determined for every first-countable* σ*-compact topological group* G *and every* $X \subseteq G$*;* (essentially, Scheepers [**38**])
(7′) *the game* $RB(X, \mathbb{R})$ *is determined for every* $X \subseteq \mathbb{R}$*;* (Galvin–Mycielski–Solovay [**12**])
(8) *on every first-countable* σ*-compact topological group* G*, the game* $\mathsf{S}_1(\mathcal{O}_{\mathrm{nbhd}}, \mathcal{O}_X)$ *is determined for every* $X \subseteq G$*;* (essentially, Scheepers [**38**])
(8′) *the game* $\mathsf{S}_1(\mathcal{O}_{\mathrm{nbhd}}, \mathcal{O}_X)$ *on* $\mathbb{R}$ *is determined for every* $X \subseteq \mathbb{R}$. (Galvin–Mycielski–Solovay [**12**])

PROOF. We already have:

- (1) $\leftrightarrow$ (2) $\leftrightarrow$ (2′) by Corollary 2.4;
- (3) $\leftrightarrow$ (4) and (3′) $\leftrightarrow$ (4′) by Theorem 3.6;
- (1) $\leftrightarrow$ (3′) by Corollary 3.10;
- (2′) $\leftrightarrow$ (5′) $\leftrightarrow$ (6′) by Corollary 4.7;
- (1) $\leftrightarrow$ (5) $\leftrightarrow$ (6) by Corollary 4.8;
- (1) $\leftrightarrow$ (7) $\leftrightarrow$ (8) by Corollary 5.5.

Now it is clear that (3) $\to$ (3′), (6′) $\to$ (6″) and (7′) $\leftrightarrow$ (8′), and it is easy to see that (5′) $\leftrightarrow$ (5″). Note also that, for every $X \subseteq \mathbb{R}$, the games $MS(X)$ and $RB(X, \mathbb{R})$

are equivalent; this yields $(7') \leftrightarrow (2')$. Thus, we will be done once we verify that $(1) \to (3)$ and $(6'') \to (1)$.

$(1) \to (3)$. Assume (1), and let X be a metric space such that ALICE does not have a winning strategy in the game $\mathsf{G}_1(\mathcal{O},\mathcal{O})$ on X. Thus X is Rothberger, hence has strong measure zero. By Theorem 1.2, X is countable, which implies that BOB has a winning strategy in $\mathsf{G}_1(\mathcal{O},\mathcal{O})$ on X.

$(6'') \to (1)$. Let X be a strong measure zero subset of $\mathbb{R}$. Then $X \cup \mathbb{Q} \subseteq \mathbb{R}$ also has strong measure zero, and it follows from Theorem 2.3 and Corollary 4.7 that ALICE does not have a winning strategy in the game $\mathsf{G}_1(\mathcal{O},\mathcal{D})$ played on $T(X\cup\mathbb{Q},\mathbb{R})$. By $(6'')$, BOB must then have a winning strategy in $\mathsf{G}_1(\mathcal{O},\mathcal{D})$ played on $T(X\cup\mathbb{Q},\mathbb{R})$, and hence in $MS(X\cup\mathbb{Q})$ — again by Corollary 4.7. Thus, in view of Theorem 2.2, $X\cup\mathbb{Q}$ is countable; therefore, X is countable. □

7. The cut-and-choose game

In this section we will focus on a game that deals with partitions of a cardinal κ. There are several variations of this game (see e.g. [**14**]), but we will be interested here only in some specific cases of the game. Given a cardinal κ, the **cut-and-choose game** $\mathsf{CC}(\kappa)$ is played between ALICE and BOB as follows: in each inning $n \in \omega$, ALICE chooses a finite partition $\mathcal{P}_n$ of κ, and then BOB chooses $P_n \in \mathcal{P}_n$. BOB is declared the winner if $\bigcup_{n\in\omega} P_n = \kappa$, and ALICE is declared the winner otherwise.

The Borel Conjecture says that the strong measure zero sets of reals are the countable ones; in case of its failure, the least cardinality of a non–strong-measure-zero set may vary. So, let $\mathrm{non}(\mathcal{SMZ})$ be the least cardinal λ such that the there is a subset of the real line with cardinality λ that does not have strong measure zero. Of course, $\mathrm{non}(\mathcal{SMZ}) \geq \aleph_1$, and BC implies $\mathrm{non}(\mathcal{SMZ}) = \aleph_1$. A strict inequality is consistent with ZFC; see e.g. [**28**].

The proof presented here for the following theorem makes use of the relation between the strong measure zero property and properties related to the Rothberger property discussed in previous sections — for a more direct proof, we refer the reader to the original paper [**33**].

THEOREM 7.1 (Scheepers [**33**]). *Let κ be an infinite cardinal. The following are equivalent:*

(1) $\kappa < \mathrm{non}(\mathcal{SMZ})$;
(2) ALICE *does not have a winning strategy in the game* $\mathsf{CC}(\kappa)$;
(3) *for every sequence $\langle \mathcal{P}_n : n\in\omega\rangle$ of finite partitions of κ, there is a sequence $\langle P_n : n\in\omega\rangle$ such that each $P_n \in \mathcal{P}_n$ and $\bigcup_{n\in\omega} P_n = \kappa$.*

PROOF. $(1) \to (2)$. Let θ be a strategy for ALICE in $\mathsf{CC}(\kappa)$; we must prove that θ is not winning. By considering common refinements, we may assume that the partitions played according to θ depend only on the inning number — i.e. they do not depend upon the history of the play so far[10]. Let $g(n)$ be the size of the partition given by θ in the inning n, and let $\{P^n_i : i < g(n)\}$ be an enumeration of

[10]This argument is very similar to the one in the proof of Lemma 4.3. If θ is a strategy for ALICE in $\mathsf{CC}(\kappa)$, we may define $\tilde{\theta}(n) = \bigcap_{s\in S_n} \theta(s)$ for each $n\in\omega$, where S_n is the collection of all the (finitely many) $s \in \wp(\kappa)^n$ such that $\langle \theta(s\restriction 0), s(0), \theta(s\restriction 1), s(1), \dots, \theta(s\restriction (n-1)), s(n-1)\rangle$ is a partial play of $\mathsf{CC}(\kappa)$. If there is a play of $\mathsf{CC}(\kappa)$ in which ALICE follows θ and wins, then playing the partition $\tilde{\theta}(n)$ in each inning n of $\mathsf{CC}(\kappa)$ constitutes a winning strategy for ALICE.

this partition. Let $T = \{f \in \omega^\omega : f(n) < g(n) \text{ for all } n \in \omega\} = \prod_{n\in\omega} g(n)$, which is a compact subset of ω^ω. For each $\xi < \kappa$, let $f_\xi \in T$ be such that $\xi \in P^n_{f_\xi(n)}$ for every $n \in \omega$. Now define $X = \{f_\xi : \xi < \kappa\}$ and consider the strategy for ALICE in the game $\mathsf{G}_1(\mathcal{O}_T, \mathcal{O}_X)$ in which, in the inning $n \in \omega$, ALICE plays the cover $\{V^n_i : i < g(n)\}$, where $V^n_i = \{f \in T : f(n) = i\}$. By Theorem 4.1, this strategy is not winning; therefore, there is an $h : \omega \to \omega$ such that $\bigcup_{n\in\omega} V^n_{h(n)} \supseteq X$. We now claim that $\kappa \subseteq \bigcup_{n\in\omega} P^n_{h(n)}$, whence θ is not a winning strategy. Indeed, given $\xi < \kappa$, let $n \in \omega$ be such that $f_\xi \in V^n_{h(n)}$; then $\xi \in P^n_{h(n)}$ by the definition of f_ξ.

(2) $\to$ (3). This is clear.

(3) $\to$ (1). Let $X \subseteq \mathbb{R}$ be a space such that $|X| = \kappa$. Without loss of generality we may assume that $X \subseteq K \subseteq \mathbb{R}$ with K compact. Therefore, by Theorem 4.1, we only need to prove that $\mathsf{S}_1(\mathcal{O}_K, \mathcal{O}_X)$ holds. Let $\langle \mathcal{V}_n : n \in \omega\rangle$ be a sequence of finite open covers for K. Write, for each n, $\mathcal{V}_n = \{V^n_0, \ldots, V^n_{g(n)}\}$. For each $n \in \omega$ and $k \leq g(n)$, let $P^n_k = V^n_k \setminus \bigcup_{j<k} V^n_j$. Note that for each n, $\{P^n_0, \ldots, P^n_{g(n)}\}$ is a partition for X. By hypothesis, there is a function f such that $X \subseteq \bigcup_{n\in\omega} P^n_{f(n)} \subseteq \bigcup_{n\in\omega} V^n_{f(n)}$. □

8. Strong measure zero in every finite power

In this section, we will deal with the property of a (subspace of a) metric space having strong measure zero in all of its finite powers. As strong measure zero is not a topological property — rather, it depends on the metric itself —, we must make clear in advance which, among the various (reasonable) metrics one may consider on products, we will be referring to.

If $n \in \omega$ and $\langle X_0, d_0\rangle, \ldots, \langle X_n, d_n\rangle$ are metric spaces, we will consider on the product $X = \prod_{i=0}^n X_i$ the metric d defined as

$$d(\langle x_0, \ldots, x_n\rangle, \langle y_0, \ldots, y_n\rangle) = \max\{d_0(x_0, y_0), \ldots, d_n(x_n, y_n)\}.$$

It is clear that a subset of X has strong measure zero in the metric d if and only if it has strong measure zero in the (perhaps more customary) Euclidean metric d' defined as

$$d'(\langle x_0, \ldots, x_n\rangle, \langle y_0, \ldots, y_n\rangle) = \sqrt{\sum_{i=0}^{n} (d_i(x_i, y_i))^2},$$

since $d(\bar{x}, \bar{y}) \leq d'(\bar{x}, \bar{y}) \leq \sqrt{n+1} \cdot d(\bar{x}, \bar{y})$ for all $\bar{x}, \bar{y} \in \prod_{i=0}^n X_i$.

Of course, it is consistent that a metric space has strong measure zero if and only if it has strong measure zero in every finite power, since both of these classes equal the class of countable metric spaces under the Borel Conjecture. However, it it also consistent that the property "strong measure zero in every finite power" is strictly stronger than strong measure zero:

THEOREM 8.1 (Sierpiński [**41**]). (1) *If $A \subseteq \mathbb{R}$ is such that $A + A = \{a + b : a, b \in A\} \subseteq \mathbb{R}$ does not have strong measure zero, then $A^2 \subseteq \mathbb{R}^2$ does not either.*

(2) CH *implies the existence of a Luzin set $L \subseteq \mathbb{R}$ such that $L + L = \mathbb{R}$.*

PROOF. (1) This follows directly from the fact that $A + A$ is the image of A^2 under the uniformly continuous function $\langle x, y\rangle \mapsto x + y$, together with the (quite straightforward) observation that every uniformly continuous image of a strong measure zero set has strong measure zero.

(2) Using CH, enumerate $\mathbb{R} = \{r_\alpha : \alpha \in \omega_1\}$ and $\{F \subseteq \mathbb{R} : F$ is a closed nowhere dense set$\} = \{F_\alpha : \alpha \in \omega_1\}$ bijectively.[11] Now, recursively for $\alpha \in \omega_1$, pick x_α in the complement of the meagre set

$$\Big(\bigcup_{\beta<\alpha}(F_\beta \cup \{x_\beta, y_\beta\})\Big) \cup \Big(r_\alpha - \bigcup_{\beta<\alpha}(F_\beta \cup \{x_\beta, y_\beta\})\Big),$$

and then define $y_\alpha = r_\alpha - x_\alpha$; note that, as $x_\alpha \notin r_\alpha - \bigcup_{\beta<\alpha}(F_\beta \cup \{x_\beta, y_\beta\})$, it follows that $y_\alpha = r_\alpha - x_\alpha \notin \bigcup_{\beta<\alpha}(F_\beta \cup \{x_\beta, y_\beta\})$.

Now let $L = \{x_\alpha : \alpha \in \omega_1\} \cup \{y_\alpha : \alpha \in \omega_1\}$ — note that $|L| = \aleph_1$ by construction. Since every $\alpha \in \omega_1$ satisfies $r_\alpha = x_\alpha + y_\alpha$, we have $L + L = \mathbb{R}$. Furthermore, as $x_\alpha, y_\alpha \in \mathbb{R} \setminus (\bigcup_{\beta<\alpha}(F_\beta \cup \{x_\beta, y_\beta\}))$ for each α, it follows that $L \cap F_\alpha \subseteq \{x_\beta : \beta \leq \alpha\} \cup \{y_\beta : \beta \leq \alpha\}$ — therefore, $|L \cap F_\alpha| \leq \aleph_0$ for each $\alpha \in \omega_1$. Hence, L is a Luzin set. □

This suggests the following question — which, to the best of our knowledge, is open:

PROBLEM 8.2. Is it consistent with ZFC that BC fails and yet every subset of $\mathbb{R}$ that has strong measure zero in every finite power is countable?

We will now turn to game-theoretical characterizations of the property of all finite powers having strong measure zero.

Let Y be a topological space and $C(Y) = \{f \in \mathbb{R}^Y : f$ is continuous$\}$. For each nonempty $X \subseteq Y$, we denote by $C_p^X(Y)$ the set $C(Y)$ endowed with the topology generated by the basic open sets

$$V(f, F, \varepsilon) = \{g \in C(Y) : \forall x \in F\ (|g(x) - f(x)| < \varepsilon)\},$$

where $f \in C(Y)$, $F \in [X]^{<\aleph_0}$ and $\varepsilon \in \mathbb{R}_{>0}$. If $X = Y$, this is the usual space $C_p(Y)$, i.e. $C(Y)$ endowed with the pointwise convergence topology — which is the same as the subspace topology inherited from the product $\mathbb{R}^Y$. For each $f \in C(Y)$, we write

$$\Omega_f^X = \{A \subseteq C(Y) : f \in \overline{A} \text{ in } C_p^X(Y)\}.$$

As usual, we omit the superscript if $X = Y$. We also denote by $\underline{0}$ the function in $C(Y)$ satisfying $\underline{0}[Y] = \{0\}$.

The following theorem and its corollary follow from Theorem 3 of [**36**] and Theorem 12 of [**37**]. Here, for a topological space Y, we denote by Ω the set of all **ω-covers** of Y, i.e. $\Omega = \{\mathcal{U} \in \mathcal{O}(Y) : \forall F \in [Y]^{<\aleph_0}\ \exists U \in \mathcal{U}\ (F \subseteq U)\}$. For a subset X of Y, we also write $\Omega_X = \{\mathcal{U} \in \mathcal{O}(Y) : \forall F \in [X]^{<\aleph_0}\ \exists U \in \mathcal{U}\ (F \subseteq U)\}$.

THEOREM 8.3 (Scheepers [**36, 37**]). *Let Y be a compact topological space and $X \subseteq Y$. The following conditions are equivalent:*

(1) ALICE *does not have a winning strategy in the game* $\mathsf{G}_1(\Omega_{\underline{0}}, \Omega_{\underline{0}}^X)$ *played on* $C(Y)$*;*
(2) $C(Y)$ *satisfies* $\mathsf{S}_1(\Omega_{\underline{0}}, \Omega_{\underline{0}}^X)$*;*
(3) ALICE *does not have a winning strategy in the game* $\mathsf{G}_1(\Omega, \Omega_X)$ *played on* Y*;*
(4) Y *satisfies* $\mathsf{S}_1(\Omega, \Omega_X)$*;*

[11] As $\mathbb{R}$ is second-countable, there are $\mathfrak{c}$ many open subsets of $\mathbb{R}$, hence $\mathfrak{c}$ many closed subsets of $\mathbb{R}$. Since every singleton is nowhere dense in $\mathbb{R}$, it follows that there are $\mathfrak{c}$ many closed nowhere dense subsets of $\mathbb{R}$.

(5) *for every $k \in \omega$, the product Y^k satisfies* $\mathsf{S}_1(\mathcal{O}, \mathcal{O}_{X^k})$.

PROOF. The implications (1) $\to$ (2) and (3) $\to$ (4) are immediate.

(1) $\to$ (3). First, for each finite set $F \subseteq Y$ and each open set $U \subseteq Y$ with $F \subseteq U$, fix $h_{F,U} \in C(Y)$ such that $F \subseteq h_{F,U}^{-1}[\{0\}]$ and $Y \setminus U \subseteq h_{F,U}^{-1}[\{1\}]$.

Let $\theta : \tau^{<\omega} \to \Omega$ be a strategy for ALICE in $\mathsf{G}_1(\Omega, \Omega_X)$ played on Y. We will use θ to define a strategy $\varrho : C(Y)^{<\omega} \to \Omega_{\underline{0}}$ for ALICE in the game $\mathsf{G}_1(\Omega_{\underline{0}}, \Omega_{\underline{0}}^X)$ played on $C(Y)$, by proceeding as follows.

ALICE starts the play with

$$\varrho(\langle\rangle) = \{h_{F,U} : F \in [Y]^{<\aleph_0} \text{ and } U \in \theta(\langle\rangle) \text{ and } F \subseteq U\} \in \Omega_{\underline{0}}.$$

If BOB responds by choosing $f_0 = h_{F_0,U_0} \in \varrho(\langle\rangle)$, ALICE's move in the next inning is

$$\varrho(\langle f_0\rangle) = \{h_{F,U} : F \in [Y]^{<\aleph_0} \text{ and } U \in \theta(\langle U_0\rangle) \text{ and } F \subseteq U\} \in \Omega_{\underline{0}}.$$

Similarly, if BOB chooses $f_1 = h_{F_1,U_1} \in \varrho\langle f_0\rangle$, ALICE's next move is

$$\varrho(\langle f_0, f_1\rangle) = \{h_{F,U} : F \in [Y]^{<\aleph_0} \text{ and } U \in \theta(\langle U_0, U_1\rangle) \text{ and } F \subseteq U\} \in \Omega_{\underline{0}};$$

and so on. In general, we define recursively

$$\varrho(\langle h_{F_j,U_j} : j < n\rangle) = \{h_{F,U} : F \in [Y]^{<\aleph_0} \text{ and } U \in \theta(\langle U_j : j < n\rangle) \text{ and } F \subseteq U\} \in \Omega_{\underline{0}}.$$

By (1), ϱ is not a winning strategy; there must then exist a play

$$\langle \varrho(\langle\rangle), f_0, \varrho(\langle f_0\rangle), f_1, \varrho(\langle f_0, f_1\rangle), f_2, \ldots\rangle$$

of $\mathsf{G}_1(\Omega_{\underline{0}}, \Omega_{\underline{0}}^X)$ played on $C(Y)$ in which ALICE makes use of the strategy ϱ and loses. We will conclude that θ is not a winning strategy for ALICE by showing that the corresponding play

$$\langle \theta(\langle\rangle), U_0, \theta(\langle U_0\rangle), U_1, \theta(\langle U_0, U_1\rangle), U_2, \ldots\rangle$$

of $\mathsf{G}_1(\Omega, \Omega_X)$ played on Y is won by BOB — where, for each $n \in \omega$, the open set $U_n \in \theta(\langle U_0, \ldots, U_{n-1}\rangle)$ is such that $f_n = h_{F_n,U_n}$ for some finite set $F_n \subseteq Y$.

Let $F \subseteq X$ be finite. As $\{f_n : n \in \omega\} \in \Omega_{\underline{0}}^X$, there is $n \in \omega$ such that $f_n \in V(\underline{0}, F, 1)$ — that is, such that $|f_n(x)| < 1$ for all $x \in F$. Then $F \subseteq f_n^{-1}[(-1,1)] \subseteq f_n^{-1}[(-\infty,1)]$; now, as $Y \setminus U_n \subseteq h_{F_n,U_n}^{-1}[\{1\}] = f_n^{-1}[\{1\}] \subseteq f_n^{-1}[[1,\infty)]$, it follows that $F \subseteq f_n^{-1}[(-\infty,1)] \subseteq U_n$. Thus $\{U_n : n \in \omega\} \in \Omega_X$, whence BOB wins.

(2) $\to$ (4). Left to the reader.

(4) $\to$ (5). Let $k \in \omega$ be fixed (which we may assume to be non-zero). By Theorem 4.1, it suffices to show that Y^k satisfies $\mathsf{S}_1(\mathcal{O}, \mathcal{O}_{X^k})$ — which, by a standard argument (see Theorem 17 of [**32**]), is equivalent to $\mathsf{S}_1(\Omega, \mathcal{O}_{X^k})$.

Let then $\langle \mathcal{V}_n : n \in \omega\rangle$ be a sequence of ω-covers of Y^k. For each $n \in \omega$ and each $F \in [Y]^{<\aleph_0}$, fix $V_n(F) \in \mathcal{V}_n$ with $F^k \subseteq V_n(F)$; as F^k is finite, we may assign to each $z \in F$ an open neighbourhood $W_n^F(z)$ of z in Y in such a way that

$$V_n(F) \supseteq \bigcup\Big\{\prod_{i=1}^{k} W_n^F(z_i) : \langle z_1, \ldots, z_k\rangle \in F^k\Big\}$$

— that is, $V_n(F) \supseteq (U_n(F))^k$, where $U_n(F) = \bigcup_{z\in F} W_n^F(z) \subseteq Y$.

As $\{U_n(F) : F \in [Y]^{<\aleph_0}\}$ is an ω-cover of Y for each $n \in \omega$, it follows from (4) that there is a sequence $\langle F_n : n \in \omega\rangle$ of elements of $[Y]^{<\aleph_0}$ that satisfies $\{U_n(F_n) : n \in \omega\} \in \Omega_X$. Note that $X^k \subseteq \bigcup_{n\in\omega} V_n(F_n)$: given $\langle x_1, \ldots, x_k\rangle \in X^k$, if $n \in \omega$ is such that $\{x_1,\ldots,x_k\} \subseteq U_n(F_n)$, then $\langle x_1,\ldots,x_k\rangle \in \{x_1,\ldots,x_k\}^k \subseteq (U_n(F_n))^k \subseteq V_n(F_n) \in \mathcal{V}_n$.

(5) $\to$ (1). First, write $\omega = \bigcup_{m,k\in\omega} N^k_m$ where each N^k_m is infinite and $N^k_m \cap N^{k'}_{m'} = \emptyset$ whenever $\langle m,k\rangle, \langle m',k'\rangle \in \omega\times\omega$ are distinct. For each $\langle m,k\rangle \in \omega\times\omega$ and each $f \in C(Y)$, define

$$V^k_m(f) = \left(f^{-1}\left[\left(-\frac{1}{2^m},\frac{1}{2^m}\right)\right]\right)^k$$

— which is an open subset of Y^k.

Let $\theta : C(Y)^{<\omega} \to \Omega_{\underline{0}}$ be a strategy for ALICE in $\mathsf{G}_1(\Omega_{\underline{0}},\Omega^X_{\underline{0}})$ played on $C(Y)$. We will now make use of θ to define, for each $\langle m,k\rangle \in \omega\times\omega$, a sequence $\langle \mathcal{U}_n : n \in N^k_m\rangle$ of open covers of Y^k.

Let $\langle m_0,k_0\rangle \in \omega\times\omega$ be such that $0 \in N^{k_0}_{m_0}$. As $\theta(\langle\rangle) \in \Omega_{\underline{0}}$, for each $\bar{y} = \langle y_1,\ldots,y_{k_0}\rangle \in Y^{k_0}$ we may choose $f_{\langle\bar{y}\rangle} \in \theta(\langle\rangle)$ such that, for all $i \in \{1,\ldots,k_0\}$, we have $|f_{\langle\bar{y}\rangle}(y_i)| < \frac{1}{2^{m_0}}$; define then $W(\langle\bar{y}\rangle) = V^{k_0}_{m_0}(f_{\langle\bar{y}\rangle})$ — which is an open neighbourhood of $\bar{y}$ in Y^{k_0}. Now Y^{k_0} is compact, hence there is a finite set $F_{\langle\rangle} \subseteq Y^{k_0}$ with $Y^{k_0} \subseteq \bigcup_{\bar{y}\in F_{\langle\rangle}} W(\langle\bar{y}\rangle)$; let then $\mathcal{U}_0 = \{W(\langle\bar{y}\rangle) : \bar{y} \in F_{\langle\rangle}\}$.

Next, let $\langle m_1,k_1\rangle \in \omega\times\omega$ be such that $1 \in N^{k_1}_{m_1}$. For each fixed $\bar{y}_0 \in F_{\langle\rangle}$, we have $\theta\left(\langle f_{\langle\bar{y}_0\rangle}\rangle\right) \in \Omega_{\underline{0}}$; thus, for every $\bar{y} = \langle y_1,\ldots,y_{k_1}\rangle \in Y^{k_1}$ we may fix $f_{\langle\bar{y}_0,\bar{y}\rangle} \in \theta\left(\langle f_{\langle\bar{y}_0\rangle}\rangle\right)$ such that, for all $i \in \{1,\ldots,k_1\}$, we have $|f_{\langle\bar{y}_0,\bar{y}\rangle}(y_i)| < \frac{1}{2^{m_1}}$. Let then $W_1(\langle\bar{y}_0,\bar{y}\rangle) = V^{k_1}_{m_1}(f_{\langle\bar{y}_0,\bar{y}\rangle})$, an open neighbourhood of $\bar{y}$ in Y^{k_1}. As Y^{k_1} is compact, there is a finite set $F_{\langle\bar{y}_0\rangle} \subseteq Y^{k_1}$ with $Y^{k_1} \subseteq \bigcup_{\bar{y}\in F_{\langle\bar{y}_0\rangle}} W(\langle\bar{y}_0,\bar{y}\rangle)$. Having done this for every $\bar{y}_0 \in F_{\langle\rangle}$, let then $\mathcal{U}_1 = \bigwedge_{\bar{y}_0\in F_{\langle\rangle}} \{W(\langle\bar{y}_0,\bar{y}\rangle) : \bar{y} \in F_{\langle\bar{y}_0\rangle}\}$.

In general, given $n \in \omega$, let $\langle m_n,k_n\rangle \in \omega\times\omega$ be such that $n \in N^{k_n}_{m_n}$. For each finite sequence $\langle\bar{y}_0,\ldots,\bar{y}_{n-1}\rangle$ such that $\bar{y}_j \in F_{\langle\bar{y}_0,\ldots,\bar{y}_{j-1}\rangle}$ for all $j < n$, one has $\theta\left(\langle f_{\langle\bar{y}_0\rangle}, f_{\langle\bar{y}_0,\bar{y}_1\rangle}, f_{\langle\bar{y}_0,\bar{y}_1,\bar{y}_2\rangle},\ldots, f_{\langle\bar{y}_0,\ldots,\bar{y}_{n-1}\rangle}\rangle\right) \in \Omega_{\underline{0}}$; therefore, for every $\bar{y} = \langle y_1,\ldots,y_{k_n}\rangle \in Y^{k_n}$ we may fix

$$f_{\langle\bar{y}_0,\ldots,\bar{y}_{n-1},\bar{y}\rangle} \in \theta\left(\langle f_{\langle\bar{y}_0\rangle}, f_{\langle\bar{y}_0,\bar{y}_1\rangle}, f_{\langle\bar{y}_0,\bar{y}_1,\bar{y}_2\rangle},\ldots, f_{\langle\bar{y}_0,\ldots,\bar{y}_{n-1}\rangle}\rangle\right)$$

such that, for all $i \in \{1,\ldots,k_n\}$, we have $|f_{\langle\bar{y}_0,\ldots,\bar{y}_{n-1},\bar{y}\rangle}(y_i)| < \frac{1}{2^{m_n}}$. Define

$$W(\langle\bar{y}_0,\ldots,\bar{y}_{n-1},\bar{y}\rangle) = V^{k_n}_{m_n}(f_{\langle\bar{y}_0,\ldots,\bar{y}_{n-1},\bar{y}\rangle}),$$

which is an open neighbourhood of $\bar{y}$ in Y^{k_n}. Since Y^{k_n} is compact, there is a finite set $F_{\langle\bar{y}_0,\ldots,\bar{y}_{n-1}\rangle} \subseteq Y^{k_n}$ such that

$$Y^{k_n} \subseteq \bigcup_{\bar{y}\in F_{\langle\bar{y}_0,\ldots,\bar{y}_{n-1}\rangle}} W(\langle\bar{y}_0,\ldots,\bar{y}_{n-1},\bar{y}\rangle).$$

Finally, define

$$\mathcal{U}_n = \bigwedge_{\bar{y}_0\in F_{\langle\rangle}} \bigwedge_{\bar{y}_1\in F_{\langle\bar{y}_0\rangle}} \cdots \bigwedge_{\bar{y}_{n-1}\in F_{\langle\bar{y}_0,\ldots,\bar{y}_{n-2}\rangle}} \{W(\langle\bar{y}_0,\ldots,\bar{y}_{n-1},\bar{y}\rangle) : \bar{y} \in F_{\langle\bar{y}_0,\ldots,\bar{y}_{n-1}\rangle}\}.$$

It now follows from (5) that, for each $\langle m,k\rangle \in \omega\times\omega$, there is a sequence $\langle U_n : n \in N^k_m\rangle$ such that $U_n \in \mathcal{U}_n$ for all $n \in N^k_m$ and $X^k \subseteq \bigcup_{n\in N^k_m} U_n$. We can now recursively pick $\bar{z}_n \in F_{\langle\bar{z}_0,\ldots,\bar{z}_{n-1}\rangle}$ for $n \in \omega$ in such a way that

$$U_n \subseteq W(\langle\bar{z}_0,\ldots,\bar{z}_{n-1},\bar{z}_n\rangle)$$

for all $n \in \omega$. We claim that $\{f_{\langle \bar{z}_0,\dots,\bar{z}_n\rangle} : n \in \omega\} \in \Omega_{\underline{0}}^X$ — which implies that θ is not a winning strategy.

Indeed, let $F \in [X]^{<\aleph_0}$ and $\varepsilon \in \mathbb{R}_{>0}$ be arbitrary. Let $F = \{x_1, \dots, x_k\}$, where $k = |F|$, and let also $m \in \omega$ be such that $\frac{1}{2^m} < \varepsilon$. Then there is $n \in N_m^k$ such that $\langle x_1, \dots, x_k\rangle \in U_n$, so

$$\begin{aligned}\langle x_1, \dots, x_k\rangle \in W(\langle \bar{z}_0, \dots, \bar{z}_{n-1}, \bar{z}_n\rangle) &= V_m^k(f_{\langle \bar{z}_0,\dots,\bar{z}_n\rangle}) \\ &= \left((f_{\langle \bar{z}_0,\dots,\bar{z}_n\rangle})^{-1}\left[\left(-\frac{1}{2^m}, \frac{1}{2^m}\right)\right]\right)^k.\end{aligned}$$

Thus $f_{\langle \bar{z}_0,\dots,\bar{z}_n\rangle}(x_i) \in \left(-\frac{1}{2^m}, \frac{1}{2^m}\right) \subseteq (-\varepsilon, \varepsilon)$ for each $i \in \{1, \dots, k\}$, whence $f_{\langle \bar{z}_0,\dots,\bar{z}_n\rangle} \in V(\underline{0}, F, \varepsilon)$. □

From Theorems 4.1 and 8.3, we obtain:

COROLLARY 8.4 (Scheepers [**36, 37**]). *Let Y be a compact metric space and $X \subseteq Y$. The following conditions are equivalent:*

(1) ALICE *does not have a winning strategy in the game* $\mathsf{G}_1(\Omega_{\underline{0}}, \Omega_{\underline{0}}^X)$ *played on* $C(Y)$*;*
(2) $C(Y)$ *satisfies* $\mathsf{S}_1(\Omega_{\underline{0}}, \Omega_{\underline{0}}^X)$*;*
(3) ALICE *does not have a winning strategy in the game* $\mathsf{G}_1(\Omega, \Omega_X)$ *played on* Y*;*
(4) Y *satisfies* $\mathsf{S}_1(\Omega, \Omega_X)$*;*
(5) *every finite power of X has strong measure zero.*

In addition to the above corollary, we mention (without proof) Theorems 5.1 and 5.3 of [**23**], in which the property of having strong measure zero in every finite power is characterized in terms of games played on (function spaces defined on) the same subspace of the Alexandroff double considered in Corollary 6.1(4):

THEOREM 8.5 (Kočinac–Scheepers [**23**]). *Let X be a dense subset of* $[0, 1]$. *The following conditions are equivalent:*

(1) ALICE *does not have a winning strategy in the game* $\mathsf{G}_1(\Omega_{\underline{0}}, \Omega_{\underline{0}}^{X\times\{1\}})$ *played on* $C(T(X, [0, 1]))$*;*
(2) $C(T(X, [0, 1]))$ *satisfies* $\mathsf{S}_1(\Omega_{\underline{0}}, \Omega_{\underline{0}}^{X\times\{1\}})$*;*
(3) ALICE *does not have a winning strategy in the game* $\mathsf{G}_1(\Omega, \Omega_{X\times\{1\}})$ *played on* $T(X, [0, 1])$*;*
(4) $T(X, [0, 1])$ *satisfies* $\mathsf{S}_1(\Omega, \Omega_{X\times\{1\}})$*;*
(5) *every finite power of X has strong measure zero.*

We also mention (again without proof) an "every finite power" counterpart of Theorem 4.9:

THEOREM 8.6 (Zindulka [**48**]). *Let X be a σ–totally-bounded metric space. The following conditions are equivalent:*

(1) ALICE *does not have a winning strategy in the game* $\mathsf{G}_1(\langle\mathcal{O}_n^{\mathrm{unif}}\rangle, \Omega)$ *played on* X*;*
(2) X *satisfies* $\mathsf{S}_1(\langle\mathcal{O}_n^{\mathrm{unif}}\rangle, \Omega)$*;*
(3) ALICE *does not have a winning strategy in the game* $\mathsf{G}_1(\Omega^{\mathrm{unif}}, \Omega)$ *played on* X*;*
(4) X *satisfies* $\mathsf{S}_1(\Omega^{\mathrm{unif}}, \Omega)$*;*
(5) *every finite power of X has strong measure zero.*

9. Higher-cardinal versions of the Borel Conjecture

In this section, we will briefly discuss versions of the Borel Conjecture for uncountable cardinals appearing in the literature — some of which are formulated (or at least have an equivalent formulation) in terms of topological games. Although some of them can be defined for each infinite cardinal κ, our main interest will be the case $\kappa = \aleph_1$.

9.1. Halko's GBC. The first version we mention is due to Halko [**17**]. In order to state it, consider, on the set ${\omega_1}^{\omega_1}$, the topology generated by the basic open sets $[p] = \{f \in {\omega_1}^{\omega_1} : p \subseteq f\}$ for $p \in {\omega_1}^{<\omega_1} = \bigcup_{\alpha<\omega_1} {\omega_1}^{\alpha}$.

DEFINITION 9.1 (Halko [**17**]). Let $X \subseteq {\omega_1}^{\omega_1}$. We say that

- X is small if, for every countable $C \subseteq \omega_1$, there is $\langle p_\alpha : \alpha \in C\rangle$ such that $p_\alpha \in {\omega_1}^{\alpha}$ for each $\alpha \in C$ and $X \subseteq \bigcup_{\alpha\in C}[p_\alpha]$ — equivalently, if, for every $\alpha \in \omega_1$, there is a sequence $\langle q_n : n \in \omega\rangle$ of elements of ${\omega_1}^{\alpha}$ such that $X \subseteq \bigcup_{n\in\omega}[q_n]$;
- X has strong measure zero if, for every uncountable $A \subseteq \omega_1$, there is $\langle p_\alpha : \alpha \in A\rangle$ such that $p_\alpha \in {\omega_1}^{\alpha}$ for each $\alpha \in A$ and $X \subseteq \bigcup_{\alpha\in A}[p_\alpha]$.

DEFINITION 9.2 (Halko [**17**]).

- The Weak Generalized Borel Conjecture (wGBC) is the statement that every small subset of ${\omega_1}^{\omega_1}$ has cardinality at most $\aleph_1$.
- The Generalized Borel Conjecture (GBC) is the statement that every strong measure zero subset of ${\omega_1}^{\omega_1}$ has cardinality at most $\aleph_1$.

Note that GBC $\to$ wGBC, since every small subset of ${\omega_1}^{\omega_1}$ has strong measure zero.

For the first theorem of this section, we recall that an ω_1-tree is a tree of height ω_1 all of whose levels are countable, and that a **Kurepa tree** is an ω_1-tree with at least $\aleph_2$ cofinal branches. The Kurepa Hypothesis (KH) is the statement "There is a Kurepa tree" — equivalently (see e.g. [**24**, Theorem II.5.18]), "There is a family $\mathcal{A} \subseteq \wp(\omega_1)$ of cardinality greater than $\aleph_1$ such that $\{A \cap C : A \in \mathcal{A}\}$ is countable for every countable $C \subseteq \omega_1$," i.e. "There is a **Kurepa family**." It is well-known (see e.g. the section "Trees in L" in Chapter 27 of [**18**]) that $\neg$KH and the existence of an inaccessible cardinal are equiconsistent relative to ZFC.

THEOREM 9.3 (Halko [**17**]). wGBC *is equivalent to* $\neg$KH.

PROOF. Note that every tree of height ω_1 and cardinality at most $\aleph_1$ can be assumed to be a subtree of $\langle {\omega_1}^{<\omega_1}, \subseteq\rangle$. The result then follows from the observation that, if T is such a tree and $B(T) \subseteq {\omega_1}^{\omega_1}$ is the set of all of the cofinal branches of T, then $B(T)$ is small if and only if all levels of T are countable. $\square$

COROLLARY 9.4 (Halko [**17**]). *The statements "There is an inaccessible cardinal" and* wGBC *are equiconsistent relative to* ZFC.

The main question concerning Halko's GBC is still open:

PROBLEM 9.5 (Halko [**17**]). Is GBC consistent with ZFC relative to the existence of some large cardinal?

As GBC $\to$ wGBC, Corollary 9.4 tells us that, in order to possibly achieve relative consistency of GBC, one must necessarily appeal to large cardinal axioms.

9.2. Tall and Usuba's $\aleph_1$-Borel Conjecture. The next version we consider is due to Tall and Usuba [**44**]. Let us start the motivation behind it with the following definition:

DEFINITION 9.6. A topological space X is **projectively countable** if $f[X]$ is countable whenever f is a continuous function from X to a separable metrizable space.

Note that one can replace "a separable metrizable space" by $[0,1]^\omega$ in the above definition, since the separable metrizable spaces are (up to homeomorphism) exactly the subspaces of $[0,1]^\omega$.

By combining Theorem 2.6 of [**22**] with Rothberger's Theorem 3.9 (see also [**5**, Theorem 37]), one has:

THEOREM 9.7 (Kočinac [**22**], Rothberger [**31**]). *The Borel Conjecture is equivalent to the statement "A Lindelöf regular space is Rothberger if and only if it is projectively countable."*

PROOF. The converse implication is immediate in view of Theorem 3.9: if the Borel Conjecture fails, then so does the R-Borel Conjecture, whence there is an uncountable Rothberger subset of the real line; this fails the "only if" part of the statement.

For the direct implication, assume BC — equivalently, R-BC.

Let X be a Rothberger space, and let $f : X \to [0,1]^\omega$ be continuous. It follows from a classical theorem due to Szpilrajn-Marczewski [**42**, Theorem 2] that every strong measure zero metric space is zero-dimensional; thus, $f[X]$ (being Rothberger) is a zero-dimensional second-countable space, hence is homeomorphic to a subspace of 2^ω. This implies that $f[X]$ is homeomorphic to a subspace of $\mathbb{R}$; therefore, $f[X]$ is countable by R-BC.

Finally, let X be a Lindelöf regular space that is projectively countable, and fix a sequence $\langle \mathcal{U}_n : n \in \omega \rangle$ of open covers of X. As Lindelöf regular spaces are paracompact, it follows from Katětov's characterization of paracompactness (see e.g. Problem 209 of Chapter 5 in [**1**]), for each $n \in \omega$ there exist a separable metrizable space Y_n, a continuous function $f_n : X \to Y_n$ and an open cover $\mathcal{V}_n$ of Y_n such that $\{f_n^{-1}[V] : V \in \mathcal{V}_n\}$ is an open refinement of $\mathcal{U}_n$. Then, since

$$\begin{array}{rccl} f & : & X & \to & \prod_{n\in\omega} Y_n \\ & & x & \mapsto & \langle f_n(x) : n \in \omega \rangle \end{array}$$

is a continuous function from X to a separable metrizable space, it follows that $f[X]$ is countable. Let then $\{x_k : k \in \omega\} \subseteq X$ be such that $f[X] = \{f(x_n) : n \in \omega\}$. If, for each $n \in \omega$, we pick $V_n \in \mathcal{V}_n$ such that $f_n(x_n) \in V_n$, we get

$$X = \bigcup_{n\in\omega} f^{-1}[\{f(x_n)\}] \subseteq \bigcup_{n\in\omega} f_n^{-1}[\{f_n(x_n)\}] \subseteq \bigcup_{n\in\omega} f_n^{-1}[V_n];$$

hence, choosing $U_n \in \mathcal{U}_n$ with $f_n^{-1}[V_n] \subseteq U_n$ for every $n \in \omega$, we obtain $X \subseteq \bigcup_{n\in\omega} U_n$. □

In view of Theorem 3.5, the above result becomes:

COROLLARY 9.8. *The Borel Conjecture is equivalent to the statement that, for every Lindelöf regular space X,* ALICE *does not have a winning strategy in the game $\mathsf{G}_1(\mathcal{O},\mathcal{O})$ played on X if and only if X is projectively countable.*

Let us now consider a natural extension of the game $\mathsf{G}_1(\mathcal{O},\mathcal{O})$ to arbitrary lengths:

DEFINITION 9.9 (Scheepers–Tall [**39**]). Let η be an ordinal, X be a topological space, and $\mathcal{O}$ be the set of all of the open covers of X. The game $\mathsf{G}_1^{\eta}(\mathcal{O},\mathcal{O})$ is played according to the following rules. In each inning $\alpha \in \eta$, ALICE chooses an open cover $\mathcal{U}_\alpha$ of X, and then BOB answers with $U_\alpha \in \mathcal{U}_\alpha$. The winner is BOB if $\bigcup_{\alpha\in\eta} U_\alpha = X$, and ALICE otherwise.

In connection with the problem of the cardinality of Lindelöf spaces with points G_δ, Tall introduced the following concept:

DEFINITION 9.10 (Tall [**43**]). A Lindelöf space is **indestructibly Lindelöf** if it remains Lindelöf in every countably closed forcing extension.

These two last definitions are connected by the next result, which is Theorem 1 of [**39**].

THEOREM 9.11 (Scheepers–Tall [**39**]). *A Lindelöf space X is indestructibly Lindelöf if and only if* ALICE *does not have a winning strategy in the game* $\mathsf{G}_1^{\omega_1}(\mathcal{O},\mathcal{O})$ *played on X.*

With this last theorem in mind, and inspired by Corollary 9.8, Tall and Usuba formulated the following "one-step-above" version of the Borel Conjecture:

DEFINITION 9.12 (Tall–Usuba [**44**]). The $\aleph_1$-Borel Conjecture is the statement "A Lindelöf regular space X is indestructibly Lindelöf if and only if every continuous image of X in $[0,1]^{\omega_1}$ has cardinality at most $\aleph_1$."

As it turns out, the $\aleph_1$-Borel Conjecture is related to large cardinal axioms. More specifically, a combination of results of Dias, Tall and Usuba yields:

THEOREM 9.13 (Dias–Tall–Usuba [**9**,**44**]). *The following statements are equiconsistent relative to* ZFC*:*

(1) *the $\aleph_1$-Borel Conjecture;*
(2) *there is a strongly inaccessible cardinal.*

COROLLARY 9.14. *The following statements are equiconsistent relative to* ZFC*:*

(1) *the $\aleph_1$-Borel Conjecture;*
(2) wGBC*;*
(3) *there is a strongly inaccessible cardinal.*

PROOF. By Corollary 9.4 and Theorem 9.13. □

We remark that the implication $\neg(2) \to \neg(1)$ in Theorem 9.13 is done in [**9**] by showing that, if there is a Kurepa tree, then there is an indestructibly Lindelöf regular space of weight $\aleph_1$ (hence it is embeddable in $[0,1]^{\omega_1}$) and size greater than $\aleph_1$. Thus, in view of Theorem 9.3, we actually have a direct implication from statement (1) to statement (2) in Corollary 9.14:

COROLLARY 9.15. *The $\aleph_1$-Borel Conjecture implies* wGBC.

We also note that the $\aleph_1$-Borel Conjecture is incompatible with BC. This is mainly because of the following result — which is essentially a consequence of the simple observation that, whenever ALICE loses a play of $\mathsf{G}_1^{\omega_1}(\mathcal{O},\mathcal{O})$ on a Lindelöf space, she loses after countably many innings.

THEOREM 9.16 (Tall [**43**]). *Every hereditarily Lindelöf space is indestructibly Lindelöf.*

PROOF. Let X be a hereditarily Lindelöf space. Suppose, by way of contradiction, that there is a winning strategy θ for ALICE in $\mathsf{G}_1^{\omega_1}(\mathcal{O}, \mathcal{O})$ on X. Let BOB play against θ by choosing, in each inning $\alpha \in \omega_1$, a point $x_\alpha \in X$ that has not yet been covered by the open sets played by BOB thus far, and then covering x_α with some open set $U_\alpha \in \mathcal{U}_\alpha$, where $\mathcal{U}_\alpha$ is the open cover provided by θ as ALICE's α-th move. (As θ is a winning strategy for ALICE, BOB can always do this.)

At the end of the play, we will have covered the set $Y = \{x_\alpha : \alpha \in \omega_1\}$ with the open sets $\{U_\alpha : \alpha \in \omega_1\}$. As Y is Lindelöf, there is $\gamma \in \omega_1$ such that $Y \subseteq \bigcup_{\alpha\in\gamma} U_\alpha$ — thus contradicting the choice of x_γ. $\square$

COROLLARY 9.17. *The $\aleph_1$-Borel Conjecture implies* CH*; hence, in particular, it implies the failure of* BC.

PROOF. By Theorem 9.16, the Cantor space 2^ω is indestructibly Lindelöf. Since 2^ω is embeddable in $[0,1]^{\omega_1}$, the $\aleph_1$-Borel Conjecture implies that $\mathfrak{c} = |2^\omega| \leq \aleph_1$. $\square$

9.3. Galvin and Scheepers's $\mathsf{BC}_{\aleph_1}$. The third and last version is due to Galvin and Scheepers [**13**]. We start by recalling the following definition:

DEFINITION 9.18 (Guran [**16**]). A topological group $\langle G, *\rangle$ with identity element e is **$\aleph_0$-bounded** if, for every open neighbourhood U of e in G, there is a countable subset $E \subseteq G$ such that $E * U = G$ — that is, if G can be covered by countably many left translates of any open neighbourhood of e given in advance.

Inspired by the equivalence (1) $\leftrightarrow$ (2) in Theorem 5.4, Galvin and Scheepers introduced the following general version of the Borel Conjecture:

DEFINITION 9.19 (Galvin–Scheepers [**13**]). Let κ be an infinite cardinal.

- For a topological group G, $\mathsf{BC}(G, \kappa)$ is the statement:
  every Rothberger-bounded subset of G has cardinality at most κ.
- BC_κ is the statement:
  $\mathsf{BC}(G, \kappa)$ holds for every $\aleph_0$-bounded topological group G of weight κ.

Note that, as every second-countable topological group is obviously $\aleph_0$-bounded, Theorem 5.4 yields that $\mathsf{BC}_{\aleph_0}$ is equivalent to BC.

The next result — in which the set 2^{ω_1} is endowed with the group operation $\oplus$ defined by the sum modulo 2 coordinate-wise — is part of Theorem 11 of [**13**].[12]

THEOREM 9.20 (Galvin–Scheepers [**13**]). *Consider the following statements:*

(1) $\mathsf{BC}_{\aleph_1}$;
(2) $\mathsf{BC}(2^{\omega_1}, \aleph_1)$;
(3) $\neg\mathsf{KH}$.

Then (1) $\to$ (2) $\to$ (3). *Furthermore, the three statements are equivalent under the Borel Conjecture.*

[12] Although we have restricted ourselves to the case $\kappa = \aleph_1$, the original theorem is done for any uncountable cardinal.

PROOF. Clearly, (1) $\to$ (2).

$\neg(3) \to \neg(2)$. Let $\mathcal{F} \subseteq \wp(\omega_1)$ be a Kurepa family. For each $A \in \mathcal{F}$, let $\chi_A \in 2^{\omega_1}$ be the characteristic function of A — that is, $\chi_A(\xi) = 1$ if and only if $\xi \in A$. We will show that $X = \{\chi_A : A \in \mathcal{F}\} \subseteq 2^{\omega_1}$ is Rothberger-bounded, which will imply $\neg(2)$ since $|X| = |\mathcal{F}| > \aleph_1$.

Suppose not, and let $\langle U_n : n \in \omega\rangle$ be a sequence of basic open neighbourhoods of $\bar{0} \in 2^{\omega_1}$ witnessing this. For each $n \in \omega$, let $F_n \in [\omega_1]^{<\aleph_0}$ be such that $U_n = \{f \in 2^{\omega_1} : \forall \xi \in F_n\ (f(\xi) = 0)\}$. Now, since $C = \bigcup_{n\in\omega} F_n$ is a countable subset of ω_1, it follows from the choice of $\mathcal{F}$ that the set $\{A \cap C : A \in \mathcal{F}\}$ is countable; let then $\langle A_n : n \in \omega\rangle$ be a sequence of elements of $\mathcal{F}$ satisfying $\{A_n \cap C : n \in \omega\} = \{A \cap C : A \in \mathcal{F}\}$. We claim that $X \subseteq \bigcup_{n\in\omega} \chi_{A_n} \oplus U_n$.

Indeed, let $A \in \mathcal{F}$ be arbitrary. Pick $n \in \omega$ such that $A \cap C = A_n \cap C$. As $F_n \subseteq C$, this implies $A \cap F_n = A_n \cap F_n$. But then, by the choice of F_n, it follows that $\chi_A \in \chi_A \oplus U_n = \chi_{A_n} \oplus U_n$.

$((3) + \mathsf{BC}) \to (1)$. Let G be an $\aleph_0$-bounded group of weight $\aleph_1$, and let $X \subseteq G$ be Rothberger-bounded. A theorem of Guran [**16**] now tells us that, as G is $\aleph_0$-bounded, it is embeddable in a product of second-countable topological groups — and furthermore that, as G has weight $\aleph_1$, such an embedding can be done in a product of $\aleph_1$ many second-countable groups. So we can assume that $G \subseteq \prod_{\alpha\in\omega_1} G_\alpha$ where each G_α is a second-countable topological group.

For each $A \subseteq \omega_1$, let $\pi_A : \prod_{\alpha\in\omega_1} G_\alpha \to \prod_{\alpha\in A} G_\alpha$ be the natural projection. As every π_A is a continuous homomorphism, it follows that $\pi_A[X]$ is Rothberger-bounded for each $A \subseteq \omega_1$. In particular, for each countable $C \subseteq \omega_1$, the set $\pi_C[X]$ is a Rothberger-bounded subset of the second-countable group $\prod_{\alpha\in C} G_\alpha$; hence, as we are assuming BC, it follows from Theorem 5.4 that

$$\pi_C[X] = \{f \restriction C : f \in X\} \text{ is countable for every countable } C \subseteq \omega_1. \ (\ddagger)$$

Note that each $f \in X$ is a subset of $S = \bigcup_{\alpha\in\omega_1} \left(\{\alpha\} \times \pi_{\{\alpha\}}[X]\right)$ — which is a set of cardinality $\aleph_1$, since $(\ddagger)$ implies that $\pi_{\{\alpha\}}[X]$ is countable for each $\alpha \in \omega_1$. We also have that $\{f \cap W : f \in X\}$ is countable for every countable $W \subseteq S$, as $\{f \cap W : f \in X\} \subseteq \{f \restriction \mathrm{dom}(W)\} = \pi_{\mathrm{dom}(W)}[X]$ is countable by $(\ddagger)$. Therefore, it must be the case that $|X| \leq \aleph_1$ — for otherwise X would be a Kurepa family on the set S, thus contradicting (3). □

In view of Theorem 9.3, this yields:

COROLLARY 9.21. $\mathsf{BC}_{\aleph_1}$ *implies* wGBC.

Consistency of $\mathsf{BC}_{\aleph_1}$ (together with BC) relative to the existence of a large cardinal was obtained in Corollary 25 of [**13**], which we quote below. Recall that a cardinal κ is said to be **1-inaccessible** if κ is the κ-th strongly inaccessible cardinal.

THEOREM 9.22 (Galvin–Scheepers [**13**]). *If the existence of a 1-inaccessible cardinal is consistent with* ZFC, *then the conjunction of the statements* BC *and* $\mathsf{BC}_{\aleph_1}$ *is consistent with* ZFC.

Acknowledgments

We would like to thank Marion Scheepers for many interesting discussions on generalizations of the Borel Conjecture to higher cardinals, to which we owe most of the material included in Section 9. We also wish to thank the referee for a careful reading of the first version of this work and for many valuable suggestions that helped improve the text.

References

[1] A. V. Arkhangel′skiĭ and V. I. Ponomarev, *Fundamentals of general topology*, Mathematics and its Applications, D. Reidel Publishing Co., Dordrecht, 1984. Problems and exercises; Translated from the Russian by V. K. Jain; With a foreword by P. Alexandroff [P. S. Aleksandrov]. MR785749

[2] L. Babinkostova, Lj. D. R. Kočinac, and M. Scheepers, *Combinatorics of open covers. XI. Menger- and Rothberger-bounded groups*, Topology Appl. **154** (2007), no. 7, 1269–1280, DOI 10.1016/j.topol.2005.09.013. MR2310460

[3] L. Babinkostova, B. A. Pansera, and M. Scheepers, *Weak covering properties and infinite games*, Topology Appl. **159** (2012), no. 17, 3644–3657, DOI 10.1016/j.topol.2012.09.009. MR2973387

[4] G. Birkhoff, *A note on topological groups*, Compositio Math. **3** (1936), 427–430. MR1556955

[5] M. Bonanzinga, F. Cammaroto, and M. Matveev, *Projective versions of selection principles*, Topology Appl. **157** (2010), no. 5, 874–893, DOI 10.1016/j.topol.2009.12.004. MR2593701

[6] E. Borel, *Sur la classification des ensembles de mesure nulle* (French), Bull. Soc. Math. France **47** (1919), 97–125. MR1504785

[7] T. J. Carlson, *Strong measure zero and strongly meager sets*, Proc. Amer. Math. Soc. **118** (1993), no. 2, 577–586, DOI 10.2307/2160341. MR1139474

[8] P. Daniels, *Pixley-Roy spaces over subsets of the reals*, Topology Appl. **29** (1988), no. 1, 93–106, DOI 10.1016/0166-8641(88)90061-2. MR944073

[9] R. R. Dias and F. D. Tall, *Indestructibility of compact spaces*, Topology Appl. **160** (2013), no. 18, 2411–2426, DOI 10.1016/j.topol.2013.07.036. MR3120656

[10] A. W. Miller and D. H. Fremlin, *On some properties of Hurewicz, Menger, and Rothberger*, Fund. Math. **129** (1988), no. 1, 17–33, DOI 10.4064/fm-129-1-17-33. MR954892

[11] F. Galvin, *Indeterminacy of point-open games* (English, with Russian summary), Bull. Acad. Polon. Sci. Sér. Sci. Math. Astronom. Phys. **26** (1978), no. 5, 445–449. MR493925

[12] F. Galvin, J. Mycielski, and R. M. Solovay, *Strong measure zero and infinite games*, Arch. Math. Logic **56** (2017), no. 7-8, 725–732, DOI 10.1007/s00153-017-0541-z. MR3696064

[13] F. Galvin and M. Scheepers, *Borel's conjecture in topological groups*, J. Symbolic Logic **78** (2013), no. 1, 168–184, DOI 10.2178/jsl.7801110. MR3087068

[14] F. Galvin and M. Scheepers, *Baire spaces and infinite games*, Arch. Math. Logic **55** (2016), no. 1-2, 85–104, DOI 10.1007/s00153-015-0461-8. MR3453580

[15] K. Gödel, *The Consistency of the Continuum Hypothesis*, Annals of Mathematics Studies, no. 3, Princeton University Press, Princeton, N. J., 1940. MR0002514

[16] I. I. Guran, *Topological groups similar to Lindelöf groups* (Russian), Dokl. Akad. Nauk SSSR **256** (1981), no. 6, 1305–1307. MR606469

[17] A. Halko, *Negligible subsets of the generalized Baire space* $\omega_1^{\omega_1}$, Ann. Acad. Sci. Fenn. Math. Diss. **107** (1996), 38. MR1413840

[18] T. Jech, *Set theory*. Springer, Berlin–Heidelberg–New York, 2002. xiii+769 pp.

[19] W. Just, A. W. Miller, M. Scheepers, and P. J. Szeptycki, *The combinatorics of open covers. II*, Topology Appl. **73** (1996), no. 3, 241–266, DOI 10.1016/S0166-8641(96)00075-2. MR1419798

[20] S. Kakutani, *Über die Metrisation der topologischen Gruppen* (German), Proc. Imp. Acad. Tokyo **12** (1936), no. 4, 82–84. MR1568424

[21] J. L. Kelley, *General topology*, D. Van Nostrand Company, Inc., Toronto-New York-London, 1955. MR0070144

[22] L. D. R. Kočinac, *Selection principles and continuous images* (English, with English and Spanish summaries), Cubo **8** (2006), no. 2, 23–31. MR2244614

[23] Lj. D. Kočinac and M. Scheepers, *Function spaces and strong measure zero sets*, Acta Math. Hungar. **82** (1999), no. 4, 341–351, DOI 10.1023/A:1006600627664. MR1675622

[24] K. Kunen, *Set theory*, Studies in Logic and the Foundations of Mathematics, vol. 102, North-Holland Publishing Co., Amsterdam-New York, 1980. An introduction to independence proofs. MR597342

[25] R. Laver, *On the consistency of Borel's conjecture*, Acta Math. **137** (1976), no. 3-4, 151–169, DOI 10.1007/BF02392416. MR422027

[26] N. Lusin, *Sur un problème de M. Baire*, Comptes Rendus Hebdomadaires des Séances de l'Académie des Sciences **158** (1914), 1258–1261.

[27] A. W. Miller, *The γ-Borel conjecture*, Arch. Math. Logic **44** (2005), no. 4, 425–434, DOI 10.1007/s00153-004-0260-0. MR2139770

[28] J. Pawlikowski, *Powers of transitive bases of measure and category*, Proc. Amer. Math. Soc. **93** (1985), no. 4, 719–729, DOI 10.2307/2045552. MR776210

[29] J. Pawlikowski, *Undetermined sets of point-open games*, Fund. Math. **144** (1994), no. 3, 279–285. MR1279482

[30] F. Rothberger, *Eine Verschärfung der Eigenschaft* **C**, Fundamenta Mathematicae **30** (1938), 50–55.

[31] F. Rothberger, *Sur les familles indénombrables de suites de nombres naturels et les problèmes concernant la propriété C* (French), Proc. Cambridge Philos. Soc. **37** (1941), 109–126. MR4281

[32] M. Scheepers, *Combinatorics of open covers. I. Ramsey theory*, Topology Appl. **69** (1996), no. 1, 31–62, DOI 10.1016/0166-8641(95)00067-4. MR1378387

[33] M. Scheepers, *Strong measure zero subsets of the real line and an infinite game*, Proceedings of the II Mathematical Conference in Priština (1996), Univ. Priština, Priština, 1997, pp. 61–65. MR1466232

[34] M. Scheepers, *Combinatorics of open covers. III. Games,* $\mathsf{C_p}(X)$, Fund. Math. **152** (1997), no. 3, 231–254. MR1444715

[35] M. Scheepers, *Combinatorics of open covers. IV. Subspaces of the Alexandroff double of the unit interval*, Topology Appl. **83** (1998), no. 1, 63–75, DOI 10.1016/S0166-8641(97)00073-4. MR1601646

[36] M. Scheepers, *Strong measure zero sets, filters, and pointwise convergence*, East-West J. Math. **1** (1998), no. 1, 109–116. MR1663012

[37] M. Scheepers, *Finite powers of strong measure zero sets*, J. Symbolic Logic **64** (1999), no. 3, 1295–1306, DOI 10.2307/2586631. MR1779763

[38] M. Scheepers, *Rothberger bounded groups and Ramsey theory*, Topology Appl. **158** (2011), no. 13, 1575–1583, DOI 10.1016/j.topol.2011.05.025. MR2812467

[39] M. Scheepers and F. D. Tall, *Lindelöf indestructibility, topological games and selection principles*, Fund. Math. **210** (2010), no. 1, 1–46, DOI 10.4064/fm210-1-1. MR2720214

[40] W. Sierpiński, *Sur un ensemble non dénombrable, dont toute image continue est de mesure nulle*, Fundamenta Mathematicae **11** (1928), 302–304.

[41] W. Sierpiński, *Sur le produit combinatoire de deux ensembles jouissant de la propriété C*, Fundamenta Mathematicae **24** (1935), 48–50.

[42] E. Szpilrajn, *La dimension et la mesure*, Fundamenta Mathematicae **28** (1937), 81–89.

[43] F. D. Tall, *On the cardinality of Lindelöf spaces with points G_δ*, Topology Appl. **63** (1995), no. 1, 21–38, DOI 10.1016/0166-8641(95)90002-0. MR1328616

[44] F. D. Tall and T. Usuba, *Lindelöf spaces with small pseudocharacter and an analog of Borel's conjecture for subsets of* $[0,1]^{\aleph_1}$, Houston J. Math. **40** (2014), no. 4, 1299–1309. MR3298750

[45] R. Telgársky, *Spaces defined by topological games*, Fund. Math. **88** (1975), no. 3, 193–223, DOI 10.4064/fm-88-3-193-223. MR380708

[46] V. V. Tkachuk, *Some new versions of an old game*, Comment. Math. Univ. Carolin. **36** (1995), no. 1, 177–196. MR1334425

[47] A. Weil, *Sur les espaces à structure uniforme et sur la topologie générale*. Publications de l'Institut Mathématique de l'Université de Strasbourg, 1. Actualités Scientifiques et Industrielles, 551. Hermann & Cie, Paris, 1937. 39 pp.

[48] O. Zindulka, *Combinatorics of uniform covers*, preprint, 2019.

Instituto de Ciências Matemáticas e de Computação, Universidade de São Paulo, Caixa Postal 668, São Carlos, SP, 13560-970, Brazil
Email address: aurichi@icmc.usp.br

Centro de Matemática, Computação e Cognição, Universidade Federal do ABC, Avenida dos Estados, 5001, Santo André, SP, 09210-580, Brazil
Email address: rodrigo.dias@ufabc.edu.br

Contemporary Mathematics
Volume **755**, 2020
https://doi.org/10.1090/conm/755/15184

Strong measure zero in Polish groups

Michael Hrušák and Ondřej Zindulka

ABSTRACT. The notion of strong measure zero is studied in the context of Polish groups. In particular, the extent to which the theorem of Galvin, Mycielski and Solovay holds in an arbitrary Polish group is studied. Hausdorff measure and dimension is used to characterize strong measure zero. The products of strong measure zero sets are examined. Sharp measure zero, a notion stronger that strong measure zero, is shown to be related to meager-additive sets in the Cantor set and Polish groups by a theorem very similar to the theorem of Galvin, Mycielski and Solovay.

1. Introduction

All spaces and topological groups considered are separable and metrizable.

A natural extension of a definition due to Borel (1919) [**6**] asserts that a metric space X has *strong measure zero* (**Smz**) if for any sequence $\langle \varepsilon_n : n \in \omega \rangle$ of positive real numbers there is a cover $\{U_n : n \in \omega\}$ of X such that $\operatorname{diam} U_n \leqslant \varepsilon_n$ for all n.

In the same paper Borel conjectured that every strong measure zero set of reals is countable. This was shown to be independent of the usual axioms of set theory by Sierpiński (1928) [**44**] and Laver (1976) [**28**]. Later it was observed by Carlson [**9**] that the *Borel Conjecture* actually implies a formally stronger statement that all separable **Smz** metric spaces are countable.

We shall investigate the behaviour of strong measure zero sets in arbitrary Polish groups. In a sense we shall investigate the world, where the Borel conjecture fails, as most if not all of our results are trivial if the Borel Conjecture holds.

The subject of inquiry of this work starts with the theorem of Galvin, Mycielski, and Solovay [**13, 14**] who, confirming a conjecture of Prikry, proved that *a set $A \subseteq \mathbb{R}$ is of strong measure zero if and only if $A + M \neq \mathbb{R}$ for every meager set $M \subseteq \mathbb{R}$.*

Relatively recently Kysiak [**26**] and Fremlin [**12**], independently, showed that an analogous theorem is true for all locally compact metrizable groups (see also [**50**]). We present a proof of Kysiak and Fremlin's result based on [**19**] and consider the natural question as to how far the result can be extended. The fact that the theorem does not in general hold for all Polish groups was established in [**19**] and

2010 *Mathematics Subject Classification.* Primary 03E17, 22B05, 22A10, 54E52.

Key words and phrases. Strong measure zero, Polish group, Galvin-Mycielski-Solovay theorem.

The first author gratefully acknowledges support from PAPIIT grant IN 100317. The second author was supported from European Regional Development Fund-Project "Center for Advanced Applied Science" (No. CZ.02.1.01/0.0/0.0/16_019/0000778).

[**50**] and extended in [**20**]. This depends on further set-theoretic axioms, as the result obviously holds for all Polish groups assuming, e.g., the Borel Conjecture.

Cardinal invariants associated with strong measure zero sets on $\mathbb{R}$, ω^ω, and 2^ω have been studied rather extensively in recent decades [**1,8,15,51**]. We review some of these, concentrating on the *uniformity* invariant of the σ-ideal $\mathsf{Smz}(\mathbb{G})$ of strong measure zero subsets of a Polish group $\mathbb{G}$. A version of the Galvin-Mycielski-Solovay theorem links this study to the investigation of the so-called *transitive coefficient* $\mathsf{cov}^*(\mathcal{M})$ in Polish groups [**1,11,31**].

It was probably the aforementioned result of Prikry, Galvin, Mycielski and Solovay that inspired a few notions of smallness on the real line and Cantor set akin to strong measure zero. E.g., a set $S \subseteq \mathbb{R}$ is *strongly meager* if $S + N \neq \mathbb{R}$ for each Lebesgue null set N; it is *null-additive* if $S + N$ is Lebesgue null for each Lebesgue null set N; and it is *meager-additive* if $S + M$ is meager for each meager set M. These notions easily extend to other Polish groups.

We will study the latter notion, which is obviously a strengthening of strong measure zero. Since the early nineties, meager-additive sets in the Cantor set receive quite some attention. Let us single out the remarkable paper of Shelah [**42**] that provides a proof that each null-additive set in the Cantor set 2^ω is meager-additive and also the underlying combinatorial characterizations of null-additive and meager-additive sets in 2^ω (cf. 7.7 below), and Pawlikowski's paper [**36**] providing fine combinatorics and study of the so called transitive coefficients mentioned above that are actually cardinal invariants of strong measure zero, meager-additive and null-additive sets, and of course Bartoszyński's book [**1**]. However, all nontrivial results on meager-additive sets depended heavily on the combinatorial and group structure of 2^ω. In 2009 Weiss [**47, 48**] found a method that made the theory transferrable to the real line. Only very recently in [**52, 54**] it was noted that there is a description of meager-additive sets that resembles very much the Borel's definition of strong measure zero. Metric spaces having this property were termed to have *sharp measure zero*. This allowed for the theory of meager-additive sets to extend to other Polish groups. We provide some highlights of the rather new theory of sharp measure in metric spaces and meager-additive sets, and sharp measure zero on 2^ω and on Polish groups, including calculation of the uniformity number of sharp measure zero and meager-additive sets.

Our set-theoretic notation is standard and follows e.g. [**21, 25**]. In particular, the set of finite ordinals is identified with the set of non-negative integers and denoted interchangeably by ω and $\mathbb{N}$. In the same vein, the non-negative integers themselves are identified with the set of smaller non-negative integers, in particular $2 = \{0, 1\}$.

All spaces considered are separable and metrizable, often endowed with a compatible metric denoted d. We denote by $B(x, \varepsilon)$ the closed ball with radius ε centered at x, the corresponding open ball will be denoted by $B^\circ(x, \varepsilon)$.

The product spaces of the type A^ω for some finite or countable set A are considered with the *metric of least difference* defined by $d(f, g) = 2^{-|f \wedge g|}$, where $f \wedge g = f{\restriction}n$ for $n = \min\{k : f(k) \neq g(k)\}$. The clopen balls in the space A^ω are represented by nodes of the tree $A^{<\omega}$, given $s \in A^{<\omega}$, we let $\langle s \rangle = \{f \in A^\omega : s \subseteq f\}$. Given a subtree T of $A^{<\omega}$, we let $[T] = \{f \in A^\omega : \forall n \in \omega\ f{\restriction}n \in T\}$ be the (closed) set of branches of T. A metric space is *analytic* if it is a continuous image of ω^ω, and it is *Borel* (*absolutely* G_δ, resp.) if it is Borel (G_δ, resp.) in its completion.

A *Polish group* is a separable, completely metrizable topological group. A compatible metric d on a separable metrizable group $\mathbb{G}$ is *left-invariant* if $d(zx, zy) = d(x, y)$ for any $x, y, z \in \mathbb{G}$.

A separable group $\mathbb{G}$ is a CLI *group* if it admits a complete left-invariant compatible metric. Abelian and locally compact Polish groups are CLI, while, e.g., the group S_∞ of all permutations of ω is not.

A separable group $\mathbb{G}$ is a TSI *group* if it admits a (both-sided) invariant compatible metric. Not every Polish group admits an invariant metric, but if it is compact or abelian, then it does. Also, any invariant metric on a Polish group is complete.

2. Strong measure zero in Polish groups

The notion of strong measure zero is in general neither a topological nor a metric property, but a *uniform* property; in particular, a uniformly continuous image of a Smz set is Smz, and if X uniformly embeds into Y, then any set $A \subseteq X$ that is not Smz in X is not Smz in Y either.

As all left-invariant (equiv right-invariant) metrics on a separable metrizable group are uniformly equivalent the notion of strong measure zero becomes seemingly "topological": a subset S of a topological group $\mathbb{G}$ is *Rothberger bounded* if for every sequence $\langle U_n : n \in \omega \rangle$ of neighbourhoods of $1_{\mathbb{G}}$ there is a sequence $\langle g_n : n \in \omega \rangle$ of elements of the group $\mathbb{G}$ such that the family $\langle g_n \cdot U_n : n \in \omega \rangle$ covers S. It follows [**12**] that a subset of a Polish group $\mathbb{G}$ is Rothberger bounded if and only if it is strong measure zero w.r.t. some (any) left-invariant metric on $\mathbb{G}$.

Many of the results stated here could be phrased in the language of uniformities and/or in terms of the property of being Rothberger bounded (see [**12**] for such treatment).

Whenever $\mathbb{G}$ is a Polish group, $\mathsf{Smz}(\mathbb{G})$ denotes the family of strong measure zero sets with respect to any left-invariant metric (i.e., the (left) Rothberger bounded sets as described above).

Of course, the choice of left-invariant over right-invariant is arbitrary, one being isomorphic to the other via the inverse map of the group in question. In fact, both the left Rothberger bounded and right Rothberger bounded set form a σ-ideal which is invariant under both left and right translations.

PROPOSITION 2.1. *$\mathsf{Smz}(\mathbb{G})$ is a bi-invariant σ-ideal.*

PROOF. To see that $\mathsf{Smz}(\mathbb{G})$ is a σ-ideal let $\{X_n : n \in \omega\} \subseteq \mathsf{Smz}(\mathbb{G})$ and a sequence $\{U_n : n \in \omega\}$ of open subsets of $\mathbb{G}$ be given. Let $\{I_n : n \in \omega\}$ be a partition of ω into infinite sets. As each X_n is of strong measure zero, there is a sequence $\{g_i : i \in I_n\} \subseteq \mathbb{G}$ such that $X_n \subseteq \bigcup_{i \in I_n} g_i \cdot U_i$. Then $\bigcup_{n \in \omega} X_n \subseteq \bigcup_{i \in \omega} g_i \cdot U_i$.

Now, let $X \in \mathsf{Smz}(\mathbb{G})$ and $g \in \mathbb{G}$ be given.

To see that $g \cdot X \in \mathsf{Smz}(\mathbb{G})$, note that if $\{U_n : n \in \omega\}$ is a sequence of open subsets of $\mathbb{G}$ and $\{g_i : i \in \omega\} \subseteq \mathbb{G}$ is such that $X \subseteq \bigcup_{n \in \omega} g_n \cdot U_n$, then $g \cdot X \subseteq \bigcup_{n \in \omega} g \cdot g_n \cdot U_n$.

To show that $X \cdot g \in \mathsf{Smz}(\mathbb{G})$, let $\{U_n : n \in \omega\}$ be a sequence of open subsets of $\mathbb{G}$. Consider the open sets $\{U_n \cdot g^{-1} : n \in \omega\}$. As $X \in \mathsf{Smz}(\mathbb{G})$ here is a sequence $\{g_n : n \in \omega\} \subseteq \mathbb{G}$ such that $X \subseteq \bigcup_{n \in \omega} g_n \cdot (U_n \cdot g^{-1})$. Then $X \cdot g \subseteq \bigcup_{n \in \omega} g_n \cdot U_n$. □

Now, assuming Borel conjecture, or assuming that the group $\mathbb{G}$ has an invariant metric, the left and right Rothberger bounded sets coincide. This is not true in general, though:

EXAMPLE 2.2. Assuming CH, there is a left Rothberger bounded subset of the group of permutations S_∞ of ω which is not right Rothberger bounded.

PROOF. Denote by Ω the set of all finite partial injective functions from some $n \in \omega$ to ω. Enumerate all sequences of elements of S_∞ as $\{y_\alpha : \alpha < \omega_1\}$, and all increasing functions from ω to ω as $\{f_\alpha : \alpha < \omega_1\}$.

We shall recursively construct $\{g_\alpha : \alpha < \omega_1\} \subseteq S_\infty$ and $\{z_\alpha : \alpha < \omega_1\} \subseteq S_\infty^\omega$ so that

(1) $\forall \beta < \alpha < \omega_1 \ \exists n \in \omega \ g_\alpha \restriction f_\beta(n) = z_\beta(n) \restriction f_\beta(n)$, while
(2) $\forall \beta < \alpha < \omega_1 \ \forall n \in \omega \ g_\alpha^{-1} \restriction n+1 \neq y_\alpha^{-1}(n) \restriction n+1$, and
(3) $\forall s \in \Omega$ and $m_0 < m_1$ the first two elements of $\omega \setminus \mathrm{rng}(s)$ $\exists a \in [\omega]^{m_1+1}$
  (a) $\forall n \in a \ s \subseteq z_\alpha(n) \restriction f_\alpha(n)$,
  (b) $\forall n \in a \ m_0 \in \mathrm{rng}(z_\alpha(n) \restriction f_\alpha(n))$,
  (c) $\forall n \in a \ m_1 \notin \mathrm{rng}(z_\alpha(n) \restriction f_\alpha(n))$, and
  (d) $\forall i \neq j \in a \ z_\alpha(i)^{-1}(m_0) \neq z_\alpha(j)^{-1}(m_0)$.

It should be clear, that if this can be accomplished then (1) guaranties that the set $X = \{g_\alpha : \alpha < \omega_1\}$ is of strong measure zero, while (2) makes sure that X^{-1} is not. The condition (3) is there for the construction not to prematurely terminate.

Assume that g_β, z_β for $\beta < \alpha$ have been constructed. First choose $z_\alpha \in S_\infty^\omega$ satisfying (3). Then enumerate $\alpha = \{\beta_i : i \in \omega\}$ and recursively find $\{n_i : i \in \omega\}$ so that $s_i = z_{\beta_i}(n_i) \restriction f_{\beta_i}(n_i)$ satisfy

(i) $s_i \subseteq s_{i+1}$,
(ii) if $m_i = \min(\omega \setminus \mathrm{rng}(s_i))$ then $m_i \in \mathrm{rng}(s_{i+1})$,
(iii) $\forall n \leqslant m_i \ \exists k_n \leqslant n \ k_n \in \mathrm{rng}(s_i) \ s_i^{-1}(k_n) \neq y_\alpha^{-1}(n)(k_n)$.

Then let $g_\alpha = \bigcup_{i \in \omega} s_i$. Then $g_\alpha \in S_\infty$ satisfying (1) by (i) and (ii), and (2) by (iii).

To construct the sequence $\langle s_n : n \in \omega \rangle$ start with $s_{-1} = \emptyset$. Having found s_i, let $m < k$ be the first two elements of $\omega \setminus \mathrm{rng}(s_i)$. By (3), there is $n_{i+1} \in \omega$ such that $s_{i+1} = z_{\beta_{i+1}}(n_{i+1}) \restriction f_{\beta_{i+1}}(n_{i+1})$ is such that $\{m, k\} \cap \mathrm{rng}(s_{i+1}) = \{m\}$, and $s_{i+1}^{-1}(n) \neq y_\alpha(n)^{-1}(n)$ for every $m \leqslant n < k$. □

There is a close relation between strong measure zero and Geometric measure theory which shall be explored later on in the text, in section 5. The first result in this direction is due to Besicovitch [**4, 5**] who showed that a set X of reals has strong measure zero if and only if every uniformly continuous image of X has Hausdorff dimension 0.

Here we shall characterize strong measure sets in Polish groups as exactly the sets of *universal invariant submeasure zero*, a result due to J. Grebík.

It is a classical result of Haar [**16**] that every locally compact Polish group admits an (essentially unique) left-invariant, countably additive, outer regular Borel measure. In a similar vein, we shall prove here that every Polish group admits a non-trivial countably subadditive, outer regular, left-invariant diffuse submeasure, a result used in the next section.

Recall that a function $\mu : \mathcal{P}(\mathbb{G}) \to \mathbb{R}^+ \cup \{\infty\}$ is a *submeasure* if $\mu(\emptyset) = 0$, and $\mu(A \cup B) \leqslant \mu(A) + \mu(B)$ whenever A, B are subsets of $\mathbb{G}$. A submeasure μ on $\mathbb{G}$ is

- *σ-subadditive* if $\mu(\bigcup_{n\in\omega} A_n) \leqslant \sum_{n\in\omega} \mu(A_n)$, for any $\{A_n : n \in \omega\} \subseteq \mathcal{P}(\mathbb{G})$,
- *outer regular* if $\mu(A) = \inf\{\mu(U) : A \subseteq U, U \text{ open in } X\}$, for any $A \subseteq \mathbb{G}$,

- *left-invariant* if $\mu(A) = \mu(g \cdot A)$, for any $A \subseteq \mathbb{G}$ and $g \in \mathbb{G}$,
- *non-atomic* or *diffuse* if $\mu(\{x\}) = 0$ for every $x \in \mathbb{G}$, and
- *non-trivial* if $\mu(\mathbb{G}) > 0$.

LEMMA 2.3. *In every Polish group $\mathbb{G}$ there is a decreasing local basis $\{U_n : n \in \omega\}$ of open sets at $1_{\mathbb{G}}$ such that for every $m \in \omega$ and $\{a_n : n > m\} \subseteq \mathcal{P}(\mathbb{G})$ such that $|a_n| = n$ for every $n > m$, $U_m \not\subseteq \bigcup_{n>m} a_n \cdot U_n$.*

PROOF. Let d be a left invariant compatible metric on $\mathbb{G}$, and let e be a complete metric on $\mathbb{G}$. Recursively choose the open sets U_n, $n \in \omega$, together with finite sets $b_n \subseteq U_n$ of size $n+1$ so that:

In d: The points of b_n are $3 \operatorname{diam} U_{n+1}$ apart, and also $3 \operatorname{diam} U_{n+1}$ apart from the complement of U_n, while

In e: $\forall m < n \, \forall \{g_i : m < i < n\}$ with $g_i \in b_i$ $\operatorname{diam} \prod_{m<i<n} g_i \cdot U_n < \frac{1}{n}$.

To verify that the sequence $\{U_n : n \in \omega\}$ has the desired property assume that $m \in \omega$ and a sequence $\{a_n : n > m\} \subseteq \mathcal{P}(\mathbb{G})$ such that $|a_n| = n$ for every $n > m$ are given. Recursively choose $g_n \in b_n$ so that the set $\prod_{m<i<n} g_i \cdot U_n \cap a_n \cdot U_n = \emptyset$. Such g_n exists as d is left invariant, hence for every $g \in a_n$ the set $g \cdot U_n$ intersects at most one of the sets $\prod_{m<i<n} g_i \cdot h \cdot U_n$ for $h \in b_n$, and $|b_n| = |a_n| + 1$. The closures of the sets $\prod_{m<i<n} g_i \cdot U_n$, for $n > m$ form a decreasing sequence of sets of e-diameter converging to 0, hence by completeness of e their intersection is a singleton $x \in U_m$ which is not in $\bigcup_{n>m} a_n \cdot U_n$. □

THEOREM 2.4 ([**20**]). *There is a non-trivial, left-invariant, outer regular, σ-subadditive diffuse submeasure on every Polish group.*

PROOF. Fix a sequence $\{U_n : n \in \omega\}$ as in Lemma 2.3 and define for $A \subseteq \mathbb{G}$:

$$\mu(A) = \inf \left\{ \sum_{i\in\omega} \frac{1}{n_i} : A \subseteq \bigcup_{i\in\omega} g_i \cdot U_{n_i} \right\}.$$

It is immediate from the definition that μ is a diffuse σ-additive, left invariant, outer regular submeasure on $\mathbb{G}$. To see that μ non-trivial it suffices to note that $\mu(U_m) = \frac{1}{m}$. To see that $\mu(U_m)$ is not less than $\frac{1}{m}$, note that by the key property of $\{U_n : n \in \omega\}$, if $U_m \subseteq \bigcup_{i\in\omega} g_i \cdot U_{n_i}$ then $\sum_{i\in\omega} \frac{1}{n_i} \geqslant \frac{1}{m}$. □

The promised characterization is the following:

THEOREM 2.5 (J. Grebík, see [**20**]). *A subset A of a Polish group $\mathbb{G}$ is of left strong measure zero if and only if $\mu(A) = 0$ for every left-invariant, outer regular, countably additive diffuse submeasure on $\mathbb{G}$.*

PROOF. Assume first that $X \in \mathbf{Smz}(\mathbb{G})$, let μ be a left-invariant, outer regular, countably additive diffuse submeasure on $\mathbb{G}$, and let $\varepsilon > 0$ be arbitrary. As μ is non-atomic and outer regular, there is a sequence $\{U_n : n \in \omega\}$ of neighborhoods of $1_{\mathbb{G}}$ such that $\sum_{n\in\omega} \mu(U_n) < \varepsilon$. Now, as $X \in \mathbf{Smz}(\mathbb{G})$, there is a sequence $\{g_n : n \in \omega\} \subseteq \mathbb{G}$ such that $X \subseteq \bigcup_{n\in\omega} g_n \cdot U_n$. By left invariance of μ, $\mu(X) \leqslant \sum_{n\in\omega} \mu(U_n) < \varepsilon$. Hence $\mu(X) = 0$.

On the other hand, assume that $X \subseteq \mathbb{G}$ has $\mu(X) = 0$ for every invariant, non-atomic, outer regular submeasure μ on $\mathbb{G}$, and let $\{V_n : n \in \omega\}$ be a sequence of open neighbourhoods of $1_{\mathbb{G}}$ in $\mathbb{G}$. Let $\{U_n : n \in \omega\}$ be a decreasing local basis as in Lemma 2.3, that is such that for every $m \in \omega$ and $\{a_n : n > m\}$ such that

$|a_n| = n$ for every $n > m$, $U_m \not\subseteq \bigcup_{n>m} a_n \cdot U_n$, by passing on to a subsequence, we may assume that $U_n \subseteq V_n$ for every $n \in \omega$. Let $\{n_i : i \in \omega\} \subseteq \omega$ be such that $n_{i+1} > n_i$ for every $i \in \omega$, and let $\mathcal{W} = \{U_{n_{i+1}} : i \in \omega\}$, and $w(U_{n_{i+1}}) = \frac{1}{n_i}$. Then define a submeasure μ by putting for $A \subseteq \mathbb{G}$

$$\mu(A) = \inf \left\{ \sum_{i\in\omega} w(W_i) : A \subseteq \bigcup_{i\in\omega} g_i \cdot W_i \right\}$$

with each $W_i \in \mathcal{W}$ and $g_i \in \mathbb{G}$. This is again a left-invariant, σ-subadditive, non-atomic, outer regular submeasure, with $\mu(U_{n_{i+1}}) = \frac{1}{n_i}$. Hence $\mu(X) = 0$, in particular, there is a sequence $\{W_j : j \in \omega\}$ and a sequence $\{q_j : j \in \omega\}$ such that $X \subseteq \bigcup_{j\in\omega} g_i \cdot W_j$ and $\sum_{j\in\omega} w(W_j) < 1$. This means that every $U_{n_{i+1}}$ appears fewer that n_i-many times as one of the W_j, so there is permutation $\pi \in S_\infty$ such that $W_{\pi(n)} \subseteq V_n$ for every $n \in \omega$, hence $X \subseteq \bigcup_{n\in\omega} g_{\pi(n)} \cdot V_n$. Hence $X \in \mathsf{Smz}(\mathbb{G})$. □

In the general context of a metric space, Szpilrajn [**45**] proved that every **Smz** set X has *universal measure zero*, i.e. has measure zero for every finite diffuse Borel measure on X. It should be noted that unlike strong measure zero sets, uncountable universal measure zero sets exist in ZFC as shown by Sierpiński and Szpilrajn [**43**].

PROPOSITION 2.6 (Szpilrajn [**45**]). *Strong measure zero sets in separable metric spaces have universal measure zero.*

PROOF. Aiming towards contradiction, suppose that X is **Smz** yet there is a diffused Borel measure μ on X such that $\mu(X) = 1$. Define a function $f : (0, \infty) \to [0, 1]$ by

$$f(r) = \sup\{\mu(E) : \operatorname{diam} E \leqslant r\}.$$

We claim that $\lim_{r\to 0} f(r) = 0$. Otherwise there is $\varepsilon > 0$ and a sequence of sets E_n such that $\operatorname{diam} E_n \searrow 0$ and $\mu(E_n) \geqslant \varepsilon$. Let $E = \bigcap_{n\in\omega} \bigcup_{m\geqslant n} E_n$. Then clearly $\mu(E) \geqslant \varepsilon > 0$. In particular $E \neq \emptyset$, i.e., there is $I \in [\omega]^\omega$ such that $\bigcap_{n\in I} E_n \neq \emptyset$. Suppose without loss of generality that $I = \omega$. Since any two sets E_n, E_m have a common point, we have $\operatorname{diam}(\bigcup_{m\geqslant n} E_n) \leqslant 2 \operatorname{diam} E_n$. Therefore $\operatorname{diam} E \leqslant 2 \lim_{n\to 0} \operatorname{diam} E_n = 0$, which contradicts $\mu(E) > 0$. We proved that $\lim_{r\to 0} f(r) = 0$. Therefore there is, for each $n \in \omega$, $\varepsilon_n > 0$ such that $\sum_n f(\varepsilon_n) < 1$. Since X is **Smz**, there are sets U_n such that $\operatorname{diam} U_n < \varepsilon_n$ that cover X. It follows that

$$1 = \mu(X) \leqslant \sum_n \mu(U_n) \leqslant \sum_n f(\operatorname{diam} U_n) \leqslant \sum_n f(\varepsilon_n) < 1,$$

the desired contradiction. □

3. The Galvin-Mycielski-Solovay Theorem in Polish groups

In this section we study the Galvin-Mycielski-Solovay theorem in the context of an arbitrary Polish group $\mathbb{G}$. We denote by $\mathcal{M}(\mathbb{G})$, or simply by $\mathcal{M}$ if there is no danger of confusion, the ideal of meager subsets of $\mathbb{G}$. Much of this section exists thanks to the following simple yet crucial observation due to Prikry:

PROPOSITION 3.1 (Prikry [**38**]). *Let $\mathbb{G}$ be a separable group, and let $S \subseteq \mathbb{G}$ be such that $S \cdot M \neq \mathbb{G}$ for all $M \in \mathcal{M}(\mathbb{G})$. Then $S \in \mathsf{Smz}(\mathbb{G})$.*

PROOF. Let S be as above, and let $\{U_n : n \in \omega\}$ be a family of open neighbourhoods of 1 in $\mathbb{G}$. Let $\{g_n : n \in \omega\} \subseteq \mathbb{G}$ be such that $U = \bigcup_{n\in\omega} g_n \cdot U_n$ is dense open in $\mathbb{G}$. Then U^{-1} is dense open in $\mathbb{G}$, the inverse being a homeomorphism, so $M = \mathbb{G} \setminus U^{-1}$ is nowhere dense in $\mathbb{G}$. As $S \cdot M \neq \mathbb{G}$, there is $x \in \mathbb{G} \setminus S \cdot M$, that is $S \subseteq x \cdot U = \bigcup_{n\in\omega} x \cdot g_n \cdot U_n$. Hence, $S \in \mathsf{Smz}(\mathbb{G})$. □

As mentioned in the introduction, Galvin, Mycielski, and Solovay [**13**, **14**] answered Prikry's question by showing that the reverse inclusion holds for $\mathbb{R}$. The same was recently proved for all locally compact groups by Kysiak [**26**] and Fremlin [**12**], independently. We shall present a proof of their theorem (the converse of Prikry's result for locally compact groups) here. Our proof follows [**19**].

Call a subset N of a topological group $\mathbb{G}$ *uniformly nowhere dense* if for every neighborhood U of 1 there is a neighborhood V of 1 such that for every $x \in \mathbb{G}$ there is a $g \in \mathbb{G}$ such that $g \cdot V \subseteq x \cdot U \setminus N$. A set $M \subseteq \mathbb{G}$ is *uniformly meager* if it can be written as a union of countably many uniformly nowhere dense sets. We denote the family of all uniformly meager subsets of $\mathbb{G}$ by $\mathcal{UM}(\mathbb{G})$ (or simply $\mathcal{UM}$). The following generalizes [**14**, Theorem 4].

PROPOSITION 3.2 ([**19**]). *Let $\mathbb{G}$ be a Polish group which is either locally compact or* TSI, *and let $S \in \mathsf{Smz}(\mathbb{G})$. Then $S \cdot M \neq \mathbb{G}$ for all $M \in \mathcal{UM}(\mathbb{G})$.*

PROOF. Assume first that $\mathbb{G}$ admits a invariant metric d. Recall that every invariant metric on a Polish group is complete. Let N be uniformly nowhere dense subset of $\mathbb{G}$. Note that for every $y \in \mathbb{G}$ and an open set $U \subseteq \mathbb{G}$, $y \cdot U \cdot N = U \cdot y \cdot N$, and $y \cdot N$ is uniformly nowhere dense. It follows that for every uniformly nowhere dense $N \subseteq \mathbb{G}$

$$\forall U \text{open } \exists V \text{open } \forall x, y \in \mathbb{G}\ \exists z \in \mathbb{G}\quad z \cdot V \subseteq x \cdot U \setminus V \cdot y \cdot N. \tag{1}$$

Now, fix a Smz set S and a uniformly meager set M written as the union of an increasing sequence $\langle N_n : n \in \omega\rangle$ of uniformly nowhere dense sets. Then there is a sequence $\langle U_n : n \in \omega\rangle$ of open subsets of $\mathbb{G}$, such that for every $n > 0$

$$\forall x, y \in \mathbb{G}\ \exists z \in \mathbb{G}\quad z \cdot U_n \subseteq x \cdot U_{n-1} \setminus U_n \cdot y \cdot N_n. \tag{2}$$

As S is Smz, for every sequence $\{U_n : n \in \omega\}$ of open sets (of diameter converging to 0) there is a sequence $\{g_n : n \in \omega\} \subseteq \mathbb{G}$ such that each $s \in S$ is contained in infinitely many of the sets $g_n \cdot U_n$. Applying (2) recursively there is a sequence $\langle x_n : n \in \omega\rangle$ of elements of $\mathbb{G}$ such that for every $n \in \omega$

$$x_{n+1} \cdot g_{n+1} \cdot U_{n+1} \subseteq x_n \cdot g_n \cdot U_n \setminus (g_{n+1} \cdot U_{n+1} \cdot N_{n+1}).$$

The sequence $\langle x_n \cdot g_n : n \in \omega\rangle$ is Cauchy, let x be its limit, i.e., $\{x\} = \bigcap_{n\in\omega} x_n \cdot g_n \cdot U_n$. Then $x \notin \bigcup_{n\in\omega} g_n \cdot U_n \cdot N_n \supseteq S \cdot M$, as the sequence $\langle N_n : n \in \omega\rangle$ is increasing and every element of S is contained in infinitely many of the $g_n \cdot U_n$ (for every $(s, m) \in S \times M$ there is an $n \in \omega$ such that $s \in g_n \cdot U_n$ and $m \in N_n$).

Now if $\mathbb{G}$ is locally compact, the proof proceeds along similar lines. Only (1) is replaced by the following lemma.

LEMMA 3.3. *Let $\mathbb{G}$ be a locally compact Polish group equipped with a complete metric d, and let $U \subseteq \mathbb{G}$ be an open set with compact closure $C = \overline{U}$ and $P \subseteq \mathbb{G}$ be compact nowhere dense. Then*

$$\forall \varepsilon > 0\ \exists \delta > 0\ \forall x \in C\ \forall y \in K\ \exists z \in C\quad B(z, \delta) \subseteq B(x, \varepsilon) \setminus (B(y, \delta) \cdot P).$$

PROOF. Fix $\varepsilon > 0$ and define $f : C \times K \to \mathbb{R}$ by

$$f(x, y) = \sup\{t : \exists z \in C\ B(z, t) \subseteq B^\circ(x, \varepsilon) \setminus y \cdot P)\}.$$

Then f is positive on $C \times K$ and attains its (positive) minimum.

To see that, consider, for each $z \in C$, the functions

$$g_z(x) = \underline{d}(z, X \setminus B^\circ(x, \varepsilon)), \quad x \in C$$
$$h_z(y) = \underline{d}(z, y \cdot P), \quad y \in K$$

and note that

$$f(x, y) = \sup_{z \in C} \min(g_z(x), h_z(y)). \tag{3}$$

Using compactness it is easy to see that, for each $z \in C$, the function h_z is lower semicontinuous and that while g_z does not have to be, it has the following lower-semicontinuity property: if $x_n \to x$ and $g_z(x_n) \to 0$, then $g_z(x) = 0$.

Now suppose that there are $(x_n, y_n) \in C \times K$ such that $f(x_n, y_n) \to 0$. Since C, K are compact, passing to subsequences we may assume $(x_n, y_n) \to (x, y) \in C \times K$. Use (3) and the semicontinuity properties of g_z and h_z to conclude that since $f(x_n, y_n) \to 0$, for any z either $g_z(x_n) \to 0$ and then $g_z(x) = 0$, or else $h_z(y_n) \to 0$ and then $h_z(y) = 0$. Use (3) again to conclude that $f(x, y) = 0$, the desired contradiction proving that there is $\eta > 0$ such that $f(x, y) > \eta$ for all x, y. It follows that

$$\forall x \in C\ \forall y \in K\ \exists z \in C \quad B(z, \eta) \subseteq B(x, \varepsilon)\ \wedge B(z, \eta) \cap y \cdot P = \emptyset.$$

The latter of course yields $B(z, \frac{\eta}{2}) \cap B(y \cdot P, \frac{\eta}{2}) = \emptyset$. On the other hand, there is $\xi > 0$ such that

$$\forall y \in K \quad B(y, \xi) \cdot P \subseteq B(y \cdot P, \tfrac{\eta}{2}).$$

It follows that $B(z, \frac{\eta}{2}) \cap B(y, \xi) \cdot P = \emptyset$. Thus letting $\delta = \min\{\frac{\eta}{2}, \xi\}$ yields the lemma. $\square$

To conclude, write $\mathbb{G}$ as the union of an increasing sequence of open sets with compact closures K_n, and write a meager set M as the union of an increasing sequence of compact nowhere dense sets P_n. Choose $x_0 \in \mathbb{G}$ and $\varepsilon_0 > 0$ such that $B(x_0, \varepsilon_0)$ is compact. Let $C = B(x_0, \varepsilon_0)$. By the above lemma there is a sequence $\langle \varepsilon_n : n \in \omega \rangle \in (0, \infty)^\omega$ such that for every $n > 0$

$$\forall x \in C\ \forall y \in K_n\ \exists z \in C \quad B(z, \varepsilon_n) \subseteq B(x, \varepsilon_{n-1}) \setminus B(y, \varepsilon_n) \cap K_n) \cdot P_n. \tag{4}$$

We may of course suppose that $\varepsilon_n \to 0$. Since S is **Smz**, there is a cover $\{E_n\}$ of S such that diam $E_n < \varepsilon_n$ for all n such that each point of S is covered by infinitely many E_n's).

For each n there is y such that $E_n \subseteq B(y, \varepsilon_n)$. Therefore, using repeatedly (4), there is a sequence $\langle x_n : n \in \omega \rangle$ in C such that for all $n \in \omega$

$$B(x_{n+1}, \varepsilon_{n+1}) \subseteq B(x_n, \varepsilon_n) \setminus (E_{n+1} \cap K_{n+1}) \cdot P_{n+1}.$$

Let x be the unique point of $\bigcap_{n \in \omega} B(x_n, \varepsilon_n)$ (there is one, since $B(x_0, \varepsilon_0)$ is compact and is unique as $\varepsilon_n \to 0$). Then $x \notin \bigcup_{n \in \omega} (E_n \cap K_n) \cdot P_n$.

Thus, to prove that x is not covered by $S \cdot M$ it suffices to show that $S \times M \subseteq \bigcup_{n \in \omega} (E_n \cap K_n) \times P_n$. Let $(s, m) \in S \times M$. There is k such that $(s, m) \in K_k \times P_k$. Since there are infinitely many n such that $s \in E_n$, there is $n \geqslant k$ such that

$s \in E_n$, hence $s \in E_n \cap K_k \subseteq E_n \cap K_n$. Also $m \in P_k \subseteq P_n$. Therefore $(s, m) \in (E_n \cap K_n) \times P_n$. The desired inclusion is proved. □

The Galvin-Mycielski-Solovay/Fremlin/Kysiak result follows from the fact that in a locally compact group every meager set is uniformly meager:

PROPOSITION 3.4 ([**19**]). *A Polish group* $\mathbb{G}$ *is locally compact if and only if* $\mathcal{M}(\mathbb{G}) = \mathcal{UM}(\mathbb{G})$.

PROOF. We shall see first that $\mathcal{M}(X) = \mathcal{UM}(X)$ *for every locally compact metric space* X.

To that end it suffices to see that every nowhere dense subset of a compact space is, in fact, uniformly nowhere dense. Let N be a nowhere dense subset of a compact space X, and let $\varepsilon > 0$. Let F be a finite subset of X such that $Z = \bigcup_{x \in F} B(x, \frac{\varepsilon}{2})$. For every $x \in F$ let $y_x \in B(x, \frac{\varepsilon}{2})$ and $\delta_x > 0$ be such that $B(y_x, \delta_x) \subseteq B(x, \frac{\varepsilon}{2}) \setminus N$. Then $\delta = \min\{\delta_x : x \in F\}$ works as $B(x, \frac{\varepsilon}{2}) \subseteq B(z, \varepsilon)$ whenever $z \in B(x, \frac{\varepsilon}{2})$.

On the other hand, $\mathcal{M}(X) \neq \mathcal{UM}(X)$ *for every nowhere locally compact complete metric space* X.

To see this let X be nowhere locally compact with a complete metric d. Then for every U with non-empty interior there is an $\varepsilon^U > 0$ and a pairwise disjoint family $\{V_k^U : k \in \omega\}$ of open balls of radius ε^U contained in U.

We shall construct a nowhere dense set N which is not uniformly meager. To do that we recursively construct a family $\{U_s : s \in \omega^{<\omega}\}$ of non-empty regular closed[1] sets so that

(1) $\operatorname{diam} U_s \leqslant 2^{-|s|}$ for every $s \in \omega^{<\omega}$,
(2) $\bigcup_{n \in \omega} U_{s^\frown n} \subseteq U_s$ for every $s \in \omega^{<\omega}$,
(3) $U_{s^\frown n} \cap U_{s^\frown m} = \emptyset$ for every $s \in \omega^{<\omega}$ and any two distinct $m, n \in \omega$,
(4) $\operatorname{int}(U_s \setminus \bigcup_{n \in \omega} U_{s^\frown n}) \neq \emptyset$ for every $s \in \omega^{<\omega}$,
(5) for all $s \in \omega^{<\omega}$ and $k \in \omega$ and $x \in V_k^{U_s}$ there is a $n \in \omega$ such that $U_{s^\frown n} \subseteq B(x, 2^{-k})$.

To do this is straightforward.

Having constructed such a family, let $N = \bigcap_{j \in \omega} \bigcup_{|s|=j} U_s$. This is the required set:

It is nowhere dense as a non-empty open set U is either disjoint from N, or contains U_s for some $s \in \omega^{<\omega}$. Then, however, $\emptyset \neq \operatorname{int}(U_s \setminus \bigcup_{n \in \omega} U_{s^\frown n}) \subseteq U \setminus N$ by the property (4) above.

Now we will prove that N is not uniformly meager in X. The set N is naturally homeomorphic to ω^ω (see properties (1)–(3) above), hence satisfies the Baire Category Theorem. Aiming toward a contradiction assume that $N \subseteq \bigcup_{l \in \omega} N_l$, where each N_l is a closed uniformly nowhere dense subset of X. By the Baire Category Theorem applied to N there is an $s \in \omega^{<\omega}$ and an $l \in \omega$ such that $U_s \cap N \subseteq N_l$, hence $U_s \cap N$ is uniformly nowhere dense. So, there is a $\delta > 0$ as in the definition of uniformly nowhere dense corresponding to ε^{U_s}. Consider $V_k^{U_s}$, for $2^{-k} < \delta$. Then, on the one hand there is an $x \in V_k^{U_s}$ such that $B(x, 2^{-k}) \subseteq V_k^{U_s} \setminus N$, and on the other hand, there is (see property (5) above) an $n \in \omega$ such that $\emptyset \neq N \cap U_{s^\frown n} \subseteq B(x, 2^{-k})$, which is a contradiction.

The result follows as every Polish group is either locally compact or nowhere locally compact. □

[1]Recall that a set U is *regular closed* if U is the closure of the interior of U.

And finally:

THEOREM 3.5 (Fremlin [**12**], Kysiak [**26**]). *Let $\mathbb{G}$ be a locally compact Polish group. A set $A \subseteq \mathbb{G}$ is of strong measure zero if and only if $A \cdot M \neq \mathbb{G}$ for every meager set $M \subseteq \mathbb{G}$*

PROOF. The theorem follows directly from Propositions 3.1, 3.2 and 3.4. □

Next we shall discuss the possibility of extending the Galvin-Mycielski-Solovay theorem to a larger class of Polish groups. First, one needs to realize that assuming the Borel conjecture, the theorem holds trivially for every Polish group $\mathbb{G}$, as strong measure zero sets in all Polish groups are exactly the countable subsets ([**9**]), hence $S \cdot M$ is meager for every strong measure zero set S and meager set M, hence $S \cdot M \neq \mathbb{G}$.

On the other hand, it was shown in [**19**], that the Galvin-Mycielski-Solovay theorem fails for the Baer-Specker group $\mathbb{Z}^\omega$, assuming $\mathsf{cov}(\mathcal{M}) = \mathfrak{c}$. We conjecture that assuming a strong failure of the Borel conjecture the locally compact Polish groups are exactly the ones for which the theorem holds.

CONJECTURE 3.6 (CH). *The Galvin-Mycielski-Solovay theorem holds in a Polish group $\mathbb{G}$ if and only if $\mathbb{G}$ is locally compact.*

That the Continuum Hypothesis is optimal in the sense that under CH the Galvin-Mycielski-Solovay Theorem fails for as many Polish groups as possible follows from the logical complexity of the problem. The statement $\mathbb{G}$ *satisfies the Galvin-Mycielski-Solovay Theorem* is a Π^2_1-statement with $\mathbb{G}$ as a parameter, and hence is decided by the Ω-logic under the Continuum hypothesis. Moreover, if the statement is true assuming CH it is true in ZFC (see e.g. [**27**]).

We shall verify (following [**20**]) that the conjecture is true for Abelian Polish groups, in fact, it is true for all groups with a complete (both-sided)-invariant metric, and also for closed subgroups of the permutation group S_∞. Whether it is true in general remains open.

There is a, perhaps an even more interesting, stronger ZFC conjecture on the structure of Polish groups. The following concept was introduced in [**19**] and the term coined in [**20**]: A nonempty subset C of a Polish group $\mathbb{G}$ is said to be *anti-GMS* if it is nowhere dense and for every sequence $\{U_n : n \in \omega\}$ of open neighborhoods of 1 there is a sequence $\{g_n : n \in \omega\}$ of elements of $\mathbb{G}$ such that for every $g \in \mathbb{G}$, the set $g \cdot \bigcup_{n\in\omega} g_n \cdot U_n$ is dense in C.

The reason for introducing anti-GMS sets is the following:

PROPOSITION 3.7 ([**19**]). *Assuming $\mathsf{cov}(\mathcal{M}) = \mathfrak{c}$, if $C \subseteq \mathbb{G}$ is anti-GMS, then there is a strong measure zero set S such that $S \cdot C = \mathbb{G}$.*

PROOF. Enumerate $\mathbb{G} = \{g_\alpha : \alpha < \mathfrak{c}\}$ and enumerate all sequences of open sets in $\mathbb{G}$ as $\{\langle U^\alpha_n : n \in \omega\rangle : \alpha < \mathfrak{c}\}$. Let $C \subseteq \mathbb{G}$ be anti-GMS, and for every $\alpha < \mathfrak{c}$ let $\langle g^\alpha_n : n \in \omega\rangle$ be such that for all $g \in \mathbb{G}$, $(g \cdot \bigcup_{n\in\omega} g^\alpha_n \cdot U^\alpha_n) \cap M$ is comeager in M. Let $U_\alpha = \bigcup_{n\in\omega} g^\alpha_n \cdot U^\alpha_n$.

As $\mathsf{cov}(\mathcal{M}) = \mathfrak{c}$, the intersection of fewer than $\mathfrak{c}$ relatively dense open subsets of M is not empty. In particular, for every $\alpha < \mathfrak{c}$, there is an

$$m_\alpha \in M \cap (g_\alpha^{-1} \cdot \bigcap_{\beta\leqslant\alpha} U_\beta).$$

There is then an $x_\alpha \in \bigcap_{\beta\leqslant\alpha} U_\beta$ such that $m_\alpha = g_\alpha^{-1} \cdot x_\alpha$. That is $g_\alpha = x_\alpha \cdot m_\alpha^{-1}$.

Let $X = \{x_\alpha : \alpha < \mathfrak{c}\}$. Then $\mathbb{G} = X \cdot M^{-1}$. Let us see that $X \in \mathsf{Smz}(\mathbb{G})$: Given a sequence $\{U_n : n \in \omega\}$ of neighbourhoods of $1_\mathbb{G}$, consider the subsequence $\{U_{2n} : n \in \omega\}$. It is listed as $\langle U_n^\alpha : n \in \omega\rangle$ for some $\alpha < \mathfrak{c}$. By the construction, every $x_\gamma \in U_\alpha$ for $\gamma \geqslant \alpha$. On the other hand, $X \setminus U_\alpha \subseteq \{x_\beta : \beta < \alpha\}$, hence has size less than $\mathsf{cov}(\mathcal{M}) = \mathfrak{c}$. Committing the sin of forward reference, by Theorem 4.2(i), $X \setminus U_\alpha$ is a **Smz** set, hence can be covered by the sets $\{U_{2n+1} : n \in \omega\}$. Hence, X has strong measure zero. □

The anti-GMS sets are our only tool for disproving the Galvin-Mycielski-Solovay Theorem in non-locally compact groups. Hence the *Strong Conjecture* is:

CONJECTURE 3.8. *Exactly one of the following holds for any Polish group $\mathbb{G}$: Either $\mathbb{G}$ is locally compact, or it contains an anti-GMS set.*

It is immediate from Proposition 3.7 that the Strong conjecture, indeed, solves the GMS conjecture stated above. It is, in fact the Strong conjecture we have verified for the aforementioned classes of groups:

THEOREM 3.9 ([**20**]). *Let $\mathbb{G}$ be a non-locally compact* TSI *Polish group. Then $\mathbb{G}$ contains an anti-GMS set.*

PROOF. Let μ be a non-trivial left-invariant countably subadditive, diffuse outer regular submeasure given by Theorem 2.4.

CLAIM 3.10. There is a nowhere dense set $C \subseteq \mathbb{G}$ such that for every open set O intersecting C there is an open set U and $\{g_m : m \in \omega\} \subseteq \mathbb{G}$ such that for every $m \in \omega$ $g_m \cdot U \subseteq O$ and $\lim_{m\in\omega} \mu(g_m \cdot U \setminus C) = 0$.

PROOF. To construct the set C let $\{B_n : n \in \omega\}$ be a basis for the topology of $\mathbb{G}$. Recursively construct an increasing sequence of open sets $\{W_k : k \in \omega\}$ and a sequence $\{A_k : k \in \omega\}$ of countable sets of pairs of the form $\langle U, \varepsilon\rangle$, where U is an open subset of $\mathbb{G}$ and $\varepsilon > 0$ such that

(1) $B_k \cap W_{k+1} \neq \emptyset$,
(2) $\forall g \in \mathbb{G}\ |\{\langle U, \varepsilon\rangle \in A_k : g \in U\}| \leqslant k$,
(3) $\forall \langle U, \varepsilon\rangle \in A_k\ \mu(\overline{W_{k+1}} \cap U \setminus \overline{W}_k) < \frac{\varepsilon}{2^k}$, and if k is the least such that $\langle U, \varepsilon\rangle \in A_k$ then $\overline{W}_k \cap U = \emptyset$, and
(4) (a) either $B_k \subseteq W_{k+1}$,
 (b) or, there is an open neighborhood V of $1_\mathbb{G}$ and there are distinct $\{g_i : i \in \omega\} \subseteq \mathbb{G}$ and $\{\varepsilon_i : i \in \omega\}$ such that for all $i \in \omega$ $g_i \cdot V \subseteq B_n$, $\langle g_i \cdot V, \varepsilon_i\rangle \in A_k$ and $\lim_{i\in\omega} \varepsilon_i = 0$.

To carry out the construction, put first $W_0 = A_0 = \emptyset$. Having constructed W_k and A_k, see first whether $B_k \subseteq \overline{W}_k$. If so, let $A_{k+1} = A_k$ and let $W_{k+1} = W_k \cup B_k$. If $B_k \not\subseteq \overline{W}_k$, let U_0, U_1 be disjoint open subsets of $B_k \setminus \overline{W}_k$. By (2) there is an open set $U_2 \subseteq U_0$ contained or disjoint from every U appearing in A_k, by non-atomicity of μ we may require $\mu(U_2)$ to be so small that $W_{k+1} = W_k \cup U_2$ satisfies (3) for all $\langle U, \varepsilon\rangle \in A_k$ such that $U_2 \subseteq U$. Finally, as U_1 is not compact, there are infinitely many balls of the same diameter (i.e. translates of the same open set) with disjoint closures contained in U_1, add them to A_{k+1} paired with some real numbers converging to 0. It is clear that (1)-(4) are satisfied.

Now, let $C = \mathbb{G} \setminus \bigcup_{k\in\omega} W_k$. Then C is a closed nowhere dense set by (1). By (3), $\mu(U \setminus C) < \varepsilon$ for every $\langle U, \varepsilon\rangle \in \bigcup_{k\in\omega} A_k$, and if $B_k \cap C \neq \emptyset$ then by (4) B_k contains infinitely many sets with the required properties. □

The set C from the claim is anti-GMS. In order to verify this let a sequence $\{U_m : m \in \omega\}$ of open neighborhoods of $1_{\mathbb{G}}$ be given, without loss of generality of diameter shrinking to 0. Let again $\{B_n : n \in \omega\}$ be a basis for the topology of $\mathbb{G}$, and let $\{A_n : n \in \omega\}$ be a partition of ω into infinite sets. For every B_n such that $B_n \cap C \neq \emptyset$ let W_n be an open neighborhood of $1_{\mathbb{G}}$ such that there are distinct $\{g_i : i \in \omega\} \subseteq \mathbb{G}$ such that $g_i \cdot W_n \subseteq B_n$ and $\lim_{i \in \omega} \mu(g_i \cdot W_n \setminus C) = 0$.

Now, let V_n be an open neighborhood of $1_{\mathbb{G}}$ such that $V_n \cdot U_j \subseteq W_n$ for all but finitely many $j \in A_n$, and let $g_j, h_j \in \mathbb{G}$ be such that $h_j \cdot W_n \subseteq B_n$, $\mu(h_j \cdot W_n \setminus C) < \mu(U_j)$, and $G = \bigcup_{j \in A_n} h_j \cdot V_n \cdot g_j^{-1}$ (here is where we use the invariance of the metric). The sequence $\{g_i : i \in \omega\} \subseteq \mathbb{G}$ witnesses (for the sequence $\{U_i : i \in \omega\}$ that C is anti-GMS. Indeed, of $g \in \mathbb{G}$ and $B_n \cap C \neq \emptyset$ then there is a $j \in A_n$ such that $g \in h_j \cdot V_n \cdot g_j^{-1}$, i.e. $g \cdot g_j \in h_j \cdot V_n$, hence

$$g \cdot g_j \cdot U_j \subseteq h_j \cdot V_n \cdot U_j \subseteq h_j \cdot W_n \subseteq B_n.$$

As $\mu(h_j \cdot W_n \setminus C) < \mu(U_j)$, we get $C \cap B_n \cap g \cdot g_j \cdot U_j \neq \emptyset$, as required. □

COROLLARY 3.11. *The strong conjecture is true for Abelian groups.*

We do not know, whether the strong conjecture is true for all Polish groups, but we can confirm it for another important class of groups – the automorphism groups of countable structures:

THEOREM 3.12 ([**20**]). *Let $\mathbb{G}$ be a non-locally compact closed subgroup of S_∞. Then $\mathbb{G}$ contains an anti-GMS set.*

PROOF. As $\mathbb{G}$ is not locally compact, there is an infinite $A \subseteq \omega$ such that for all $n \in A$ there are infinitely many $m \in \omega$ for which there are $g \in G$ such that $g \restriction n = 1_{\mathbb{G}} \restriction n$ and $g(n) = m$. Given $n, m \in A$, $n < m$ define

$$R_{n,m} = \{(g,h) \in \mathbb{G} \times \mathbb{G} : g(n) \notin \operatorname{rng}(h \restriction m) \text{ and } h(n) \notin \operatorname{rng}(g \restriction m)\}.$$

Note that the relation is left-invariant. Let $B = \{u \in \omega^{<\omega} : U$ is one-to-one and $\operatorname{dom}(u) \in A$ and $u \subseteq g$ for some $g \in \mathbb{G}\}$.

CLAIM 3.13. For every $n < m \in A$ and $u \in B$ such that $|u| = n$ there are $g, h \in \mathbb{G}$ both extending u such that $(g,h) \in R_{n,m}$.

PROOF. By the left-invariance of $R_{n,m}$, we may assume that u is the identity on $n = \operatorname{dom}(u)$. Let $\{g_k : k \in \omega\} \subseteq \mathbb{G}$ be such that $g_k \restriction n = u$ and $\{g_k(n)) : k \in \omega\}$ are pairwise distinct, by further shrinking we may assume that for every $l \in [n,m)$ either $\{g_k(n)) : k \in \omega\}$ are pairwise distinct or all equal. In particular, for every $k_0 \in \omega$ the set $\{k_1 \in \omega : g_{k_0}(n) \in \operatorname{rng}(g_{k_1} \restriction [n,m)\}$ is finite, hence $(g_0, g_k) \in R_{n,m}$ for almost all $k \in \omega$. □

Given a subset H of B let $T(H) = \{u \in B : \forall k \in \omega\ u \restriction k \notin H\}$. Construct $D \subseteq B$ such that

(1) $\forall u \in B\ \exists v \in D\ u \subseteq v$,
(2) $\forall v, v' \in D\ \operatorname{rng}(v) \subseteq \operatorname{rng}(v')$ or $\operatorname{rng}(v') \subseteq \operatorname{rng}(v)$, and
(3) $\forall u \in T(D)\ \forall m \in \omega\ \exists v \in T(D)\ m \in \operatorname{rng}(v)$.

It is easy to construct such a set using a simple bookkeeping argument. Having done so, let $C = [T(D)]$.

To see that C is anti-GMS, consider an infinite set $Z \subseteq A$. By the Claim there is a sequence $\{g_n : n \in Z\} \subseteq \mathbb{G}$ such that

$$\forall u \in B\ \exists n_o < n_1 \in Z\ u \subseteq g_{n_0} \cap g_{n_1} \text{ and } (g_{n_o}, g_{n_1}) \in R_{|u|,n_0}.$$

To finish the proof it suffices to show that $g \cdot \bigcup_{n \in Z} \langle g_n \restriction n \rangle \cap C$ is dense in C for every $g \in \mathbb{G}$. To that end fix $g \in \mathbb{G}$ and $v \in B$ such that $\langle v \rangle \cap C \neq \emptyset$, and let $k = \operatorname{dom}(v)$ and $u = g^{-1} \cdot v$. Choose $n_0 < n_1 \in Z$ such that $u \subseteq g_{n_0} \cap g_{n_1}$ and $(g_{n_o}, g_{n_1}) \in R_{k,n_1}$. Then either $\langle g \cdot g_{n_0} \restriction n_0 \rangle \cap C \neq \emptyset$ or $\langle g \cdot G_{n_1} \restriction n_1 \rangle \cap C \neq \emptyset$: If not, then there are $s_0, s_1 \in D$ such that $s_0 \subseteq g \cdot g_{n_0} \restriction n_0$ and $s_1 \subseteq g \cdot g_{n_1} \restriction n_1$. Neither $s_0 \subseteq v$ nor $s_1 \subseteq v$ as $C \cap \langle v \rangle \neq \emptyset$, so $g \cdot g_{n_0}(k) \in \operatorname{rng}(s_1)$ and $g \cdot g_{n_1}(k) \in \operatorname{rng}(s_0)$. This, however, contradicts the assumption that $\operatorname{rng}(s_0) \subseteq \operatorname{rng}(s_1)$ or $\operatorname{rng}(s_1) \subseteq \operatorname{rng}(s_0)$. □

4. Cardinal invariants of Smz in Polish groups

Given an ideal $\mathcal{I}$ of subsets of a set X the following are the standard cardinal invariants associated with $\mathcal{I}$:

$$\begin{aligned}
\mathsf{non}(\mathcal{I}) &= \min\{|Y| : Y \subseteq X \wedge Y \notin \mathcal{I}\},\\
\mathsf{add}(\mathcal{I}) &= \min\{|\mathcal{A}| : \mathcal{A} \subseteq \mathcal{I} \wedge \bigcup \mathcal{A} \notin \mathcal{I}\},\\
\mathsf{cov}(\mathcal{I}) &= \min\{|\mathcal{A}| : \mathcal{A} \subseteq \mathcal{I} \wedge \bigcup \mathcal{A} = X\},\\
\mathsf{cof}(\mathcal{I}) &= \min\{|\mathcal{A}| : \mathcal{A} \subseteq \mathcal{I} \wedge (\forall I \in \mathcal{I})(\exists A \in \mathcal{A})(I \subseteq A)\}.
\end{aligned}$$

We denote by $\mathcal{M}, \mathcal{N}$ the ideals of meager and Lebesgue null subsets of 2^ω, respectively. For $f, g \in \omega^\omega$, we say that $f \leqslant^* g$ if $f(n) \leqslant g(n)$ for all but finitely many $n \in \omega$ (the order of *eventual dominance*). A family $F \subseteq \omega^\omega$ is *bounded* if there is an $h \in \omega^\omega$ such that $f \leqslant^* h$ for all $f \in F$; and F is *dominating* if for any $g \in \omega^\omega$ there is $f \in F$ such that $g \leqslant^* f$. The cardinal invariants related to eventual dominance are $\mathfrak{b}$ (the minimal cardinality of an unbounded family) and $\mathfrak{d}$ (the minimal cardinality of a dominating family).

We shall briefly review the results (not necessarily in the chronological order) concerning other cardinal invariants of $\mathsf{Smz}(\mathbb{G})$, after which we give a more detailed account of the $\mathsf{non}(\mathsf{Smz})$ in Polish groups. We shall denote by Smz the ideal of strong measure zero subsets of $\mathbb{R}$. Concerning additivity of Smz, Carlson [**9**] in effect showed that $\mathsf{add}(\mathcal{N}) \leqslant \mathsf{add}(\mathsf{Smz})$, that $\mathsf{add}(\mathsf{Smz}) \leqslant \mathsf{non}(\mathcal{N})$ is a triviality, while Goldstern, Judah and Shelah [**15**] showed that consistently $\mathsf{cof}(\mathcal{M}) < \mathsf{add}(\mathsf{Smz})$, and, of course, Laver [**28**] that consistently $\mathsf{add}(\mathsf{Smz}) < \mathfrak{b}$ and Baumgartner [**2**] that consistently $\mathsf{add}(\mathsf{Smz}) < \mathsf{non}(\mathcal{N})$.

For cofinality of Smz, there are lower bounds $\mathsf{cov}(\mathcal{N})$ and $\mathsf{cov}(\mathcal{M})$ (see below), and it is folklore fact that assuming CH $\mathsf{cof}(\mathsf{Smz}) > \mathfrak{c}$, while the Borel conjecture produces models where $\mathsf{cof}(\mathsf{Smz}) = \mathfrak{c}$. Yorioka [**51**] and more recently Cardona [**7**] produced models of ZFC where $\mathsf{cof}(\mathsf{Smz}) < \mathfrak{c}$. According to our knowledge, it has not been subject to study if $\mathsf{add}(\mathsf{Smz}(\mathbb{G}))$, and/or $\mathsf{cof}(\mathsf{Smz}(\mathbb{G}))$ may vary depending on the Polish group in question.

It is, however, known that the uniformity numbers may differ depending on the group. There is also a surprising asymmetry between $\mathsf{cov}(\mathsf{Smz})$ and $\mathsf{non}(\mathsf{Smz})$. The trivial lower bound for $\mathsf{cov}(\mathsf{Smz})$ is $\mathsf{cov}(\mathcal{N})$ (every Smz-set has Lebesgue measure zero) and it also seems to be the best one. Pawlikowski [**37**] showed that $\mathsf{cov}(\mathsf{Smz}) < \mathsf{add}(\mathcal{M})$ is consistent. On the other hand, Cardona, Mejía and Rivera-Madrid [**8**] recently showed that $\mathsf{cov}(\mathsf{Smz}) = \omega_2 = \mathfrak{c}$ in the iterated Sacks model, hence, there seems to be very little in terms of upper bounds on $\mathsf{cov}(\mathbf{Smz})$, in particular, consistently $\mathsf{cof}(\mathcal{N}) < \mathsf{cov}(\mathbf{Smz})$.

Before moving on we shall mention some of the open problems concerning these invariants:

QUESTION 4.1. (1) ([**8**]) Is it consistent that all four of the cardinal invariants corresponding to **Smz** have simultaneously different values?
(2) ([**8**]) Is it consistent that $\mathsf{add}(\mathbf{Smz}) < \min\{\mathsf{cov}(\mathbf{Smz}), \mathsf{non}(\mathbf{Smz})\}$?
(3) Do any of $\mathsf{add}(\mathbf{Smz}(\mathbb{G}))$, $\mathsf{cov}(\mathbf{Smz}(\mathbb{G}))$, $\mathsf{cof}\,\mathbf{Smz}(\mathbb{G}))$ depend on which Polish group one considers?

Finally, we arrive at the uniformity of $\mathbf{Smz}(\mathbb{G})$ which we shall discuss in considerably more detail. Two more invariants are required here (see [**1, 19, 29, 31**]):

$$\begin{aligned}\mathfrak{eq} &= \min\{|F| : F \subseteq \omega^\omega \text{ bounded}, \forall g \in \omega^\omega\ \exists f \in F\ \forall n \in \omega\ f(n) \neq g(n)\}.\\ \mathfrak{eq}^* &= \min\{|F| : F \subseteq \omega^\omega \text{ bounded},\\ &\qquad \forall g, h \in \omega^\omega\ \exists f \in F\ \exists^\infty n\ \forall k \in [h(n), h(n+1))\ f(n) \neq g(n)\}.\end{aligned}$$

It is a theorem of Bartoszyński [**1**, 2.4.1] that omitting "bounded" from the definition of $\mathfrak{eq}$ yields $\mathsf{cov}(\mathcal{M})$, i.e.

$$\mathsf{cov}(\mathcal{M}) = \min\{|F| : F \subseteq \omega^\omega\ \forall g \in \omega^\omega\ \exists f \in F\ \forall n \in \omega\ f(n) \neq g(n)\}.$$

The following diagram (see [**1, 18**] for proofs) summarizes the provable inequalities between the cardinal invariants mentioned (the arrows point from the smaller to the larger cardinal).

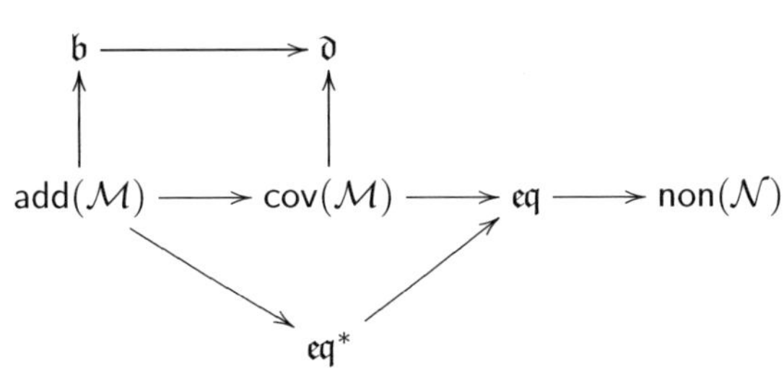

In addition, $\mathsf{add}(\mathcal{M}) = \min\{\mathfrak{b}, \mathfrak{eq}\} = \min\{\mathfrak{b}, \mathfrak{eq}^*\}$, while $\mathsf{cov}(\mathcal{M}) < \min\{\mathfrak{d}, \mathfrak{eq}\}$ is consistent with ZFC by a theorem of Goldstern, Judah and Shelah [**15**]. For any separable metric space X there are upper and lower bounds for $\mathsf{non}(\mathbf{Smz}(X))$ given by Rothberger [**40**] and Szpilrajn [**46**], respectively. The uniformity invariant $\mathsf{non}(\mathbf{Smz}(X))$ for $X = 2^\omega$ and $X = \omega^\omega$ was calculated by Bartoszyński [**1**], and Fremlin and Miller [**30**], respectively.

THEOREM 4.2. *Let X be a separable metric space that is not* **Smz**.
(i) ([**40**]) $\mathsf{cov}(\mathcal{M}) \leqslant \mathsf{non}(\mathbf{Smz}(X))$,
(ii) ([**46**]) *if X is not of universal measure zero, then* $\mathsf{non}(\mathbf{Smz}(X)) \leqslant \mathsf{non}(\mathcal{N})$,
(iii) ([**1**]) $\mathsf{non}(\mathbf{Smz}(\omega^\omega)) = \mathsf{cov}(\mathcal{M})$ *and*
(iv) ([**30**]) $\mathsf{non}(\mathbf{Smz}(2^\omega)) = \mathfrak{eq}$.

PROOF. (i) by now is standard: Given a separable metric space X and a sequence $\{\varepsilon_n : n \in \omega\}$, pick a countable dense set $\{d_n : n \in \omega\} \subseteq X$. For every one to one function $f \in \omega^\omega$ let $U_f = \bigcup_{n\in\omega} B(x_n, \varepsilon_{f_n})$. Now, assume that $|X| < \mathsf{cov}(\mathcal{M})$. To finish the proof it suffices to note that for every $x \in X$ the set $N_x = \{f \in \omega^\omega : f$

is one-to-one and $x \notin U_f\}$ is nowhere dense in the (closed) subspace of ω^ω consisting of one-to-one functions.

To see (ii) recall that every **Smz** set is of universal measure zero by Proposition 2.6, so it suffices to show that $\mathsf{non}(\mathcal{N})$ is the minimal size of a space which is not of universal measure zero. To see this note that any diffuse Borel probability measure μ on a separable metric space X extends to a finite Borel diffuse probability measure $\overline{\mu}$ on its completion $\hat{X}$ by putting $\overline{\mu}(A) = \mu(A \cap X)$. Now, by a theorem of Parthasarathy [**35**] there is a measure preserving Borel isomorphism between $\hat{X}$ with $\overline{\mu}$ and $[0, 1]$ equipped with the Lebesgue measure, hence $|X| \geqslant \mathsf{non}(\mathcal{N})$.

For (iii) it suffices to see that $\mathsf{non}(\mathbf{Smz}(\omega^\omega)) \leqslant \mathsf{cov}(\mathcal{M})$. Let $F \subseteq \omega^\omega$ be such that $|F| = \mathsf{cov}(\mathcal{M})$ and $\forall g \in \omega^\omega\ \exists f \in F\ \forall n \in \omega\ f(n) \neq g(n)$. We shall show that $F \notin \mathbf{Smz}(\omega^\omega)$. Let $\langle s_n : n \in \omega \rangle$ be a sequence of elements of $2^{<\omega}$ such that $|s_n| = n + 1$. Define $g \in \omega^\omega$ by putting $g(n) = s_n(n)$. Then there is an $f \in F$ such that $f(n) \neq g(n)$ for all $n \in \omega$. This, however, means, that $s_n \not\subseteq f$ for any $n \in \omega$. That is no sequence of open sets of diameter $\frac{1}{2^{n+1}}$ covers F.

The proof of (iv) is similar. First we shall show that $\mathsf{non}(\mathbf{Smz}(2^\omega)) \leqslant \mathfrak{eq}$. To that end let $X \subseteq 2^\omega$ be of size less that $\mathfrak{eq}$, and let a sequence $\langle \varepsilon_n : n \in \omega \rangle$ of positive real numbers be given. Let $h \in \omega^\omega$ be such that $\frac{1}{2^{h(n)}} \leqslant \varepsilon_n$ for every $n \in \omega$. For each $n \in \omega$ enumerate 2^n – the set of binary sequences of length n - as $\{s^n_m : m < 2^\omega\}$. For every $x \in X$ let $f_x \in \omega^\omega$ be defined by

$$f_x(n) = m \text{ if and only if } x \restriction h(n) = s^{h(n)}_m.$$

Then $f_x(n) \leqslant 2^{h(n)}$ for every $x \in X$ and $n \in \omega$. As $|X| < \mathfrak{eq}$, there is a $g \in \omega^\omega$ such that $f_x \cap g \neq \emptyset$ for every $x \in X$, and without loss of generality, $g(n) \leqslant 2^{h(n)}$ for every $n \in \omega$ (values above are irrelevant, and can be changed). Then $\langle \langle s^{h(n)}_{g(n)} \rangle : n \in \omega \rangle$ covers X.

On the other hand, assume that $F \subseteq \omega^\omega$ is a bounded family of size less than $\mathsf{non}(\mathbf{Smz}(2^\omega))$. Let $h \in \omega^\omega$ be such that $f(n) \leqslant 2^{h(n)}$ for every $f \in F$ and $n \in \omega$. Let $\{s^n_m : m < 2^\omega\}$ enumerate 2^n as above. For every $f \in \omega^\omega$ let

$$x_f = s^{h(0)}_{f(0)}{}^\frown s^{h(1)}_{f(1)}{}^\frown s^{h(2)}_{f(2)}{}^\frown \cdots$$

and consider the set $X = \{x_f : f \in F\}$. As X is **Smz**, there is a sequence $\langle t_n : n \in \omega \rangle$ such that

(1) $t_n \in 2^{\sum_{i \leqslant n} h(n)}$, and
(2) $\forall f \in F\ \exists n \in \omega\ t_n \subseteq x_f$.

Let $g \in \omega^\omega$ be such that $g(n) \leqslant 2^{h(n)}$ for all $n \in \omega$, and $g(n) = m$ whenever $t_{n+1} = t_n{}^\frown s^{h(n)}_m$. Note that $g(n) = f(n)$ whenever $t_n \subseteq x_f$. □

In particular, the theorem evaluates $\mathsf{non}(\mathbf{Smz})$ for two groups: the compact boolean group 2^ω and the Baer-Specker group $\mathbb{Z}^\omega$. In order to extend these to a wider class of groups we shall need two easy observations:

As mentioned before, strong measure zero is a uniform property, in particular, a uniformly continuous image of a **Smz** set is **Smz** and, on the other hand, if X uniformly embeds into Y, then any set $A \subseteq X$ that is not **Smz** in X is not **Smz** in Y either. It follows that:

LEMMA 4.3. (i) *If a space Y is a uniformly continuous image of a space X then* $\mathsf{non}(\mathbf{Smz}(X)) \leqslant \mathsf{non}(\mathbf{Smz}(Y))$.
(ii) *If X uniformly embeds into Y, then* $\mathsf{non}(\mathbf{Smz}(X)) \geqslant \mathsf{non}(\mathbf{Smz}(Y))$.

LEMMA 4.4. *A* CLI *Polish group is either locally compact, or contains a uniform copy of* ω^ω.

PROOF. Let $\mathbb{G}$ be a group equipped with a complete, left-invariant metric d. Assuming $\mathbb{G}$ is not locally compact no open set is totally bounded, hence for every $\varepsilon > 0$ exists $\delta > 0$ such that the ball $B(1, \varepsilon)$ contains an infinite set of points that are pairwise at least δ apart. Using this fact construct, for each $n \in \omega$, $\varepsilon_n > 0$ and an infinite set $\{x_n^i : i \in \omega\} \subseteq B(1, \varepsilon_n)$ such that if $i \neq j$ then $d(x_n^i, x_n^j) > 5\varepsilon_{n+1}$. For $s \in \omega^n$ let $y_s = x_0^{s(0)} \cdot x_1^{s(1)} \cdot x_{n-2}^{s(n-2)} \cdot \dots \cdot x_{n-1}^{s(n-1)}$. The construction ensures that for any $f \in \omega^\omega$ the sequence $\langle y_{f\restriction n} : n \in \omega \rangle$ is Cauchy. Let z_f be its limit. It is easy to check that since d is left-invariant, the mapping $f \mapsto z_f$ is a uniform embedding of ω^ω into $\mathbb{G}$. □

THEOREM 4.5. *Let* $\mathbb{G}$ *be a* CLI *Polish group.*

(i) *If* $\mathbb{G}$ *is locally compact, then* $\mathsf{non}(\mathbf{Smz}(\mathbb{G})) = \mathfrak{eq}$,
(ii) *if* $\mathbb{G}$ *is not locally compact, then* $\mathsf{non}(\mathbf{Smz}(\mathbb{G})) = \mathsf{cov}(\mathcal{M})$.

PROOF. As every Polish group contains a (uniform) copy of 2^ω,

$$\mathsf{cov}(\mathcal{M}) \leqslant \mathsf{non}(\mathbf{Smz}(\mathbb{G})) \leqslant \mathfrak{eq}$$

for every Polish group $\mathbb{G}$. Similarly, $\mathsf{non}(\mathbf{Smz}(\mathbb{G})) \leqslant \mathsf{cov}(\mathcal{M})$ whenever $\mathbb{G}$ contains a uniform copy of ω^ω by Theorem 4.2. In particular, $\mathsf{non}(\mathbf{Smz}(\mathbb{G})) = \mathsf{cov}(\mathcal{M})$ if $\mathbb{G}$ is a non-locally compact CLI group.

So all that remains to be seen is that $\mathfrak{eq} \leqslant \mathsf{non}(\mathbf{Smz}(\mathbb{G}))$ for any locally compact Polish group $\mathbb{G}$. Write $\mathbb{G}$ as an increasing union of compact subsets K_n, $n \in \omega$. As each K_n is a uniformly continuous image of 2^ω (see e.g. [**22**, Theorem 4.18]), every subset of K_n which is not of strong measure zero has size at least $\mathfrak{eq}$ by Lemma 4.3(i) and Theorem 4.2(iii). On the other hand, as $\mathbf{Smz}(\mathbb{G})$ is a σ-ideal (Proposition 2.1), every subset of $\mathbb{G}$ which is not $\mathbf{Smz}$ has a non-$\mathbf{Smz}$ intersection with one of the k_n's, hence $\mathfrak{eq} \leqslant \mathsf{non}(\mathbf{Smz}(\mathbb{G}))$. □

It is, of course, a natural question whether the result of the Theorem (or of the preceding lemma) remains true also for Polish groups which are not CLI.

The final remark of this section deals with *transitive covering for category* (considered by Bartoszyński [**1**, 2.7] for 2^ω, and Miller and Steprāns [**31**] for general Polish group and further studied in [**19**]):

$$\mathsf{cov}^*(\mathcal{M}(\mathbb{G})) = \min\{|A| : A \subseteq \mathbb{G} \text{ and } A \cdot M = \mathbb{G} \text{ for some meager set } M \subseteq \mathbb{G}\}$$

By Theorem 3.5, for a locally compact group $\mathbb{G}$ strong measure zero sets coincide with the sets whose meager translates do not cover $\mathbb{G}$, hence, in particular, $\mathsf{non}(\mathbf{Smz}(\mathbb{G})) = \mathsf{cov}^*(\mathcal{M}(\mathbb{G}))$ for every locally compact group $\mathbb{G}$. It follows from Prikry's Proposition 3.1 that $\mathsf{cov}^*(\mathcal{M}(\mathbb{G})) \leqslant \mathsf{non}(\mathbf{Smz}(\mathbb{G}))$, for every Polish group $\mathbb{G}$, hence $\mathsf{cov}(\mathcal{M}) \leqslant \mathsf{cov}^*(\mathcal{M}(\mathbb{G})) \leqslant \mathfrak{eq}$ for any Polish group. As $\mathsf{non}(\mathbf{Smz}(\mathbb{G})) = \mathsf{cov}(\mathcal{M})$ for all CLI groups which are not locally compact, we can conclude that:

COROLLARY 4.6. $\mathsf{non}(\mathbf{Smz}(\mathbb{G})) = \mathsf{cov}^*(\mathcal{M}(\mathbb{G}))$ *for any* CLI *group.*

The conjecture is that the two numbers coincide for any Polish group.

QUESTION 4.7. Is $\mathsf{cov}^*(\mathcal{M}(\mathbb{G})) = \mathsf{cov}(\mathcal{M})$ for all Polish groups which are not locally compact?

5. Strong measure zero and Hausdorff measures

As mentioned above, there is a profound connection between strong measure zero and Hausdorff measures. The following characterizations of strong measure zero in terms of Hausdorff measures and dimensions were proved in [**54**]. They are based on a classical Besicovitch result [**4, 5**].

Hausdorff measure. Before getting any further we need to review Hausdorff measure and dimension. We set up the necessary definitions and recall relevant facts.

A non-decreasing, right-continuous function $h : [0, \infty) \to [0, \infty)$ such that $h(0) = 0$ and $h(r) > 0$ if $r > 0$ is called a *gauge*. Gauges are often ordered as follows, cf. [**39**]:

$$g \prec h \quad \stackrel{\text{def}}{\equiv} \quad \lim_{r \to 0+} \frac{h(r)}{g(r)} = 0.$$

Notice that for any sequence $\langle h_n : n \in \omega \rangle$ of gauges there is a gauge h such that $h \prec h_n$ for all n.

Given $\delta > 0$, call a cover $\mathcal{A}$ of a set $E \subseteq X$ a *δ-fine cover* if $\forall A \in \mathcal{A}$ $\operatorname{diam} A \leqslant \delta$. If h is a gauge, the *h-dimensional Hausdorff measure* $\mathcal{H}^h(E)$ of a set $E \subseteq X$ is defined thus: For each $\delta > 0$ define

$$\mathcal{H}^h_\delta(E) = \inf \left\{ \sum_{n \in \omega} h(\operatorname{diam} E_n) : \{E_n\} \text{ is a countable } \delta\text{-fine cover of } E \right\}$$

and let

$$\mathcal{H}^h(E) = \sup_{\delta > 0} \mathcal{H}^h_\delta(E).$$

In the common case of $h(r) = r^s$ for some $s > 0$, we write $\mathcal{H}^s$ for $\mathcal{H}^h$, and likewise for $\mathcal{H}^h_\delta$.

Properties of Hausdorff measures are well-known. The following, including the two lemmas, can be found e.g. in [**39**]. The restriction of $\mathcal{H}^h$ to Borel sets is a G_δ-regular Borel measure. Recall that a sequence of sets $\langle E_n : n \in \omega \rangle$ is termed a *λ-cover* of $E \subseteq X$ if every point of E is contained in infinitely many E_n's.

THEOREM 5.1 ([**4, 5**]). *A metric space X is* **Smz** *if and only if $\mathcal{H}^h(X) = 0$ for each gauge h.*

PROOF. Suppose first that X is **Smz**. Let h be a gauge. For each $\delta > 0$ pick a sequence $\langle \varepsilon_n \rangle$ such that $0 < \varepsilon_n < \delta$ and $h(\varepsilon_n) < \delta 2^{-n}$. There is a cover $\{U_n\}$ of X such that $\operatorname{diam} U_n < \varepsilon_n$ for all n. Obviously $\sum h(\operatorname{diam} U_n) < \delta$ and it follows that $\mathcal{H}^h_\delta(X) \leqslant \delta$. Let $\delta \to 0$ to get $\mathcal{H}^h(X) = 0$.

Now suppose that $\mathcal{H}^h(X) = 0$ for every gauge. Let $\langle \varepsilon_n \rangle$ be a sequence of positive numbers. Choose a gauge h such that $h(\varepsilon_n) > \frac{1}{n}$. Since $\mathcal{H}^h(X) = 0$, there is a countable cover $\{U_n\}$ such that $\sum h(\operatorname{diam} U_n) < 1$. As h is right-continuous, there are $\delta_n > \operatorname{diam} U_n$ such that $\sum h(\delta_n) < 1$. Since $\delta_n > 0$, rearranging the sequence we may suppose that δ_n decrease. Therefore $nh(\delta_n) \leqslant \sum_{i<n} h(\delta_n) < 1$. It follows that $h(\delta_n) < \frac{1}{n} < h(\varepsilon_n)$ and consequently $\delta_n < \varepsilon_n$ and in particular $\operatorname{diam} U_n < \varepsilon_n$, as required. □

We will need a cartesian product inequality. Given two metric spaces X and Y with respective metrics d_X and d_Y, provide the cartesian product $X \times Y$ with the maximum metric

$$d((x_1,y_1),(x_2,y_2)) = \max(d_X(x_1,x_2), d_Y(y_1,y_2)). \tag{5}$$

A gauge h satisfies the *doubling condition* or h is *doubling* if $\overline{\lim}_{r\to 0}\frac{h(2r)}{h(r)} < \infty$.

LEMMA 5.2 ([**17, 23**]). *Let X, Y be metric spaces, g a gauge and h a doubling gauge. Then $\mathcal{H}^h(X)\,\mathcal{H}^g(Y) \leqslant \mathcal{H}^{hg}(X\times Y)$.*

The following lemma on uniformly continuous mappings is well-known, see, e.g., [**39**, Theorem 29].

LEMMA 5.3. *Let $f:(X,d_X)\to(Y,d_Y)$ be a uniformly continuous mapping and g its modulus, i.e., $d_Y(f(x),f(y)) \leqslant g(d_X(x,y))$ for all $x,y\in X$. Then $\mathcal{H}^h(f(X)) \leqslant \mathcal{H}^{h\circ g}(X)$ for any gauge h.*

Recall that the *Hausdorff dimension* of X is defined by

$$\dim_{\mathsf{H}} X = \sup\{s>0 : \mathcal{H}^s(X)=\infty\} = \inf\{s>0 : \mathcal{H}^s(X)=0\}.$$

Properties of Hausdorff dimension are well-known. In particular, $\dim_{\mathsf{H}} X = 0$ if X is countable; and if $f: X\to Y$ is Lipschitz, then $\dim_{\mathsf{H}} f(X) \leqslant \dim_{\mathsf{H}} X$.

COROLLARY 5.4 ([**54**]). *Let X be a metric space. The following are equivalent.*

(i) *X is* **Smz**,
(ii) *$\dim_{\mathsf{H}} f(X) = 0$ for each uniformly continuous mapping f on X,*
(iii) *$\dim_{\mathsf{H}}(X,\rho) = 0$ for each uniformly equivalent metric ρ on X.*

PROOF. (i)⇒(ii) Let $s>0$ be arbitrary. Let $f: X\to Y$ be uniformly continuous and let g be the modulus of f. Define $h(x) = (g(x))^s$. Lemma 5.3(i) yields $\mathcal{H}^s(f(X)) \leqslant \mathcal{H}^h(X)$. By the above theorem $\mathcal{H}^h(X)=0$ and thus $\mathcal{H}^s(f(X))=0$. Since this holds for all $s>0$, it follows that $\dim_{\mathsf{H}} f(X)=0$.

(ii)⇒(iii) is trivial.

(iii)⇒(i) Denote by d the metric of X. Let h be a gauge. Choose a strictly increasing, convex (and in particular subadditive) gauge g such that $g \prec h$. The properties of g ensure that $\rho(x,y) = g(d(x,y))$ is a uniformly equivalent metric on X. The identity map $\mathrm{id}_X : (X,\rho)\to(X,d)$ is of course uniformly continuous and its modulus is g^{-1}, the inverse of g. Hence by Lemma 5.3(i) $\mathcal{H}^h(X,d) \leqslant \mathcal{H}^{h\circ g^{-1}}(X,\rho)$. Since $g\prec h$, we have $\mathcal{H}^{h\circ g^{-1}}(X,\rho) \leqslant \mathcal{H}^1(X,\rho)$ and $\mathcal{H}^1(X,\rho)=0$ by (ii). Thus $\mathcal{H}^h(X,d)=0$. □

Our next goal is to characterize **Smz** by behavior of cartesian products. Recall that for $p\in 2^{<\omega}$, $\langle p\rangle = \{x\in 2^\omega : p\subseteq x\}$ denotes the cone determined by p and for $T\subseteq 2^{<\omega}$ we let $\langle T\rangle = \bigcup_{p\in T}\langle p\rangle$.

The coordinatewise addition modulo 2 makes 2^ω a compact topological group. Routine proofs show that in the metric of the least difference (defined in the introduction) the 1-dimensional Hausdorff measure $\mathcal{H}^1$ coincides on Borel sets with its Haar measure, i.e., the usual product measure on 2^ω. In particular $\mathcal{H}^1(2^\omega)=1$.

We consider the important σ-ideal $\mathcal{E}$ on 2^ω generated by closed null sets, i.e., the ideal of all subsets of 2^ω that are contained in an F_σ set of Haar measure zero.

LEMMA 5.5. (i) *For each $I\in[\omega]^\omega$, the set $C_I = \{x\in 2^\omega : x{\restriction}I\equiv 0\}$ is in $\mathcal{E}$.*
(ii) *For each $h\prec 1$ there is $I\in[\omega]^\omega$ such that $\mathcal{H}^h(C_I)>0$.*

PROOF. (i) Let $I \in [\omega]^\omega$. For each $n \in \omega$, the family $\{\langle p\rangle : p \in C_I{\restriction}n\}$ is obviously a 2^{-n}-fine cover of C_I of cardinality $2^{|n\setminus I|}$. Therefore $\mathcal{H}^1_{2^{-n}}(C_I) \leqslant 2^{|n\setminus I|}2^{-n} = 2^{-|n\cap I|}$. Hence $\mathcal{H}^1(C_I) \leqslant \lim_{n\to\infty} 2^{-|n\cap I|} = 0$.

(ii) $h \prec 1$ yields $\frac{h(2^{-n})}{2^{-n}} \to \infty$. Therefore there is $I \in [\omega]^\omega$ sparse enough to satisfy $2^{|n\cap I|} \leqslant \frac{h(2^{-n})}{2^{-n}}$, i.e., $2^{-|n\setminus I|} \leqslant h(2^{-n})$ for all $n \in \omega$. Consider the product measure λ on C_I given as follows: If $p \in 2^n$ and $\langle p\rangle \cap C_I \neq \emptyset$, put $\lambda(\langle p\rangle \cap C_I) = 2^{-|n\setminus I|}$. Straightforward calculation shows that $h(\operatorname{diam} E) \geqslant \lambda(E)$ for each $E \subseteq C_I$. Hence $\sum_n h(\operatorname{diam} E_n) \geqslant \sum_n \lambda(E_n) \geqslant \lambda(C_I) = 1$ for each cover $\{E_n\}$ of C_I and $\mathcal{H}^h(C_I) \geqslant 1$ follows. □

THEOREM 5.6. *The following are equivalent.*

(i) X *is* **Smz**,
(ii) $\mathcal{H}^h(X \times Y) = 0$ *for every gauge h and every compact metric space Y such that* $\mathcal{H}^h(Y) = 0$,
(iii) $\mathcal{H}^1(X \times E) = 0$ *for every* $E \in \mathcal{E}$.

PROOF. (i)⇒(ii): Suppose X is **Smz** and let Y be compact. Let $\delta > 0$. Since $\mathcal{H}^h(Y) = 0$, for each $j \in \omega$ there is a finite cover $\mathcal{U}_j$ of Y such that $\sum_{U\in\mathcal{U}_j} h(\operatorname{diam} U) < 2^{-j-1}\delta$. We may also assume that $\operatorname{diam} U < \delta$ for all $U \in \mathcal{U}_j$.

Let $\varepsilon_j = \min\{\operatorname{diam} U : U \in \mathcal{U}_j\}$. Since X is **Smz**, there is a cover $\{V_j\}$ of X such that $\operatorname{diam} V_j \leqslant \varepsilon_j$. Define

$$\mathcal{W} = \{V_j \times U : j \in \omega,\ U \in \mathcal{U}_j\}.$$

$\mathcal{W}$ is obviously a cover of $X \times Y$. The choice of ε_j yields $\operatorname{diam}(V_j \times U) = \operatorname{diam} U$ for all j and $U \in \mathcal{U}_j$. Therefore

$$\sum_{W\in\mathcal{W}} h(\operatorname{diam} W) = \sum_{j\in\omega}\sum_{U\in\mathcal{U}_j} h(\operatorname{diam} U) < \sum_{j\in\omega} 2^{-j-1}\delta = \delta.$$

It follows that $\mathcal{H}^h_\delta(X \times Y) < \delta$, and $\mathcal{H}^h(X \times Y) = 0$ obtains by letting $\delta \to 0$.

(ii)⇒(iii)⇒(iv) is trivial.

(iv)⇒(i): Suppose X is not **Smz**. We will show that $\mathcal{H}^1(X \times E) > 0$ for some $E \in \mathcal{E}$. By assumption and Theorem 5.4 there is a gauge h such that $\mathcal{H}^h(X) > 0$. We may assume h be concave and $h(r) \geqslant \sqrt{r}$. In particular, by concavity of h the function $g(r) = r/h(r)$ is increasing. Moreover, $h(r) \geqslant \sqrt{r}$ yields $\lim_{r\to 0} g(r) = 0$, i.e., g is a gauge, and $g \prec 1$. Further, $g(2r) = 2r/h(2r) \leqslant 2r/h(r) = 2g(r)$, i.e., g is doubling.

Use Lemma 5.5(ii) to find $I \in [\omega]^\omega$ such that $\mathcal{H}^g(C_I) > 0$ and let $E = C_I$. By Lemma 5.5(i), $E \in \mathcal{E}$. Since g is doubling, Lemma 5.2 applies:

$$\mathcal{H}^1(X \times C_I) = \mathcal{H}^{h\cdot g}(X \times C_I) \geqslant \mathcal{H}^h(X) \cdot \mathcal{H}^g(C_I) > 0. \qquad \square$$

COROLLARY 5.7. *If X is* **Smz**, *then* $\dim_{\mathsf{H}} X \times Y = \dim_{\mathsf{H}} Y$ *for every compact metric space Y.*

Note that this consistently fails when we drop the assumption that Y is compact: By a classical example (cf. [**12**, 534P]), if $\mathsf{cov}(\mathcal{M}) = \mathfrak{c}$, then there is a **Smz** set $X \subseteq \mathbb{R}$ such that $X + X = \mathbb{R}$. Since $X + X$ is a Lipschitz image of $X \times X$, we have $\dim_{\mathsf{H}} X \times X \geqslant \dim_{\mathsf{H}} X + X = 1$, while X is **Smz** and $\dim_{\mathsf{H}} X = 0$.

6. Sharp measure zero

Recall that a set S in a Polish group is called *meager-additive* if $S \cdot M$ is meager for every meager set M. This is obviously a strengthening of the "algebraic" characterization of **Smz** in the Galvin-Mycielski-Solovay Theorem. Meager-additive sets, in particular in 2^ω, have received a lot of attention, e.g., in [**1, 33, 42, 47**].

Very recently it was shown that meager-additive sets are characterized by a combinatorial condition very similar to the definition of **Smz** and also in terms of Hausdorff measures. In this section we will have a look at these descriptions.

DEFINITION 6.1. A set $S \subseteq X$ in a complete metric space X has *sharp measure zero* if for every gauge h there is a σ-compact set $K \supseteq S$ such that $\mathcal{H}^h(K) = 0$.

We first work towards an intrinsic definition equivalent to the one above. The following variation of Hausdorff measure seems to be the right notion for that. Let h be a gauge. For each $\delta > 0$ define

$$\overline{\mathcal{H}}^h_\delta(E) = \inf\left\{\sum_{n=0}^{N} h(\operatorname{diam} E_n) : \{E_n : n \leqslant N\} \text{ is a } \textit{finite } \delta\text{-fine cover of } E\right\}$$

and let

$$\overline{\mathcal{H}}^h_0(E) = \sup_{\delta>0} \overline{\mathcal{H}}^h_\delta(E).$$

Note the striking similarity with the Hausdorff measure. The only difference is that only finite covers are taken into account. It is easy to check that $\overline{\mathcal{H}}^h_0$ is finitely subadditive. However, it is not a measure, since it may (due to the finite covers) lack σ-additivity. To turn it into a measure we need to apply the operation known as Munroe's *Method I construction* (cf. [**32**] or [**39**]):

$$\overline{\mathcal{H}}^h(E) = \inf\Big\{\sum_{n\in\omega} \overline{\mathcal{H}}^h_0(E_n) : E \subseteq \bigcup_{n\in\omega} E_n\Big\}.$$

Thus defined set function $\overline{\mathcal{H}}^h$ is indeed an outer measure whose restriction to Borel sets is a Borel measure. We will called it *h-dimensional upper Hausdorff measure.*

Upper Hausdorff measures behave much like Hausdorff measures. We list some important properties of upper Hausdorff measures. We refer to [**54**] for details.

Denote $\mathcal{N}_\sigma(\overline{\mathcal{H}}^h_0)$ the smallest σ-additive ideal that contains all sets E with $\overline{\mathcal{H}}^h_0(E) = 0$. Note that while $E \in \mathcal{N}_\sigma(\overline{\mathcal{H}}^h_0) \Rightarrow \overline{\mathcal{H}}^h(E) = 0$, the reverse implication in general fails. Write $E_n \nearrow E$ to denote that $\langle E_n : n \in \omega\rangle$ is an increasing sequence of sets with union E. The following lists some basic features of $\overline{\mathcal{H}}^h$ and $\overline{\mathcal{H}}^h_0$.

LEMMA 6.2 ([**54**]). *Let h be a gauge and E a set in a metric space.*

(i) *If $\overline{\mathcal{H}}^h_0(E) < \infty$, then E is totally bounded.*
(ii) $\overline{\mathcal{H}}^h_0(E) = \overline{\mathcal{H}}^h_0(\overline{E})$.
(iii) $\overline{\mathcal{H}}^h_0(E) = \mathcal{H}^h(E)$ *if E is compact.*
(iv) *If $E \in \mathcal{N}_\sigma(\overline{\mathcal{H}}^h_0)$, then $\overline{\mathcal{H}}^h(E) = 0$.*
(v) *If X is complete, $E \subseteq X$ and $E \in \mathcal{N}_\sigma(\overline{\mathcal{H}}^h_0)$, then there is a σ-compact set $K \supseteq E$ such that $\mathcal{H}^h(K) = 0$.*
(vi) *If X is complete and $E \subseteq X$, then*

$$\overline{\mathcal{H}}^h(E) = \inf\{\mathcal{H}^h(K) : K \supseteq E \text{ is } \sigma\text{-compact}\}.$$

(vii) *In particular $\overline{\mathcal{H}}^h(E) = \mathcal{H}^h(E)$ if E is σ-compact.*
(viii) *If $g \prec h$ and $\overline{\mathcal{H}}^g(E) < \infty$, then $E \in \mathcal{N}_\sigma(\overline{\mathcal{H}}^h_0)$; in particular, $\overline{\mathcal{H}}^h(E) = 0$.*

It is straightforward from Lemma 6.2 that the following intrinsic definition of sharp measure zero is consistent with the one above.

DEFINITION 6.3. A metric space X has sharp measure zero if $\overline{\mathcal{H}}^h(X) = 0$ for every gauge h. Sharp measure zero is abbreviated as $\mathsf{Smz}^\sharp$.

It is no surprise that Theorem 5.4 has a counterpart for $\mathsf{Smz}^\sharp$, with basically the same proof. Upper Hausdorff dimension is defined as expected:

$$\overline{\dim}_{\mathsf{H}} X = \sup\{s > 0 : \overline{\mathcal{H}}^s(X) = \infty\} = \inf\{s > 0 : \overline{\mathcal{H}}^s(X) = 0\},$$

see [**53, 54**] for more on the Hausdorff dimension.

THEOREM 6.4 ([**54**]). *Let X be a metric space. The following are equivalent.*

(i) *X is $\mathsf{Smz}^\sharp$,*
(ii) *$\overline{\dim}_{\mathsf{H}} f(X) = 0$ for each uniformly continuous mapping f on X,*
(iii) *$\overline{\dim}_{\mathsf{H}}(X, \rho) = 0$ for each uniformly equivalent metric ρ on X.*

It is straightforward from the definition and Theorem 6.4 that $\mathsf{Smz}^\sharp$ is a σ-additive property and that it is preserved by uniformly continuous mappings.

Sharp measure zero can be described in terms of covers. The description is strikingly similar to the Borel's definition of strong measure zero.

A countable cover $\{U_j\}$ of X is a called a *γ-cover* if each $x \in X$ belongs to all but finitely many U_j.

The following notion was studied, e.g., in [**24**]. A sequence $\langle W_n \rangle$ of sets in X is called a *γ-groupable cover* if there is a partition $\omega = I_0 \cup I_1 \cup I_2 \cup \dots$ into consecutive finite intervals (i.e. I_{j+1} is on the right of I_j for all j) such that the sequence $\langle \bigcup_{n \in I_j} W_n : j \in \omega \rangle$ is a γ-cover. The partition $\langle I_j \rangle$ will be occasionally called a *witnessing partition* and the finite families $\{U_n : n \in I_j\}$ will be occasionally called *witnessing families.*

LEMMA 6.5 ([**54**]). *$E \in \mathcal{N}_\sigma(\overline{\mathcal{H}}^h_0)$ if and only if E has a γ-groupable cover $\langle U_n : n \in \omega \rangle$ such that $\sum_{n \in \omega} h(\operatorname{diam} U_n) < \infty$.*

PROOF. Suppose that $E \in \mathcal{N}_\sigma(\overline{\mathcal{H}}^h_0)$. Let $E_n \nearrow E$ be such that $\overline{\mathcal{H}}^h_0(E_n) = 0$. For each n let $\mathcal{G}_n$ be a finite cover of E_n such that $\sum_{G \in \mathcal{G}_n} h(\operatorname{diam} G) < 2^{-n}$. Since the family $\mathcal{G} = \bigcup_n \mathcal{G}_n$ is obviously a γ-groupable cover, we are done.

In the opposite direction, suppose that $\langle U_n : n \in \omega \rangle$ is a γ-groupable cover $\langle U_n : n \in \omega \rangle$ such that $\sum_{n \in \omega} h(\operatorname{diam} U_n) < \infty$ with witnessing families $\mathcal{G}_j$. Let $E_k = \bigcap_{j \geqslant k} \bigcup \mathcal{G}_j$. Then $E = \bigcup_{k \in \omega} E_k$. For each k, the set E_k is covered by each $\mathcal{G}_j$, $j \geqslant k$, and $\sum_{G \in \mathcal{G}_j} h(\operatorname{diam} G)$ is as small as needed for j large enough. Hence $\overline{\mathcal{H}}^h_0(E_k) = 0$ and consequently $E \in \mathcal{N}_\sigma(\overline{\mathcal{H}}^h_0)$. □

THEOREM 6.6 ([**54**]). *A metric space X is $\mathsf{Smz}^\sharp$ if and only if it has the following property: for every sequence $\langle \varepsilon_n : n \in \omega \rangle$ of positive real numbers there is a γ-groupable cover $\{U_n : n \in \omega\}$ of X such that $\operatorname{diam} U_n \leqslant \varepsilon_n$ for all n.*

PROOF. The pattern of the proof is the same as that of the proof of the Besicovitch's theorem 5.1, but there are details that make it much more involved.

The forward implication is easy. Let h be a gauge. Pick $\varepsilon_n > 0$ such that $\sum_n h(\varepsilon_n) < \infty$. By assumption, there is a γ-groupable cover $\langle G_n \rangle$ such that $\operatorname{diam} G_n \leqslant \varepsilon_n$. Therefore $\sum_n h(\operatorname{diam} G_n) \leqslant \sum_n h(\varepsilon_n) < \infty$. Now apply Lemma 6.5 to conclude that $X \in \mathcal{N}_\sigma(\overline{\mathcal{H}}^h_0)$ and in particular $\overline{\mathcal{H}}^h(X) = 0$.

The reverse implication: Let $\langle \varepsilon_n \rangle \in (0,\infty)^\omega$. Choose a gauge g such that $g(\varepsilon_n) > \frac{1}{n}$ for all $n \geqslant 1$ and then a gauge $h \prec g$. Since $\overline{\mathcal{H}}^h(X) = 0$, Lemma 6.2(viii) yields $X \in \mathcal{N}_\sigma(\overline{\mathcal{H}}^g_0)$, which in turn yields, with the aid of Lemma 6.5, a γ-groupable cover $\langle G_n : n \in \omega \rangle$ such that $\sum_n g(\operatorname{diam} G_n) < \infty$. Let $\{I_j : j \in \omega\}$ be the witnessing partition and $\mathcal{G}_j = \{G_n : n \in I_j\}$ the witnessing families.

We want to permute the cover so that diameters decrease. Some of the diameters may be 0. Also, the permutation may break down the witnessing families. We thus have to exercise some care.

For each n choose $\delta_n > \operatorname{diam} G_n$ so that $\sum_n g(\delta_n) < \infty$. Next choose an increasing sequence $\langle j_k : n \in \omega \rangle$ satisfying for all $k \in \omega$

(a) $\sum\{g(\delta_n) : n \in I_{j_k}\} < 2^{-k-1}$,

(b) $\max\{\delta_n : n \in I_{j_{k+1}}\} < \min\{\delta_n : n \in I_{j_k}\}$.

Let $I = \bigcup_{k\in\omega} I_{j_k}$. Rearrange G_n's within each group $\mathcal{G}_{j_k}$ so that $\langle \delta_n : n \in I_{j_k} \rangle$ form a non-increasing sequence. Together with (b) this ensures that the sequence $\langle \delta_n : n \in I \rangle$ is non-increasing.

For each $n \in \omega$ let $\widehat{n} \in I$ be the unique index such that $n = |I \cap \widehat{n}|$ and define $H_n = G_{\widehat{n}}$. It follows, with the aid of (a) and the definition of g, that for all $n \in \omega$

$$g(\operatorname{diam} H_n) = g(\operatorname{diam} G_{\widehat{n}}) \leqslant g(\delta_{\widehat{n}}) \leqslant \frac{1}{n} \sum\{g(\delta_m) : m \in I, m \leqslant \widehat{n}\}$$
$$\leqslant \frac{1}{n} \sum\{g(\delta_m) : m \in I\} \leqslant \frac{1}{n} < g(\varepsilon_n)$$

and thus $\operatorname{diam} H_n \leqslant \varepsilon_n$ for all n. Moreover, the families $\mathcal{G}_{j_k}$, $k \in \omega$, witness that $\langle H_n : n \in \omega \rangle$ is a γ-groupable cover. □

THEOREM 6.7 ([**54**]). *Let X be a metric space. The following are equivalent.*

(i) *X is* $\mathsf{Smz}^\sharp$,

(ii) *$\overline{\mathcal{H}}^h(X \times Y) = 0$ for each gauge h and Y such that $\overline{\mathcal{H}}^h_0(Y) = 0$,*

(iii) *$\overline{\mathcal{H}}^1(X \times E) = 0$ for each $E \in \mathcal{E}$.*

PROOF. (i)⇒(ii): Suppose X is $\mathsf{Smz}^\sharp$. Let h be a gauge and $\overline{\mathcal{H}}^h_0(Y) = 0$. By Lemma 6.5 there is a γ-groupable cover $\mathcal{U}$ of Y such that $\sum_{U\in\mathcal{U}} h(\operatorname{diam} U) < \infty$. For each $U \in \mathcal{U}$ there is $\delta_U > \operatorname{diam} U$ such that $\sum_{U\in\mathcal{U}} h(\delta_U) < \infty$. Denote by $\mathcal{U}_j$ the witnessing families and let $\varepsilon_j = \min\{\delta_U : U \in \mathcal{U}_j\}$. Since X is $\mathsf{Smz}^\sharp$, Theorem 6.6 yields a γ-groupable cover $\langle V_j : n \in \omega \rangle$ of X such that $\operatorname{diam} V_j \leqslant \varepsilon_j$. Denote by $\mathcal{V}_k$ the witnessing families. Define a family of sets in $X \times Y$

$$\mathcal{W} = \{V_j \times U : j \in \omega, U \in \mathcal{U}_j\}.$$

It is routine to check that $\mathcal{W}$ is a γ-groupable cover of $F \times Y$. Since $\operatorname{diam}(V_j \times U) \leqslant \delta_U$ for all j and $U \in \mathcal{U}_j$ by the choice of ε_j, we have

$$\sum_{W\in\mathcal{W}} h(\operatorname{diam} W) \leqslant \sum_{U\in\mathcal{U}} h(\delta_U) < \infty.$$

Thus it follows from Lemma 6.5 that $X \times Y \in \mathcal{N}_\sigma(\overline{\mathcal{H}}^h_0)$ and in particular $\overline{\mathcal{H}}^h(X \times Y) = 0$.

(ii)⇒(iii) is trivial. The proof of (iii)⇒(i) is very much like that of Theorem 5.6. Suppose X is not $\mathsf{Smz}^\sharp$. We need to find $E \in \mathcal{E}$ such that $\overline{\mathcal{H}}^1(X \times E) > 0$. By assumption there is a gauge h such that $\overline{\mathcal{H}}^h(X) > 0$. We may suppose h is concave, and find a doubling gauge $g \prec 1$ such that $g(r)h(r) = r$. Then use Lemma 5.5(ii) to find $I \in [\omega]^\omega$ such that $\mathcal{H}^g(C_I) > 0$. We now need a product inequality on upper Hausdorff measures analogous to Lemma 5.2 proved in [**54**, 3.5,7.4].

LEMMA ([**54**]). *Let X, Y be metric spaces and g a gauge and h a doubling gauge. Then $\mathcal{H}^h(X)\,\overline{\mathcal{H}}^g(Y) \leqslant \overline{\mathcal{H}}^{hg}(X \times Y)$.*

Using this lemma, we get

$$\overline{\mathcal{H}}^1(X \times C_I) = \overline{\mathcal{H}}^{h\cdot g}(X \times C_I) \geqslant \overline{\mathcal{H}}^h(X) \cdot \mathcal{H}^g(C_I) > 0. \qquad \square$$

As we already mentioned, under $\mathsf{cov}(\mathcal{M}) = \mathfrak{c}$ there is an example ([**12**, 534P]) of a **Smz** set $X \subseteq \mathbb{R}$ such that $X \times X$ is not **Smz**. Scheepers [**41**] examines thoroughly conditions imposed on a **Smz** set X that would ensure that a product of X with another **Smz** set is **Smz**. A recent roofing result claims that if one of the factors is **Smz**$^\sharp$, then the product is **Smz**.

THEOREM 6.8 ([**54**]). (i) *If X and Y are* **Smz**$^\sharp$*, then $X \times Y$ is* **Smz**$^\sharp$*.*
(ii) *If X is* **Smz** *and Y is* **Smz**$^\sharp$*, then $X \times Y$ is* **Smz**.

PROOF. Suppose Y is **Smz**$^\sharp$. By Lemma 6.2(viii), $Y \in \mathcal{N}_\sigma(\overline{\mathcal{H}}_0^h)$ for all gauges h.

(i) If X is **Smz**$^\sharp$, then Theorem 6.7(ii) yields $\overline{\mathcal{H}}^h(X \times Y) = 0$ for all gauges h.

(ii) Let h be a gauge. Since Y is **Smz**$^\sharp$, it is σ-totally bounded and therefore there is a σ-compact set $K \supseteq Y$ in the completion of Y such that $\mathcal{H}^h(K) = 0$. Since X is **Smz**, Theorem 5.6(ii) yields $\mathcal{H}^h(X \times Y) = 0$, which is by Theorem 5.4(ii) enough. $\square$

This theorem, together with the above example, provides an easy argument that shows that consistently not every **Smz** set is **Smz**$^\sharp$: The **Smz** set X such that $X \times X$ is not **Smz** cannot be **Smz**$^\sharp$.

We illustrate the power of the theorem by the following

COROLLARY 6.9. *Let $X \subseteq \mathbb{R}^2$. The following are equivalent.*

(i) *X is* **Smz**$^\sharp$*,*
(ii) *all orthogonal projections of X on lines are* **Smz**$^\sharp$*,*
(iii) *at least two orthogonal projections of X on lines are* **Smz**$^\sharp$*.*

PROOF. Since orthogonal projections are uniformly continuous, (i)$\Rightarrow$(ii) from preservation of **Smz**$^\sharp$ by uniformly continuous mappings. (ii)$\Rightarrow$(iii) is trivial.
(iii)$\Rightarrow$(i): Let L_1, L_2 be two nonparallel lines and π_1, π_2 the corresponding orthogonal projections. *Mutatis mutandis* we may suppose that L_1 is the x-axis and L_2 is the y-axis. Thus $X \subseteq \pi_1 X \times \pi_2 X$. Theorem 6.8(ii) thus concludes the proof. $\square$

Theorem 6.8(ii) also raises the question whether a space whose product with any **Smz** set of reals is **Smz** has to be **Smz**$^\sharp$. As shown in [**54**], the answer is consistently no: In the forcing extension constructed by Corazza in [**10**] we have the following. A similar observation was noted without proof in [**33**] and also in [**49**].

PROPOSITION 6.10. *In the Corazza model there is a set $X \subseteq 2^\omega$ that is not* **Smz**$^\sharp$ *and yet $X \times Y$ is* **Smz** *for each* **Smz** *set $Y \subseteq 2^\omega$.*

7. Meager additive sets and sharp measure zero

We now look at the meager-additive sets in Polish groups and establish their surprising and profound connection with $\mathsf{Smz}^\sharp$ sets. The theory nicely parallels the Galvin-Mycielski-Solovay Theorem. Most of the material of this section comes from [**52**] and [**54**].

Whenever $\mathbb{G}$ is a Polish group, $\mathsf{Smz}^\sharp(\mathbb{G})$ denotes the family of sharp measure zero sets with respect to any left-invariant metric. The notion of $\mathsf{Smz}^\sharp$ is of course a uniform invariant – it is neither a topological, nor a metric property. Therefore it does not matter which left-invariant metric we choose. The same proof shows that $\mathsf{Smz}^\sharp$ sets, just like Smz sets, form a bi-invariant σ-ideal.

PROPOSITION 7.1. $\mathsf{Smz}^\sharp(\mathbb{G})$ *is a bi-invariant* σ*–ideal.*

Recall that if $\mathbb{G}$ is a Polish group, we denote by $\mathcal{M}(\mathbb{G})$, or simply by $\mathcal{M}$ if there is no danger of confusion, the ideal of meager subsets of $\mathbb{G}$.

DEFINITION 7.2. A set $S \subseteq \mathbb{G}$ is called *meager-additive* (or $\mathcal{M}$*-additive*) if $S \cdot M$ is meager for every meager set $M \subseteq \mathbb{G}$. The family of all meager-additive sets is denoted by $\mathcal{M}^*(\mathbb{G})$.

It is straightforward from the definition that

PROPOSITION 7.3. $\mathcal{M}^*(\mathbb{G})$ *is a bi-invariant* σ*–ideal.*

The hard implication of Galvin-Mycielski-Solovay Theorem claims that if S is Smz set in a locally compact group, then $S \cdot M \neq \mathbb{G}$. The analogous statement for $\mathsf{Smz}^\sharp$ and meager-additive sets is about as hard as that.

THEOREM 7.4 ([**52**]). *Let* $\mathbb{G}$ *be a locally compact Polish group. Then every* $\mathsf{Smz}^\sharp$ *set* $S \subseteq \mathbb{G}$ *is* $\mathcal{M}$*-additive, i.e.,* $\mathsf{Smz}^\sharp(\mathbb{G}) \subseteq \mathcal{M}^*(\mathbb{G})$.

PROOF. The proof utilizes Lemma 3.3. Suppose $S \subseteq \mathbb{G}$ is a $\mathsf{Smz}^\sharp$ set and let $M \subseteq \mathbb{G}$ be meager. Let K_n be compact sets in $\mathbb{G}$ with $K_n \nearrow \mathbb{G}$ and let P_n be compact nowhere dense sets with $P_n \nearrow M$. Let $\{U_k\}$ be a countable base of $\mathbb{G}$. For each k choose $x_0^k \in \mathbb{G}$ and $\varepsilon_0^k > 0$ such that $B(x_0^k, \varepsilon_0^k) \subseteq U_k$ is compact, and let $C_k = B(x_0^k, \varepsilon_0^k)$. Use Lemma 3.3 to recursively construct a sequence $\langle \varepsilon_n : n \in \omega \rangle$ of positive numbers such that

$$\begin{aligned} &\forall n\ \forall i \leqslant n\ \forall x \in C_i\ \forall y \in K_n\ \exists z \in C_i \\ &\qquad B(z, \varepsilon_n) \subseteq B(x, \varepsilon_{n-1}) \setminus ((B(y, \varepsilon_n) \cap K_n) \cdot P_n). \end{aligned} \tag{6}$$

Since S is $\mathsf{Smz}^\sharp$, there is an γ-groupable cover $\{E_n\}$ of S such that $\operatorname{diam} E_n < \varepsilon_n$ for all n. Hence for each n there is y such that $E_n \subseteq B(y, \varepsilon_n)$. Therefore we may use (6) to construct for each k a sequence $\langle x_n^k : n \in \omega \rangle$ such that

$$B(x_{n+1}^k, \varepsilon_{n+1}) \subseteq B(x_n^k, \varepsilon_n) \setminus ((E_{n+1} \cap K_{n+1}) \cdot P_{n+1}). \tag{7}$$

It is easy to check that since $\{E_n\}$ is a γ-groupable cover of S and $K_n \nearrow \mathbb{G}$, the family $\{E_n \cap K_n\}$ is also a γ-groupable cover of S. Thus we might have supposed that $E_n \subseteq K_n$, and also that all E_n's are closed. Therefore (7) simplifies to

$$B(x_{n+1}^k, \varepsilon_{n+1}) \subseteq B(x_n^k, \varepsilon_n) \setminus (E_{n+1} \cdot P_{n+1}). \tag{8}$$

In particular, $B(x^k_n,\varepsilon_n)$ is a decreasing sequence of compact balls for all k and thus there is a point $x^k\in U_k$ such that

$$x^k \notin \bigcup_{n\in\omega}(E_n\cdot P_n). \tag{9}$$

Now construct a set $\widehat{S}$ as follows: Let $\mathcal{G}_j$ be the groups of E_n's witnessing to the γ-groupability of $\{E_n\}$. Put $G_n=\bigcap_{n\in\mathcal{G}_j}E_n$ and let $F_n=\bigcap_{i<n}G_i$ and $\widehat{S}=\bigcup_{n\in\omega}F_n$. It is clear that since E_n's are closed, the set $\widehat{S}$ is F_σ, and clearly $S\subseteq\widehat{S}$. Moreover, routine calculation shows that $\widehat{S}\times M\subseteq\bigcup_n E_n\times P_n$. Therefore (9) yields $x^k\notin\widehat{S}\cdot M$ for all k. So letting $D=\{x^k:k\in\omega\}$, the set D is disjoint with $\widehat{S}\cdot M$ and it is dense in X. Since $\widehat{S}$ and M are σ-compact, so is the set $\widehat{S}\times M$. It follows that $\widehat{S}\cdot M$, being a continuous image of $\widehat{S}\times M$, is also σ-compact. In summary, $\widehat{S}\cdot M$ is an F_σ set disjoint with a dense set, and is thus meager. □

One would expect that the reverse implication that parallels the trivial Proposition 3.1 of Prikry would be also very easy. Surprisingly, it is not easy at all. Only very recently it was proved in [**52**] that it holds for Polish groups that admit a (both-sided) invariant metric.

Recall that Polish groups that admit an invariant metric are referred to as TSI groups. Compact or abelian Polish groups are TSI; any invariant metric on a TSI group is complete.

THEOREM 7.5 ([**52**]). *Let $\mathbb{G}$ be a* TSI *Polish group. If $S\subseteq\mathbb{G}$ is an $\mathcal{M}$-additive set, then S is* $\mathbf{Smz}^\sharp$ *(in any metric on $\mathbb{G}$).*

THEOREM 7.6. *Let $\mathbb{G}$ be a locally compact* TSI *Polish group. Then* $\mathbf{Smz}^\sharp(\mathbb{G})=\mathcal{M}^*(\mathbb{G})$.

It takes several pages to prove Theorem 7.5, in contrast to the five lines of the proof of Proposition 3.1. We present a proof of the particular case of $\mathbb{G}=2^\omega$ that takes advantage of the regular combinatorial structure of the Cantor set and a deep characterization of $\mathcal{M}$-additive sets by Shelah ([**42**] or [**1**, Theorem 2.7.17]) and is thus much shorter.

LEMMA 7.7 ([**42**]). *$X\subseteq 2^\omega$ is $\mathcal{M}$-additive if and only if*

$$\begin{aligned}&\forall f\in\omega^{\uparrow\omega}\ \exists g\in\omega^\omega\ \exists y\in 2^\omega\ \forall x\in X\ \forall^\infty n\ \exists k\\ &\quad g(n)\leqslant f(k)<f(k+1)\leqslant g(n+1)\ \&\ x{\restriction}[f(k),f(k+1))=y{\restriction}[f(k),f(k+1)).\end{aligned}$$

THEOREM 7.8. $\mathbf{Smz}^\sharp(2^\omega)=\mathcal{M}^*(2^\omega)$.

PROOF. Let $S\subseteq 2^\omega$ be $\mathcal{M}$-additive. Let h be a gauge. We will show that $\overline{\mathcal{H}}^h(S)=0$. Define recursively $f\in\omega^{\uparrow\omega}$ subject to

$$2^{f(k)}\cdot h\big(2^{-f(k+1)}\big)\leqslant 2^{-k},\quad k\in\omega.$$

By Lemma 7.7 there is $g\in\omega^\omega$ and $y\in 2^\omega$ such that

$$\begin{aligned}&\forall x\in X\ \forall^\infty n\ \exists k\\ &\qquad g(n)\leqslant f(k)<g(n+1)\ \&\ x{\restriction}[f(k),f(k+1))=y{\restriction}[f(k),f(k+1)).\end{aligned}\tag{10}$$

Define

$$\begin{aligned} \mathcal{B}_k &= \{\langle p^\frown y\restriction[f(k), f(k+1))\rangle : p \in 2^{f(k)}\}, && k \in \omega, \\ \mathcal{G}_n &= \bigcup\{\mathcal{B}_k : g(n) \leqslant f(k) < g(n+1)\}, && n \in \omega, \\ \mathcal{B} &= \bigcup_{k\in\omega} \mathcal{B}_k = \bigcup_{n\in\omega} \mathcal{G}_n. \end{aligned}$$

With this notation (10) reads

$$\forall x \in X\ \forall^\infty n\ \exists G \in \mathcal{G}_n\ x \in G. \tag{11}$$

Since each of the families $\mathcal{G}_n$ is finite, it follows that $\mathcal{G}_n$'s witness that $\mathcal{B}$ is a γ-groupable cover of X. Using Lemma 6.5 (and Lemma 6.2(iv)) it remains to show that the Hausdorff sum $\sum_{B\in\mathcal{B}} h(\operatorname{diam} B)$ is finite. Note that the cones forming the families $B \in \mathcal{B}_k$ are actually balls of radius $2^{-f(k+1)}$, i.e., $\operatorname{diam} B = 2^{-f(k+1)}$ for all k and all $B \in \mathcal{B}_k$. Each of the cones is determined by one $p \in 2^{f(k)}$, therefore $|\mathcal{B}_k| = 2^{f(k)}$. Overall we have

$$\sum_{B\in\mathcal{B}} h(\operatorname{diam} B) = \sum_{k\in\omega}\sum_{B\in\mathcal{B}_k} h(\operatorname{diam} B) = \sum_{k\in\omega} 2^{f(k)} \cdot h(2^{-f(k+1)}) \leqslant \sum_{k\in\omega} 2^{-k} < \infty. \qquad \square$$

The most interesting question raised by Theorems 7.4 and 7.5 is of course if the analogue of Conjecture 3.6 holds for $\mathsf{Smz}^\sharp$.

QUESTION 7.9 (CH). Is it true that if the inclusion $\mathsf{Smz}^\sharp(\mathbb{G}) \subseteq \mathcal{M}^*(\mathbb{G})$ holds for a Polish group $\mathbb{G}$, then $\mathbb{G}$ is locally compact?

Another open problem is the rôle of TSI in Theorem 7.6.

QUESTION 7.10. Can the TSI assumption in Theorem 7.6 be dropped, or weakened to CLI?

Continuous images and cartesian products of meager-additive sets. Since meager-additive sets coincide with $\mathsf{Smz}^\sharp$ sets in TSI locally compact groups, they are preserved by continuous mappings and by cartesian products.

THEOREM 7.11. *Let $\mathbb{G}_1$ be a* TSI *Polish group and $\mathbb{G}_2$ a locally compact Polish group. Let $f : \mathbb{G}_1 \to \mathbb{G}_2$ a continuous mapping. If $X \subseteq \mathbb{G}_1$ is $\mathcal{M}$-additive, then so is $f(X)$.*

THEOREM 7.12. *Let $\mathbb{G}_1, \mathbb{G}_2$ be* TSI *locally compact Polish groups. Let $X_1 \subseteq \mathbb{G}_1$, $X_2 \subseteq \mathbb{G}_2$.*

(i) *If X_1 is* Smz *and X_2 is $\mathcal{M}$-additive, then $X_1 \times X_2$ is* Smz.
(ii) *If X_1 and X_2 are $\mathcal{M}$-additive, then so is $X_1 \times X_2$.*

COROLLARY 7.13. *Let $\mathbb{G}_1, \mathbb{G}_2$ be* TSI *locally compact Polish groups. Let $X \subseteq \mathbb{G}_1 \times \mathbb{G}_2$. The following are equivalent.*

(i) *X is $\mathcal{M}$-additive,*
(ii) $\operatorname{proj}_1 X$ *and* $\operatorname{proj}_2 X$ *are $\mathcal{M}$-additive,*
(iii) $\operatorname{proj}_1 X \times \operatorname{proj}_2 X$ *is $\mathcal{M}$-additive.*

We conclude this section with a few remarks on meager-additive sets in the Cantor set 2^ω. There are a few variations of $\mathcal{M}$-additivity: A set S in 2^ω is

- *$\mathcal{M}$-additive* if $\forall M \in \mathcal{M}\quad S + M \in \mathcal{M}$,
- *sharply $\mathcal{M}$-additive* if $\forall M \in \mathcal{M}\ \exists F \supseteq S\ F_\sigma\quad F + M \in \mathcal{M}$,

- *flatly $\mathcal{M}$-additive* if $\forall M \in \mathcal{M}\ \exists F \supseteq S\ F_\sigma \quad F + M \neq 2^\omega$,
- *$\mathcal{E}$-additive* if $\forall E \in \mathcal{E}\ S + E \in \mathcal{E}$,
- *sharply $\mathcal{E}$-additive* if $\forall E \in \mathcal{E}\ \exists F \supseteq S\ F_\sigma \quad F + E \in \mathcal{E}$.

The question whether $\mathcal{E}$-additive sets are related to $\mathcal{M}$-additive sets was posed by Nowik and Weiss [**34**]. Their question was answered in [**54**] by the following theorem. Let us note that while one would expect that, e.g., the proof of (ii)$\Rightarrow$(iv) is a matter of routine, it is actually surprisingly difficult.

THEOREM 7.14 ([**54**]). *Let $S \subseteq 2^\omega$. The following properties of S are equivalent.*

(i) *S is $\mathsf{Smz}^\sharp$,*
(ii) *S is $\mathcal{M}$-additive,*
(iii) *S is sharply $\mathcal{M}$-additive,*
(iv) *S is flatly $\mathcal{M}$-additive,*
(v) *S is $\mathcal{E}$-additive,*
(vi) *S is sharply $\mathcal{E}$-additive.*

The definitions of these notions extend in a straightforward way to Polish groups (or in case $\mathcal{E}$ is considered, locally compact Polish groups). However, the proofs of the above theorem depends very much on the fine combinatorial structure of 2^ω and are thus not easily transferable to a context of a Polish group. With quite some effort the equivalence (i)$\Leftrightarrow$(ii)$\Leftrightarrow$(iii)$\Leftrightarrow$(iv) in TSI locally compact Polish groups was proved in [**52**]. (The equivalence (i)$\Leftrightarrow$(ii) is presented in this section as Theorem 7.6.) But the equivalences including $\mathcal{E}$ are still not understood.

QUESTION 7.15 ([**52**]). Let $\mathbb{G}$ be a locally compact TSI Polish group.

(i) Is every $\mathcal{M}$-additive set in $\mathbb{G}$ $\mathcal{E}$-additive?
(ii) Is every $\mathcal{E}$-additive set in $\mathbb{G}$ $\mathcal{M}$-additive?
(iii) Is every $\mathcal{E}$-additive set in $\mathbb{G}$ sharply $\mathcal{E}$-additive?

8. Uniformity number of meager-additive sets

Since the notion of sharp measure zero is rather new, not much is known about the cardinal invariants of the σ-ideal of $\mathsf{Smz}^\sharp$ sets. We will investigate only the uniformity number of $\mathsf{Smz}^\sharp$ for metric spaces and Polish groups. We refer to section 4 for the notation.

Bartoszyński [**1**] and Pawlikowski [**36**] investigated and calculated a related cardinal – the uniformity number of the ideal of meager-additive sets in 2^ω. This cardinal invariant is termed *transitive additivity of $\mathcal{M}$* and denoted by $\mathsf{add}^*(\mathcal{M})$. By [**1**, 2.7.14], $\mathsf{add}^*(\mathcal{M}) = \mathfrak{eq}^*$.

The two results on Polish groups, Theorem 8.3 and Corollary 8.4, come from [**54**]. The other results of this section are new.

THEOREM 8.1. *Let X be a separable metric space that is not $\mathsf{Smz}^\sharp$.*

(i) $\mathsf{non}(\mathsf{Smz}^\sharp(\omega^\omega)) = \mathsf{add}(\mathcal{M})$,
(ii) $\mathsf{non}(\mathsf{Smz}^\sharp(2^\omega)) = \mathfrak{eq}^*$,
(iii) $\mathsf{add}(\mathcal{M}) \leqslant \mathsf{non}(\mathsf{Smz}^\sharp(X))$,
(iv) *if X is σ-totally bounded, then* $\mathfrak{eq}^* \leqslant \mathsf{non}(\mathsf{Smz}^\sharp(X))$,
(v) *if X is not of universal measure zero, then* $\mathsf{non}(\mathsf{Smz}^\sharp(X)) \leqslant \mathsf{non}(\mathcal{N})$.

PROOF. (i) Recall that the metric on ω^ω is the least difference metric given by $d(f,g) = 2^{-n(f,g)}$, where $n(f,g) = \min\{n : f(n) \neq g(n)\}$.

Let $S \subseteq \omega^\omega$ be an unbounded set such that $|X| = \mathfrak{b}$. Since X is not bounded, it is not σ-totally bounded and in particular it is not $\mathsf{Smz}^\sharp$. It follows that $\mathsf{non}\,\mathsf{Smz}^\sharp(\omega^\omega) \leqslant \mathfrak{b}$.

Theorem 4.2(iii) yields a set $S \subseteq \omega^\omega$ that is not Smz and $|S| = \mathsf{cov}\,\mathcal{M}$. This set is clearly not $\mathsf{Smz}^\sharp$ and thus $\mathsf{non}\,\mathsf{Smz}^\sharp(\omega^\omega) \leqslant \mathsf{cov}\,\mathcal{M}$.

It follows that $\mathsf{non}\,\mathsf{Smz}^\sharp(\omega^\omega) \leqslant \min(\mathsf{cov}\,\mathcal{M}, \mathfrak{b})$. By a theorem of Miller [**29**] $\min(\mathsf{cov}\,\mathcal{M}, \mathfrak{b}) = \mathsf{add}(\mathcal{M})$ (see also [**1**, 2.2.9]). Thus we have $\mathsf{non}\,\mathsf{Smz}^\sharp(\omega^\omega) \leqslant \mathsf{add}(\mathcal{M})$.

In the other direction, suppose that $S \subseteq \omega^\omega$ is not $\mathsf{Smz}^\sharp$. By Theorem 7.5, S is not meager-additive in the group $\mathbb{Z}^\omega$. Therefore there is a meager set $M \subseteq \mathbb{Z}^\omega$ such that $S + M = \bigcup_{s\in S}(M+s)$ is not meager and hence $|S| \geqslant \mathsf{add}(\mathcal{M})$. Thus $\mathsf{non}\,\mathsf{Smz}^\sharp(\omega^\omega) \geqslant \mathsf{add}(\mathcal{M})$.

(ii) By Theorem 7.4, a set $S \subseteq 2^\omega$ is $\mathsf{Smz}^\sharp$ if and only if it is meager-additive. Thus $\mathsf{non}\,\mathsf{Smz}^\sharp(2^\omega) = \mathsf{add}^*(\mathcal{M})$ and the latter equals by the aforementioned result of Bartoszyński to $\mathfrak{eq}^*$.

(iii) Let $\{z_m : m \in \omega\} \subseteq X$ be a dense set in X. To each $x \in X$ assign a function $\widehat{x} \in \omega^\omega$ defined by

$$\widehat{x}(n) = \min\{m : d(x, z_m) < 2^{-n}\}. \tag{12}$$

We claim that the inverse map $\widehat{x} \mapsto x$ is Lipschitz. Indeed, if $d(\widehat{x},\widehat{y}) = 2^{-n}$, then $\widehat{x}(n-1) = \widehat{y}(n-1)$ and in particular $\exists m\ d(x,z_m) < 2^{-n+1}$ and $d(y,z_m) < 2^{-n+1}$. Therefore $d(x,y) < 2\cdot 2^{-n+1}$, whence $d(x,y) < 4d(\widehat{x},\widehat{y})$. In particular $\widehat{x} \mapsto x$ is uniformly continuous. So if $S \subseteq X$ is not $\mathsf{Smz}^\sharp$, then $\widehat{S} = \{\widehat{x} : x \in S\}$ is not $\mathsf{Smz}^\sharp$ as well. If follows that $\mathsf{non}(\mathsf{Smz}^\sharp(X)) \geqslant \mathsf{non}(\mathsf{Smz}^\sharp(\omega^\omega))$ and (iii) follows from (i).

(iv) Consider the completion X^* of X. Since X is σ-totally bounded, there is a σ-compact set $K \subseteq X^*$ that contains X. Let $K_n \nearrow K$ be a sequence of compact sets. Suppose that $S \subseteq X$ is a not $\mathsf{Smz}^\sharp$ set such that $|S| = \mathsf{non}(\mathsf{Smz}^\sharp(X))$. There is n such that $S \cap K_n$ is not $\mathsf{Smz}^\sharp$, therefore $|S\cap K_n| = \mathsf{non}(\mathsf{Smz}^\sharp(X))$. Since K_n is compact, it is *dyadic*: there is a uniformly continuous mapping $f : 2^\omega \to K_n$ onto K_n. For each $x \in S \cap K_n$ pick $\widehat{x} \in f^{-1}(x)$ and set $\widehat{S} = \{\widehat{x} : x \in S \cap K_n\}$. Then $f(\widehat{S}) = S \cap K_n$, hence $\widehat{S}$ is not $\mathsf{Smz}^\sharp$, and clearly $|\widehat{S}| = |S \cap K_n| = \mathsf{non}(\mathsf{Smz}^\sharp(X))$. It follows that $\mathsf{non}(\mathsf{Smz}^\sharp(2^\omega)) \leqslant \mathsf{non}(\mathsf{Smz}^\sharp(X))$ whence $\mathsf{add}^*(\mathcal{M}) \leqslant \mathsf{non}(\mathsf{Smz}^\sharp(X))$ by (ii).

(v) is a trivial consequence of Theorem 4.2(ii). □

As expected, for analytic metric spaces we can do better. Recall that a metric space has the *small ball property* if it admits a base $\{B_n\}$ such that $\operatorname{diam} B_n \to 0$. This notion is due to Behrends and Kadec [**3**]. We refer to [**19**] for more information.

THEOREM 8.2. *Let X be an uncountable analytic metric space.*

(i) $\mathsf{add}(\mathcal{M}) \leqslant \mathsf{non}(\mathsf{Smz}^\sharp(X)) \leqslant \mathfrak{eq}^*$,
(ii) *if X is σ-totally bounded, then* $\mathsf{non}(\mathsf{Smz}^\sharp(X)) = \mathfrak{eq}^*$,
(iii) *if X does not have the small ball property, then* $\mathsf{non}(\mathsf{Smz}^\sharp(X)) = \mathsf{add}(\mathcal{M})$.

PROOF. (i) The left-hand inequality is Theorem 8.1(iii). Since X contains a (uniform) copy of 2^ω, Theorem 8.1(ii) yields $\mathfrak{eq}^* = \mathsf{non}(\mathsf{Smz}^\sharp(2^\omega)) \geqslant \mathsf{non}\,\mathsf{Smz}^\sharp(\mathbb{G})$.

(ii) follows from (i) and Theorem 8.1(iv).

(iii) Consider the mapping $x \mapsto \widehat{x}$ defined by (12). Let $B \subseteq \omega^\omega$ be an unbounded set such that $|B| = \mathfrak{b}$. As shown in [**19**, 4.4], the set $\widehat{X} = \{\widehat{x} : x \in X\}$ is dominating in ω^ω. Therefore for each $f \in B$ there is $x_f \in X$ such that $\widehat{x}_f \geqslant^* f$. Set $S = \{x_g : f \in B\}$. Then $\widehat{S}$ is not bounded, because it dominates B. It follows that S is not σ-totally bounded in X, and in particular it is not $\mathsf{Smz}^\sharp$. Since clearly $|\widehat{S}| = |B| = \mathfrak{b}$, we conclude that $\mathsf{non}(\mathsf{Smz}^\sharp(X)) \leqslant \mathfrak{b}$. By (i) also $\mathsf{non}(\mathsf{Smz}^\sharp(X)) \leqslant \mathsf{add}(\mathcal{M})$, so $\mathsf{non}(\mathsf{Smz}^\sharp(X)) \leqslant \min(\mathfrak{eq}^*, \mathfrak{b}) = \mathsf{add}(\mathcal{M})$, and the reverse inequality follows from (i). □

We will now calculate uniformity numbers of $\mathsf{Smz}^\sharp(\mathbb{G})$ for CLI Polish groups and of $\mathcal{M}^*(\mathbb{G})$ for TSI Polish groups.

THEOREM 8.3 ([**54**]). *Let $\mathbb{G}$ be a* CLI *Polish group.*

(i) *If $\mathbb{G}$ is locally compact, then* $\mathsf{non}(\mathsf{Smz}^\sharp(\mathbb{G})) = \mathfrak{eq}^*$.
(ii) *If $\mathbb{G}$ is not locally compact, then* $\mathsf{non}(\mathsf{Smz}^\sharp(\mathbb{G})) = \mathsf{add}(\mathcal{M})$.

PROOF. (i) is a straightforward from Theorem 8.2(ii).

(ii) By Lemma 4.3, $\mathbb{G}$ contains a uniform copy of ω^ω. Therefore $\mathsf{non}(\mathsf{Smz}^\sharp(\mathbb{G})) \leqslant \mathsf{non}(\mathsf{Smz}^\sharp(\omega^\omega))$. Now apply Theorem 8.1(i) to get $\mathsf{non}(\mathsf{Smz}^\sharp(\mathbb{G})) \leqslant \mathsf{non}(\mathcal{M})$. The reverse inequality is straightforward from Theorem 8.2(i). □

The following is a trivial consequence of this theorem and Theorem 7.4.

COROLLARY 8.4 ([**54**]). *Let $\mathbb{G}$ be a* TSI *Polish group.*

(i) *If $\mathbb{G}$ is locally compact, then $\mathcal{M}^*(\mathbb{G}) = \mathfrak{eq}^*$.*
(ii) *If $\mathbb{G}$ is not locally compact, then $\mathcal{M}^*(\mathbb{G}) =$* $\mathsf{add}(\mathcal{M})$.

We conclude with a simple argument that shows that $\mathsf{Smz}^\sharp$ is consistently not a topological property. The Baer-Specker group $\mathbb{Z}^\omega$ is a TSI Polish group. On the other hand, it is homeomorphic to the set of irrational numbers, so regard it as a subset of $\mathbb{R}$. By Theorem 8.3(ii) there is a set $X \subseteq \mathbb{Z}^\omega$ such that $|X| = \mathsf{add}(\mathcal{M})$ and that is not $\mathsf{Smz}^\sharp$ in the invariant metric. On the other hand, if $\mathsf{add}(\mathcal{M}) < \mathsf{add}^* \mathcal{M}$, then X is by Theorem 8.3(i) $\mathsf{Smz}^\sharp$ in the metric of the real line. Since $\mathsf{add}(\mathcal{M}) < \mathsf{add}^* \mathcal{M}$ is relatively consistent with ZFC (as proved by Pawlikowski [**36**]), X is consistently a set that is $\mathsf{Smz}^\sharp$ is one metric on $\mathbb{Z}^\omega$ and not $\mathsf{Smz}^\sharp$ in another homeomorphic metric.

References

[1] T. Bartoszyński and H. Judah, *Set theory*, A K Peters, Ltd., Wellesley, MA, 1995. On the structure of the real line. MR1350295

[2] J. E. Baumgartner, *Iterated forcing*, Surveys in set theory, London Math. Soc. Lecture Note Ser., vol. 87, Cambridge Univ. Press, Cambridge, 1983, pp. 1–59, DOI 10.1017/CBO9780511758867.002. MR823775

[3] E. Behrends and V. M. Kadets, *Metric spaces with the small ball property*, Studia Math. **148** (2001), no. 3, 275–287, DOI 10.4064/sm148-3-6. MR1880727

[4] A. S. Besicovitch, *Concentrated and rarified sets of points*, Acta Math. **62** (1933), no. 1, 289–300, DOI 10.1007/BF02393607. MR1555386

[5] A. S. Besicovitch, *Correction*, Acta Math. **62** (1933), no. 1, 317–318, DOI 10.1007/BF02393610. MR1555389

[6] E. Borel, *Sur la classification des ensembles de mesure nulle* (French), Bull. Soc. Math. France **47** (1919), 97–125. MR1504785

[7] Miguel A. Cardona, *Yorioka's characterization of the cofinality of the strong measure zero ideal and its independency from the continuum*, Submitted April 2019.

[8] Miguel A. Cardona, Diego A. Mejía, and Ismael E. Rivera-Madrid, *The covering number of the strong measure zero ideal can be above almost everything else*, Submitted on 5 Feb 2019.

[9] T. J. Carlson, *Strong measure zero and strongly meager sets*, Proc. Amer. Math. Soc. **118** (1993), no. 2, 577–586, DOI 10.2307/2160341. MR1139474

[10] P. Corazza, *The generalized Borel conjecture and strongly proper orders*, Trans. Amer. Math. Soc. **316** (1989), no. 1, 115–140, DOI 10.2307/2001276. MR982239

[11] T. Dobrowolski and W. Marciszewski, *Covering a Polish group by translates of a nowhere dense set*, Topology Appl. **155** (2008), no. 11, 1221–1226, DOI 10.1016/j.topol.2008.02.010. MR2421832

[12] D. H. Fremlin, *Measure theory. Vol. 5. Set-theoretic measure theory. Part II*, Torres Fremlin, Colchester, 2015. Corrected reprint of the 2008 original. MR3723041

[13] F. Galvin, J. Mycielski, and R. M. Solovay, *Strong Measure Zero Sets*, Notices of the American Mathematical Society **26** (1979), no. 3, A-280.

[14] F. Galvin, J. Mycielski, and R. M. Solovay, *Strong measure zero and infinite games*, Arch. Math. Logic **56** (2017), no. 7-8, 725–732, DOI 10.1007/s00153-017-0541-z. MR3696064

[15] M. Goldstern, H. Judah, and S. Shelah, *Strong measure zero sets without Cohen reals*, J. Symbolic Logic **58** (1993), no. 4, 1323–1341, DOI 10.2307/2275146. MR1253925

[16] A. Haar, *Der Massbegriff in der Theorie der kontinuierlichen Gruppen* (German), Ann. of Math. (2) **34** (1933), no. 1, 147–169, DOI 10.2307/1968346. MR1503103

[17] J. D. Howroyd, *On Hausdorff and packing dimension of product spaces*, Math. Proc. Cambridge Philos. Soc. **119** (1996), no. 4, 715–727, DOI 10.1017/S0305004100074545. MR1362951

[18] M. Hrušák, D. Meza-Alcántara, and H. Minami, *Pair-splitting, pair-reaping and cardinal invariants of F_σ-ideals*, J. Symbolic Logic **75** (2010), no. 2, 661–677, DOI 10.2178/jsl/1268917498. MR2648159

[19] M. Hrušák, W. Wohofsky, and O. Zindulka, *Strong measure zero in separable metric spaces and Polish groups*, Arch. Math. Logic **55** (2016), no. 1-2, 105–131, DOI 10.1007/s00153-015-0459-2. MR3453581

[20] M. Hrušák and J. Zapletal, *Strong measure zero sets in Polish groups*, Illinois J. Math. **60** (2016), no. 3-4, 751–760. MR3707641

[21] T. Jech, *Set theory*, Springer Monographs in Mathematics, Springer-Verlag, Berlin, 2003. The third millennium edition, revised and expanded. MR1940513

[22] A. S. Kechris, *Classical descriptive set theory*, Graduate Texts in Mathematics, vol. 156, Springer-Verlag, New York, 1995. MR1321597

[23] J. D. Kelly, *A method for constructing measures appropriate for the study of Cartesian products*, Proc. London Math. Soc. (3) **26** (1973), 521–546, DOI 10.1112/plms/s3-26.3.521. MR0318427

[24] L. D. R. Kočinac and M. Scheepers, *Combinatorics of open covers. VII. Groupability*, Fund. Math. **179** (2003), no. 2, 131–155, DOI 10.4064/fm179-2-2. MR2029229

[25] K. Kunen, *Set theory*, Studies in Logic and the Foundations of Mathematics, vol. 102, North-Holland Publishing Co., Amsterdam, 1983. An introduction to independence proofs; Reprint of the 1980 original. MR756630

[26] Marcin Kysiak, *On Erdős-Sierpiński duality between Lebesgue measure and Baire category (in Polish)*, Master's thesis (2000).

[27] P. B. Larson, *The stationary tower*, University Lecture Series, vol. 32, American Mathematical Society, Providence, RI, 2004. Notes on a course by W. Hugh Woodin. MR2069032

[28] R. Laver, *On the consistency of Borel's conjecture*, Acta Math. **137** (1976), no. 3-4, 151–169, DOI 10.1007/BF02392416. MR422027

[29] A. W. Miller, *Some properties of measure and category*, Trans. Amer. Math. Soc. **266** (1981), no. 1, 93–114, DOI 10.2307/1998389. MR613787

[30] A. W. Miller and D. H. Fremlin, *On some properties of Hurewicz, Menger, and Rothberger*, Fund. Math. **129** (1988), no. 1, 17–33, DOI 10.4064/fm-129-1-17-33. MR954892

[31] A. W. Miller and J. Steprāns, *The number of translates of a closed nowhere dense set required to cover a Polish group*, Ann. Pure Appl. Logic **140** (2006), no. 1-3, 52–59, DOI 10.1016/j.apal.2005.09.010. MR2224048

[32] M. E. Munroe, *Introduction to measure and integration*, Addison-Wesley Publishing Company, Inc., Cambridge, Mass., 1953. MR0053186

[33] A. Nowik, M. Scheepers, and T. Weiss, *The algebraic sum of sets of real numbers with strong measure zero sets*, J. Symbolic Logic **63** (1998), no. 1, 301–324, DOI 10.2307/2586602. MR1610427

[34] A. Nowik and T. Weiss, *On the Ramseyan properties of some special subsets of* 2^ω *and their algebraic sums*, J. Symbolic Logic **67** (2002), no. 2, 547–556, DOI 10.2178/jsl/1190150097. MR1905154

[35] K. R. Parthasarathy, *Introduction to probability and measure*, Springer-Verlag New York Inc., New York, 1978. MR0651013

[36] J. Pawlikowski, *Powers of transitive bases of measure and category*, Proc. Amer. Math. Soc. **93** (1985), no. 4, 719–729, DOI 10.2307/2045552. MR776210

[37] ———, *Finite support iteration and strong measure zero sets*, J. Symbolic Logic **55** (1990), no. 2, 674–677. MR1056381

[38] Karel Prikry, unpublished result.

[39] C. A. Rogers, *Hausdorff measures*, Cambridge University Press, London-New York, 1970. MR0281862

[40] F. Rothberger, *Sur les familles indénombrables de suites de nombres naturels et les problèmes concernant la propriété C* (French), Proc. Cambridge Philos. Soc. **37** (1941), 109–126. MR4281

[41] M. Scheepers, *Finite powers of strong measure zero sets*, J. Symbolic Logic **64** (1999), no. 3, 1295–1306, DOI 10.2307/2586631. MR1779763

[42] S. Shelah, *Every null-additive set is meager-additive*, Israel J. Math. **89** (1995), no. 1-3, 357–376, DOI 10.1007/BF02808209. MR1324470

[43] W. Sierpiński and E. Szpilrajn, *Remarque sur le problème de la mesure*, Fund. Math. **26** (1936), 256–261.

[44] Wacław Sierpiński, *Sur un ensemble non denombrable, dont toute image continue est de mesure nulle*, Fundamenta Mathematicae **11** (1928), 302–304.

[45] E. Szpilrajn, *Remarques sur les fonctions sousharmoniques* (French), Ann. of Math. (2) **34** (1933), no. 3, 588–594, DOI 10.2307/1968179. MR1503125

[46] ———, *Remarques sur les fonctions complètement additives d'ensemble et sur les ensembles jouissant de la propriété de Baire*, Fund. Math. **22** (1934), no. 1, 303–311.

[47] T. Weiss, *On meager additive and null additive sets in the Cantor space* 2^ω *and in* $\mathbb{R}$, Bull. Pol. Acad. Sci. Math. **57** (2009), no. 2, 91–99, DOI 10.4064/ba57-2-1. MR2545840

[48] ———, *Addendum to "On meager additive and null additive sets in the Cantor space* 2^ω *and in* $\mathbb{R}$*" (Bull. Polish Acad. Sci. Math. 57 (2009), 91–99) [mr2545840]*, Bull. Pol. Acad. Sci. Math. **62** (2014), no. 1, 1–9. MR3241126

[49] T. Weiss, *Properties of the intersection ideal* $\mathcal{M} \cap \mathcal{N}$ *revisited*, Bull. Pol. Acad. Sci. Math. **65** (2017), no. 2, 107–111, DOI 10.4064/ba8098-8-2017. MR3731016

[50] Wolfgang Wohofsky, *Special sets of real numbers and variants of the Borel Conjecture*, Ph.D. thesis, Technische Universität Wien, 2013.

[51] T. Yorioka, *The cofinality of the strong measure zero ideal*, J. Symbolic Logic **67** (2002), no. 4, 1373–1384, DOI 10.2178/jsl/1190150290. MR1955243

[52] Ondřej Zindulka, *Meager-additive sets in topological groups*, Submitted.

[53] O. Zindulka, *Universal measure zero, large Hausdorff dimension, and nearly Lipschitz maps*, Fund. Math. **218** (2012), no. 2, 95–119, DOI 10.4064/fm218-2-1. MR2957686

[54] O. Zindulka, *Strong measure zero and meager-additive sets through the prism of fractal measures*, Comment. Math. Univ. Carolin. **60** (2019), no. 1, 131–155. MR3946667

Centro de Ciencias Matemáticas, Universidad Nacional Autónoma de México, Campus Morelia, 58089, Morelia, Michoacán, México.

Email address: `michael@matmor.unam.mx`

URL: `http://www.matmor.unam.mx/~michael`

Department of Mathematics, Faculty of Civil Engineering, Czech Technical University, Thákurova 7, 160 00 Prague 6, Czech Republic

Email address: `ondrej.zindulka@cvut.cz`

URL: `http://mat.fsv.cvut.cz/zindulka`

Contemporary Mathematics
Volume **755**, 2020
https://doi.org/10.1090/conm/755/15177

Ramsey theory and the Borel conjecture

Marion Scheepers

ABSTRACT. The Borel covering property, introduced a century ago by E. Borel, is intimately connected with Ramsey theory, initiated ninety years ago in an influential paper of F.P. Ramsey. The current state of knowledge about the connection between the Borel covering property and Ramsey theory is outlined in this paper. Initially the connection is established for the situation when the set with the Borel covering property is a proper subset of a σ-compact uniform space. Then the connection is explored for a stronger covering property introduced by Rothberger. After establishing the fact that in this case several landmark Ramseyan theorems are characteristic of this stronger covering property, the case when the space with this stronger covering property is in fact σ-compact is explored.

In his 1919 paper [**5**] E. Borel, studying the notion of Lebesgue measure zero, introduced a covering property for subsets of the real line. A set X of real numbers has Borel's covering property if for each sequence $(\epsilon_n : n \in \mathbb{N})$ of positive real numbers there is a sequence $(x_n : n \in \mathbb{N})$ of elements of X such that the set $\{(x_n - \epsilon_n,\ x_n + \epsilon_n) : n \in \mathbb{N}\}$ of open intervals of real numbers covers X. Here, and throughout the paper, $\mathbb{N}$ denotes the set of positive integers, also called natural numbers.

Borel's definition has been adapted in several ways to other contexts. For a subset X of a metric space (M, d) the corresponding definition is as follows: For each sequence $(\epsilon_n : n \in \mathbb{N})$ of positive real numbers there is a sequence $(x_n : n \in \mathbb{N})$ of elements of X such that the set $\{B_d(x_n,\ \epsilon_n) : n \in \mathbb{N}\}$ is a cover for X. Here, $B_d(x, \epsilon)$ denotes the set $\{y \in X : d(x, y) < \epsilon\}$. For a subset X of a topological group $(G, \odot)$ the corresponding definition is as follows: For each sequence $(U_n : n \in \mathbb{N})$ of neighborhoods of the identity element id of G, there is a sequence $(x_n : n \in \mathbb{N})$ of elements of X such that the set $\{x_n \odot U_n : n \in \mathbb{N}\}$ is a cover for X. Here the symbol $x \odot N$ denotes the set $\{x \odot y : y \in N\}$.

A remark made in [**5**] provided an initial stimulus for investigating Borel's covering property: Borel noted that every countable set of real numbers has Borel's covering property, and he speculated that the only sets of real numbers that have this covering property are the countable sets. This speculation became known as Borel's Conjecture. By 1976 it was known that Borel's conjecture is independent of

2000 *Mathematics Subject Classification.* 03E02, 03E05, 05D10, 22A05, 54D20, 54G10, 54H11 .

Key words and phrases. Ramsey theory, Rothberger bounded, strong measure zero, uniformizable space, topological group, infinite game.

the Zermelo-Fraenkel axioms, denoted ZFC. Despite the status of Borel's Conjecture as a statement undecidable in the axiomatic system ZFC, research on Borel's covering property and its relatives continues unabated and has affected numerous branches of mathematics.

In [**28**] Rothberger modified the definition of Borel's covering property for the context of general topological spaces: For a topological space (S, τ) and a subset X of S, the analogous covering property is as follows: For each sequence $(\mathcal{O}_n : n \in \mathbb{N})$ of open covers of S there is a sequence $(T_n : n \in \mathbb{N})$ of open sets such that for each n it is the case that $T_n \in \mathcal{O}_n$, and $\{T_n : n \in \mathbb{N}\}$ is a cover of X. This was a step towards a much broader framework, selection principles in mathematics, in which the Borel covering property and its modifications would be special cases. In this paper we take this broader approach, but we confine ourselves to instances of topological covering properties intimately related to the Borel covering property.

Ten years after Borel's paper [**5**] F.P. Ramsey reported [**27**] a significant extension of the classical pigeon hole principle. This work was taking place independently of considerations related to the Borel covering property, and was an initial stimulus for the development of the field now known as Ramsey theory.

The purpose of this paper is to give a coherent expository survey on the connection between covering properties related to the Borel covering property, on the one hand, and Ramsey theory on the other hand. As such the reader should not expect a large volume of new results, but instead a reorganization of several known results into one narrative. As this paper is only one among several in this volume, each dedicated to a different aspect of mathematics inspired by the Borel covering property, the reader can expect an overlap of some basic information among these papers. For example, in the first section of this paper, to set the stage, we describe the broader selection principles context for Borel's covering property. In the second section of the paper we connect this broader context with an essential tool, an infinite two-person game inspired by work of F. Galvin, for making the connection in section 3 with Ramsey theory. With this initial outline established, the rest of the paper narrates the deep connections between the covering properties under consideration, and several landmark developments in Ramsey theory. Ramseyan equivalents of several other covering properties besides the Borel covering property and its relatives have been explored. A very interesting recent investigation along these lines has been published by Tsaban in [**43**].

Without further notice all topological spaces in this paper are assumed to be Hausdorff spaces - meaning that for any two distinct points x and y, there are disjoint open sets U and V such that $x \in U$ and $y \in V$.

1. Borel's covering property as a selection principle

Rothberger's approach in [**28**] to the Borel covering property suggests a common framework for treating Borel's covering property and its variations. To describe this framework let $\mathcal{A}$ and $\mathcal{B}$ be families of sets. The symbol

$$\mathsf{S}_1(\mathcal{A}, \mathcal{B})$$

denotes the statement (called a selection principle) that there is for each sequence $(O_n : n \in \mathbb{N})$ of elements of $\mathcal{A}$ a corresponding sequence $(T_n : n \in \mathbb{N})$ such that for each $n \in \mathbb{N}$ we have $T_n \in O_n$, and $\{T_n : n \in \mathbb{N}\}$ is an element of $\mathcal{B}$. To see that the selection principle $\mathsf{S}_1(\mathcal{A},\mathcal{B})$ subsumes the already mentioned variations on Borel's covering property, consider the following.

EXAMPLE 1.1. For a metric space (M,d) define for $\epsilon>0$ the set $\mathcal{U}_\epsilon=\{B_d(x,\epsilon): x\in M\}$. Then $\mathcal{U}_\epsilon$ is an open cover of M. Use the symbol $\mathcal{O}_{diam}$ to denote the set $\{\mathcal{U}_\epsilon:\ \epsilon>0\}$ of open covers of M arising in this way. For a subset X of M, let $\mathcal{O}_X$ denote the set of $\mathcal{V}$ where $\mathcal{V}$ consists of open subsets of M and $\mathcal{V}$ covers X. Then X has the Borel covering property exactly when the selection principle $\mathsf{S}_1(\mathcal{O}_{diam},\mathcal{O}_X)$ holds.

EXAMPLE 1.2. For a topological group $(G,\odot)$ define for a open neighborhood N of the identity element i_G the set $\mathcal{U}_N=\{x\odot N:\ x\in G\}$. Then $\mathcal{U}_N$ is an open cover of G. Use the symbol $\mathcal{O}_{nbd}$ to denote the set $\{\mathcal{U}_N:\ N \text{ a neighborhood of } i_G\}$ of open covers of G arising in this way. For a subset X of G, let $\mathcal{O}_X$ denote the set of $\mathcal{V}$ where $\mathcal{V}$ consists of open subsets of G and $\mathcal{V}$ covers X. Then X has the Borel covering property exactly when the selection principle $\mathsf{S}_1(\mathcal{O}_{nbd},\mathcal{O}_X)$ holds.

Borel's covering property arose for the real line $\mathbb{R}$ with its standard algebraic operations and its Euclidean metric. For this specific example Borel's covering property can be interpreted both as a property of the metric space $(\mathbb{R},|\cdot|)$, and (independently) as a property of the topological group $(\mathbb{R},\ +)$. A natural common generalization is obtained by formulating the covering property in terms of uniform spaces. See for example [**45**], Chapter 9 for more details about uniform spaces.

DEFINITION 1.3. For a set X, a collection Ψ of subsets of $X\times X$ is said to be a uniformity on X if Ψ has the following properties:

(U1) For each $U\in\Psi$ it is the case that $\{(x,x): x\in X\}\subseteq U$
(U2) If U and V are members of Ψ, then $U\cap V$ is a member of Ψ
(U3) If U is a member of Ψ and V is a subset of $X\times X$ such that $U\subseteq V$, then V is an element of Ψ
(U4) If U is a member of Ψ, then also the set $\{(y,x):\ (x,y)\in U\}$ is a member of Ψ
(U5) For each $U\in\Psi$ there is a $V\in\Psi$ such that whenever $(x,y)\in V$ and $(y,z)\in V$, then it is the case that $(x,z)\in U$

Let Ψ is a uniformity on a set X. The pair (X,Ψ) is said to be a uniform space. An element of the uniformity Ψ is said to be an entourage. Convenient notation for the properties stated in (U4) and in (U5) is as follows: For an entourage U, U^{-1} denotes $\{(y,x):\ (x,y)\in U\}$. For a subset U of $X\times X$ the symbol $U\circ U$ denotes the set $\{(x,z)\in X\times X:\ \text{there is a } y\in X \text{ such that } (x,y)\in U \text{ and } (y,z)\in U\}$. An entourage U is said to be symmetric if $U=U^{-1}$. Properties (U1), (U2) and (U3) state that a uniformity is a filter on the set $X\times X$ with the property that each element of the filter contains the set $\Delta_X=\{(x,x): x\in X\}$, while property (U4) states that the filter is closed under the operation U^{-1}, and property (U5) states that for each element U of the filter there is an element V of the filter such that $V\circ V\subseteq U$. The blanket assumption of this paper is that topological spaces considered here are Hausdorff spaces: This requirement is imposed on the uniformities relevant to this paper by requiring that a uniformity Ψ is *separating*, which is that $\Delta_X=\bigcap\{U: U\in\Psi\}$.

A uniformity Ψ on a set X gives rise to a topology on X: For each $U\in\Psi$, and for each $x\in X$ define

$$U(x)=\{y\in X:\ (x,y)\in U\}.$$

Then for $x \in X$ the set $\mathcal{N}_\Psi(x) = \{U(x) : U \in \Psi\}$ defines a neighborhood basis of x in X, and each subset of X of the form $U(x)$ is said to be a neighborhood of x. Given a set $\mathcal{N}$ of neighborhood bases for the elements of a set X, there is a standard topology on X, denoted $\tau_\mathcal{N}$, giving rise to $\mathcal{N}$ – see for example [**45**], Section 5. Accordingly, given a uniformity Ψ on the set X, there is a naturally associated topology, denoted τ_Ψ, on X. By classical results of Tychonoff and of Weil [**44**] the topology of a topological space is uniformizable (i.e., there is a uniformity Ψ such that the topology is of the form τ_Ψ) if, and only if, the space is completely regular. Since the spaces in this paper are all assumed to be Hausdorff spaces, being uniformizable is equivalent to being a Tychonoff space.

EXAMPLE 1.4. For uniformity Ψ on the set X define a special subfamily of the family of all open covers of the space (X, τ_Ψ): For any $N \in \Psi$, the open cover $\{Int_X(N(x)) : x \in X\}$ of X is denoted $\mathcal{O}(N)$. The symbol $\mathcal{O}_\Psi$ denotes the set $\{\mathcal{O}(N) : N \in \Psi\}$ of these special open covers of X. When Y is a subset of X, then the symbol $\mathcal{O}_Y$ denotes the set of all families $\mathcal{U}$ for which each element of $\mathcal{U}$ is an open set in the topological space (X, τ_Ψ), and for which $Y \subseteq \bigcup \mathcal{U}$. Borel's covering property translates to $\mathsf{S}_1(\mathcal{O}_\Psi, \mathcal{O}_Y)$.

Generalizing further, Nachbin introduced the notion of a quasi uniformity [**24**]:

DEFINITION 1.5 (Nachbin 1948). A set $\mathcal{U} \subseteq X \times X$ is a quasi-uniformity if

(1) $\mathcal{U}$ is a filter,
(2) For all $U \in \mathcal{U}$ the set $\{(x, x) : x \in X\}$ is a subset of U,
(3) For all $U \in \mathcal{U}$ there is a $V \in \mathcal{U}$ such that $V \circ V \subseteq U$.

If $\mathcal{U}$ is a quasi-uniformity on X, then $(X, \mathcal{U})$ is said to be a quasi-uniform space.

Observe that a quasi-uniformity is a uniformity whenever for each $U \in \mathcal{U}$ it is also the case that $U^{-1} \in \mathcal{U}$. For a quasi-uniformity $\mathcal{U}$, the symbol $\mathcal{U}^*$ denotes the set $\{U^{-1} : U \in \mathcal{U}\}$. Note that $\mathcal{U}^*$ also is a quasi-uniformity and is called the conjugate quasi-uniformity of $\mathcal{U}$. As in the case of uniformities, if U is a member of a quasi-uniformity $\mathcal{U}$, then $U(x)$ denotes the set $\{y \in X : (x, y) \in U\}$. There is a natural topology associated with a quasi-uniformity: If $\mathcal{U}$ is a quasi-uniformity on the set X, define

$$\tau_\mathcal{U} = \{G \subseteq X : (\forall x \in G)(\exists U \in \mathcal{U})(U(x) \subseteq G)\}.$$

Then $\tau_\mathcal{U}$ is a topology on X. A topology τ on X is *compatible* with the quasi-uniformity $\mathcal{U}$ if $\tau = \tau_\mathcal{U}$.

EXAMPLE 1.6. For quasi uniformity $\mathcal{U}$ on the set X define a special subfamily of the family of all open covers of the space $(X, \tau_\mathcal{U})$: For an element U of $\mathcal{U}$, the open cover $\{U(x) : x \in X\}$ of X is denoted $\mathcal{O}(U)$. The symbol $\mathcal{O}_\mathcal{U}$ denotes the set $\{\mathcal{O}(N) : N \in \mathcal{U}\}$ of these special open covers of X. When Y is a subset of X, then the symbol $\mathcal{O}_Y$ denotes the set of all families $\mathcal{V}$ for which each element of $\mathcal{V}$ is an open set in the topological space $(X, \tau_\mathcal{U})$, and for which $Y \subseteq \bigcup \mathcal{V}$. In this example Borel's covering property translates to $\mathsf{S}_1(\mathcal{O}_\mathcal{U}, \mathcal{O}_Y)$.

EXAMPLE 1.7. For a general topological space (X, τ), let $\mathcal{O}$ denote the collection of all open covers of X. We say that (X, τ) has the *Rothberger covering property* if $\mathsf{S}_1(\mathcal{O}, \mathcal{O})$ holds. Rothberger introduced this covering property in [**28**]. It is evident that if τ is the topology τ_Ψ generated by the (quasi-) uniformity Ψ

on X, and if (X, τ_Ψ) satisfies $\mathsf{S}_1(\mathcal{O}, \mathcal{O})$, then it satisfies $\mathsf{S}_1(\mathcal{O}_\Psi, \mathcal{O})$. As such, the Rothberger covering property implies the Borel covering property.

2. An infinite game and covering properties

We now arrive at the crucial tool that links the considered covering properties with Ramseyan properties.

DEFINITION 2.1. Let families of sets, $\mathcal{A}$ and $\mathcal{B}$, be given. The game $\mathsf{G}_1(\mathcal{A}, \mathcal{B})$, played between players ONE and TWO, is defined as follows: The game has an inning per positive integer n. In inning n ONE first chooses an element O_n of $\mathcal{A}$, and then TWO responds by choosing an element T_n of O_n. The play

$$O_1,\ T_1,\ O_2,\ T_2,\ \cdots,\ O_n,\ T_n,\ \cdots$$

is won by player TWO if $\{T_n : n \in \mathbb{N}\}$ is a member of $\mathcal{B}$. Else, ONE wins the play.

The instance of this game when $\mathcal{A} = \mathcal{B} = \mathcal{O}$, the set of all open covers of a topological space (X, τ), was introduced by F. Galvin in [**15**]. Note that if player ONE does not have a winning strategy in the game $\mathsf{G}_1(\mathcal{A}, \mathcal{B})$, then the selection principle $\mathsf{S}_1(\mathcal{A}, \mathcal{B})$ holds: For let a sequence $(A_n : n \in \mathbb{N})$ of elements of the family $\mathcal{A}$ be given. Assign the strategy F, defined to play A_n in the n-th inning, to player ONE of the game $\mathsf{G}_1(\mathcal{A}, \mathcal{B})$. Since ONE has no winning strategy in this game, F is not a winning strategy. Consider a play of the game where ONE used the strategy F, but lost. It is of the form

$$A_1, B_1, A_2, B_2, \cdots, A_n, B_n, \cdots$$

where for each n we have $B_n \in A_n$. Since ONE lost this play, TWO won and we have $\{B_n : n \in \mathbb{N}\} \in \mathcal{B}$. But then the sequence $(B_n : n \in \mathbb{N})$ witnesses $\mathsf{S}_1(\mathcal{A}, \mathcal{B})$ for the sequence $(A_n : n \in \mathbb{N})$.

One of the fundamental questions is: When does the selection principle $\mathsf{S}_1(\mathcal{A}, \mathcal{B})$ imply that ONE has no winning strategy in the game $\mathsf{G}_1(\mathcal{A}, \mathcal{B})$? Perhaps the most delicate instance of a positive result for this question is the following theorem of Pawlikowski [**25**]:

THEOREM 2.2 (J. Pawlikowski). *For any topological space (X, τ), the selection principle $\mathsf{S}_1(\mathcal{O}, \mathcal{O})$ holds if, and only if, ONE has no winning strategy in the game $\mathsf{G}_1(\mathcal{O}, \mathcal{O})$.*

More easily proved instances of such equivalences can be obtained by imposing additional constraints on the underlying topological space, or on the objects ONE is allowed to play during the game. In the rest of this section we prove, for the convenience of the reader, some instances relevant to the Borel covering property. The proofs are not new but can be gleaned from the proof of Theorem 2 in [**13**] where it is proved for metric spaces:

THEOREM 2.3. *For (X, Ψ) a σ -totally bounded uniform space and Y a subset of X, the following are equivalent:*

(1) *(X, Ψ) satisfies $\mathsf{S}_1(\mathcal{O}_\Psi, \mathcal{O}_Y)$.*
(2) *ONE has no winning strategy in the game $\mathsf{G}_1(\mathcal{O}_\Psi, \mathcal{O}_Y)$.*

PROOF. The proof of (2) $\Rightarrow$ (1) has been explained above. We prove that (1) $\Rightarrow$ (2) : Let F be a strategy for player ONE in the game $\mathsf{G}_1(\mathcal{O}_\Psi, \mathcal{O}_Y)$ on X. Since X is σ-totally bounded, and write $X = \bigcup_{n\in\mathbb{N}} X_n$ where for each n we have

$\emptyset \neq X_n \subseteq X_{n+1}$ and each X_n is totally bounded. For each n, putting $Y_n = Y \cap X_n$, the uniform space (X, Ψ) satisfies the property $\mathsf{S}_1(\mathcal{O}_\Psi, \mathcal{O}_{Y_n})$.

To defeat ONE's strategy TWO will in specific innings m concentrate attention on specific Y_n's. To this end, partition $\mathbb{N}$ into infinitely many infinite subsets S_n, $n \in \mathbb{N}$. For innings numbered by members of S_n TWO will focus on Y_n. We now use ONE's strategy F to recursively define a sequence $\{N_k : k \in \mathbb{N}\}$ and an array of sets $\mathcal{U}(T_1, ..., T_k)$ where:

(1) For each k, N_k is a symmetric member of the uniformity Ψ;
(2) With n_1 such that $1 \in S_{n_1}$, $\mathcal{U}(\emptyset)$ is a finite subset of $F(\emptyset)$ (ONE's first move) that covers Y_1 (as X_1 is totally bounded), and N_1 is such that for each $x \in Y_1$ there is a $V \in \mathcal{U}(\emptyset)$ with $(N_1 \circ N_1)(x) \subseteq V$;
(3) For each $(T_1, \ldots, T_k)$ such that $T_1 \in \mathcal{U}(\emptyset)$, $T_2 \in \mathcal{U}(T_1), \cdots$ and $T_k \in \mathcal{U}(T_1, \cdots ., T_{k-1})$ and for n_{k+1} such that $k+1 \in S_{n_{k+1}}$ we have $\mathcal{U}(T_1, \cdots, T_k)$ a finite subset of $F(T_1, \cdots, T_k)$ that covers $Y_{n_{k+1}}$. Note that there are only finitely many such $(T_1, ..., T_k)$. N_{k+1} is a member of the uniformity such that for each such sequence $(T_1, ..., T_k)$ and for each $x \in Y_{n_{k+1}}$ there is a $U \in \mathcal{U}(T_1, ..., T_k)$ with $(N_{k+1} \circ N_{k+1})(x) \subseteq U$.

With this data available, construct a play against F won by TWO as follows:

Fix an $m \in \mathbb{N}$. Since $\mathsf{S}_1(\mathcal{O}_\Psi, \mathcal{O}_{Y_m})$ holds, select for each $k \in S_m$ an $x_k \in Y$ such that $(N_k(x_k) : k \in S_m)$ covers Y_m. We may assume each x_k is in Y_m - for suppose an x_k is not in Y_m. If $N_k(x_k) \cap Y_m = \emptyset$ we may with impunity replace this x_k by one from Y_m. However, if $N_k(x_k) \cap Y_m \neq \emptyset$, then let y be an element of this intersection. We claim that $N_k(x_k) \cap Y_m \subseteq (N_k \circ N_k)(y)$. For let $z \in N_k(x_k) \cap Y_m$ be given. Note that $(x_k, y) \in N_k$ and $(x_k, z) \in N_k$. Since N_k is symmetric we have $(y, x_k) \in N_k$ and $(x_k, z) \in N_k$. But then $(y, z) \in N_k \circ N_k$, which implies that $z \in (N_k \circ N_k)(y)$. Thus, letting this y be the new x_k we see that we may choose for each $k \in S_m$ an $x_k \in Y_k$ such that $((N_k \circ N_k)(x_k) : k \in S_m)$ covers Y_k. Finally, recursively choose a sequence $(T_k : k \in \mathbb{N})$ as follows: Choose $T_1 \in \mathcal{U}(\emptyset)$ with $(N_1 \circ N_1)(x_1) \subseteq T_1$. With $T_1, ..., T_m$ chosen, choose $T_{m+1} \in \mathcal{U}(T_1, ..., T_m)$ with $(N_{m+1} \circ N_{m+1})(x_{m+1}) \subseteq T_{m+1}$. Then the sequence $F(\emptyset), T_1, F(T_1), ..., T_k, F(T_1, ..., T_k), T_{k+1}, ...$ is an F -play lost by ONE. □

Strengthening the hypothesis on X in Theorem 2.3 from σ-totally bounded to σ-compact gives a result towards a characterization of $\mathsf{S}_1(\mathcal{O}_\Psi, \mathcal{O}_X)$ in terms of a Ramseyan property. The argument needs, as one of its components, the following Theorem 2.4 which is proven in Theorem 33 on p. 199 of [**21**]:

THEOREM 2.4. *Let (X, Ψ) be a uniform space and $K \subset X$ a subset compact in the uniform topology. For each open cover $\mathcal{U}$ of X there is a member W of the uniformity Ψ such that for each $x \in K$ there is a $U \in \mathcal{U}$ such that $W(x) \subseteq U$.*

Another component towards a Ramseyan characterization is next given in Theorem 2.5. Note that the equivalence of (2) and (3) for general topological spaces (not only σ-compact uniform spaces) holds by Pawlikowski's theorem stated above as Theorem 2.2.

THEOREM 2.5. *Let (X, Ψ) be a σ-compact uniform space and let Y be a subset of X. The following are equivalent:*

(1) *(X, Ψ) satisfies $\mathsf{S}_1(\mathcal{O}_\Psi, \mathcal{O}_Y)$.*
(2) *(X, Ψ) satisfies $\mathsf{S}_1(\mathcal{O}, \mathcal{O}_Y)$.*

(3) *ONE has no winning strategy in the game* $\mathsf{G}_1(\mathcal{O}, \mathcal{O}_Y)$.
(4) *ONE has no winning strategy in the game* $\mathsf{G}_1(\mathcal{O}_\Psi, \mathcal{O}_Y)$.

PROOF. The equivalence of (1) and (4) was proven in Theorem 2.3. We must show that (1) implies (2), (2) implies (3) and (3) implies (4). Some of these implications have been proven in prior work, as we shall point out. For the convenience of the reader we give proofs here.

Proof that (1) implies (2): Let $(\mathcal{O}_n : n \in \mathbb{N})$ be a sequence of open covers of X. Partition $\mathbb{N}$ into infinitely many infinite subsets S_n, $n \in \mathbb{N}$. Fix any $k \in \mathbb{N}$, and consider the subsequence $(\mathcal{O}_n : n \in S_k)$ of the given sequence of open covers. For a fixed $n \in S_k$, apply Theorem 2.4 to $\mathcal{O}_n$ and the compact subset K_k of X. Fix an element W_n of Ψ such that for each $x \in K_k$ there is a $U \in \mathcal{O}_n$ for which we have $W_n(x) \subseteq U$. We may assume that W_n is symmetric. Then choose $V_n \in \Psi$ symmetric such that $V_n \circ V_n \subset W_n$

Now for each $n \in S_k$ the set $\mathcal{U}_n = \{V_n(x) : x \in X\}$ is an element of $\mathcal{O}_\Psi$. Applying (1) for $\mathsf{S}_1(\mathcal{O}_\Psi, \mathcal{O}_{Y_k})$ to the sequence $(\mathcal{U}_n : n \in S_k)$, choose for each $n \in S_k$ an x_n such that $Y_k \subseteq \bigcup_{n \in S_k} V_n(x_n)$. For each $n \in S_k$ consider $V_n(x_n)$: If $V_n(x_n) \cap K_k \neq \emptyset$ but $x_n \notin K_k$, then consider $y \in K_k \cap V_n(x_n)$. Then we have $K_k \cap V_n(x_n) \subseteq W_n(y)$: For pick a $z \in K_k \cap V_n(x_n)$. We have $(z, x_n) \in V_n$, and $(x_n, y) \in V_n$. Since $V_n \circ V_n \subseteq W_n$ it follows that $(z, y) \in W_n$, and thus $z \in W_n(y)$. In this case we redefine x_n to be such a y. Then the sequence $(W_n(x_n) : n \in S_k)$ is a cover of Y_k. For each $n \in S_k$, when $x_n \in K_k$, choose a $U_n \in \mathcal{O}_n$ for which $W_n(x_n) \subseteq U_n$, and otherwise choose $U_n \in \mathcal{O}_n$ arbitrarily. Then the sequence $(U_n : n \in S_k)$ covers Y_k.

It follows that the sequence $(U_n : n \in \mathbb{N})$ where of each n we have $U_n \in \mathcal{O}_n$, is an open cover of Y. This completes the proof that (1) implies (2).

Proof that (2) implies (3): (We follow the argument in the proof of (2) $\Rightarrow$ (3) in Theorem 9 of [**36**]) To this end, write

$$X = \bigcup_{n \in \mathbb{N}} K_n$$

where for $m < n$ we have $K_m \subset K_n$ and each K_m is compact. For each n, put $Y_n = Y \cap K_n$.

Fix a partition $\mathbb{N} = \cup\{S_n : n \in \mathbb{N}\}$ so that each S_n is infinite and whenever $m \neq n$, then S_m and S_n are disjoint. Let F be a strategy of player ONE of the game $\mathsf{G}_1(\mathcal{O}, \mathcal{O}_Y)$. To find a play of $\mathsf{G}_1(\mathcal{O}, \mathcal{O}_Y)$ during which ONE used the strategy F and yet TWO won, we proceed as follows:

Recursively define a sequence $(W_n : n \in \mathbb{N})$ of elements of the uniformity Ψ, and an array $(\mathcal{U}(U_1, \cdots, U_n) : n \in \mathbb{N})$ of finite families of open sets such that:

(1) With n_1 such that $1 \in S_{n_1}$, $\mathcal{U}(\emptyset)$ is a finite subset of $F(\emptyset)$ which is an open cover of Y_{n_1}, and W_1 is a member of Ψ such that for each $x \in Y_1$ we have $W_1(x) \subset U$ for some $U \in \mathcal{U}(\emptyset)$;
(2) For each $(U_1, \cdots, U_k)$ such that $U_1 \in \mathcal{U}(\emptyset)$ and $U_{i+1} \in \mathcal{U}(U_1 ..., U_i)$, $i < k$, and for n_{k+1} such that $k+1 \in S_{n_{k+1}}$, the set $\mathcal{U}(U_1, \cdots, U_k)$ is a finite subset of $F(U_1, \cdots, U_k)$ which covers $Y_{n_{k+1}}$ and $W_{n_{k+1}} \in \Psi$ is such that $W_{n_{k+1}} \subset W_{n_k}$ and for each $x \in Y_{n_{k+1}}$ there is a $U \in \mathcal{U}(U_1, \cdots, U_k)$ with $W_{n_{k+1}}(x) \subseteq U$.

Suppose we have defined these objects up to the m-th stage. Here is how stage $m+1$'s definition is made: For each sequence $(U_1, \cdots, U_n)$ of open sets such that $U_1 \in \mathcal{U}(\emptyset)$, $U_2 \in \mathcal{U}(U_1), \cdots$. and $U_m \in \mathcal{U}(U_1, \cdots, U_{m-1})$, apply ONE's strategy F to obtain an open cover $F(U_1, \cdots, U_m)$ of X. Determine the n_{m+1} or which $m+1 \in S_{n_{m+1}}$ and then let $\mathcal{U}(U_1, \cdots, U_{m+1})$ be a finite subset of $F(U_1, \cdots, U_{m+1})$ which covers $Y_{n_{m+1}}$. For this finite cover of $Y_{n_{m+1}}$ choose a $Z_{n_{m+1}} \subset W_{n_m}$ in Ψ such that for each $x \in Y_{n_{m+1}}$ there is a $U \in \mathcal{U}(U_1, \cdots, U_{n_{m+1}})$ for which $Z_{n_{m+1}}(x) \subseteq U$. Do this for all the finitely many possible $(U_1, \cdots, U_{m+1})$, and finally let $W_{n_{m+1}}$ be an element of Ψ contained in each $Z_{n_{m+1}}$.

Note that for any fixed m the selection principle $\mathsf{S}_1(\mathcal{O}, \mathcal{O}_{Y_m})$ holds. Thus for the sequence $(W_n : n \in S_m)$ choose a corresponding sequence $(x_n : n \in S_m)$ of elements of Y_m such that Y_m is covered by the set $\{W_n(x_n) : n \in S_m)$. Then the sequence $(W_n(x_n)) : n \in \mathbb{N})$ covers Y.

Recursively choose a sequence $(U_n : n \in \mathbb{N})$ as follows: Choose $U_1 \in \mathcal{U}(\emptyset)$ with $W_1(x_1) \subset U_1$, and with $U_1, \cdots, U_m$ chosen, choose $U_{m+1} \in \mathcal{U}(U_1, \cdots, U_m)$ with $W_{n_{m+1}}(x_{n_{m+1}}) \subset U_{m+1}$. These choices are possible on account of the way we chose the sets W_{n_m}. Then

$$F(\emptyset), U_1, F(U_1), U_2, F(U_1, U_2), \cdots$$

is an F-play of $\mathsf{G}_1(\mathcal{O}, \mathcal{O}_Y)$ which is lost by ONE.

(3)⇒(4): Since $\mathcal{O}_\Psi \subset \mathcal{O}$ it follows that if ONE has no winning strategy in $\mathsf{G}_1(\mathcal{O}, \mathcal{O}_Y)$, then ONE has no winning strategy in $\mathsf{G}_1(\mathcal{O}_\Psi, \mathcal{O})$. □

EXAMPLE 2.6. The equivalence of (1) and (2) in Theorem 2.5 requires hypotheses on the underlying space. To demonstrate, consider the following example (communicated to me by Kameryn J. Williams): For any uncountable set X there is a quasi-uniformity Ψ on X such that the selection principle $\mathsf{S}_1(\mathcal{O}_\Psi, \mathcal{O})$ holds, yet the topology generated by Ψ is the discrete topology. For let an uncountable set X be given. For each countably infinite subset F of X define

$$U_F = \Delta_X \bigcup ((X \setminus F) \times X)$$

Then let Ψ be the filter generated on $X \times X$ by the family $\{U_F : F \in [X]^{\aleph_0}\}$. Then Ψ is a quasi-uniformity on X, and the topology τ_Ψ generated on X by Ψ is the set

$$\{V \subseteq X : (\forall x \in V)(\exists U \in \mathcal{U})(U(x) \subseteq V)\}.$$

To see that for each $x \in X$ the set $\{x\}$ is a member of τ_Ψ, note that for any countable set $F \subset X$ with $x \in F$ we have, for $U_F = \Delta_X \bigcup ((X \setminus F) \times X)$ that $U_F(x) = \{x\}$. To see that (X, Ψ) satisfies the selection principle $\mathsf{S}_1(\mathcal{O}_\Psi, \mathcal{O})$, let a sequence $(U_n : n \in \mathbb{N})$ of elements of $\mathcal{U}$ be given. Note that each U_n contains a set of the form U_{F_n} where F_n is a countably infinite subset of X. Choose, for each n an $x_n \notin \bigcup\{F_m : m \in \mathbb{N}\}$. Then the sequence $(U_n(x_n) : n \in \mathbb{N})$ covers X. An even stronger fact holds: TWO has a winning strategy in the game $\mathsf{G}_1(\mathcal{O}_\Psi, \mathcal{O})$: For when ONE plays the open cover $\{U(x) : x \in X\}$ for some U in Ψ, then there is an $x \in X$ for which $U(x) = X$. Yet, X does not satisfy the property $\mathsf{S}_1(\mathcal{O}, \mathcal{O})$, for take for each n the open cover $\mathcal{O}_n = \{\{x\} : x \in X\}$ of X. Since X is uncountable, no countable subset of the set $\{\{x\} : x \in X\}$ covers X.

On the other hand, the hypotheses in Theorem 2.5 are not the optimal hypotheses under which the statements of the theorem are true. For example

EXAMPLE 2.7. In [**40**] Proposition 6 it is shown that there is for each uncountable cardinal κ a T_0 topological group that has cardinality κ, does not embed as a closed subgroup into any σ-compact group, and yet TWO has a winning strategy in the game $\mathsf{G}_1(\mathcal{O}_{nbd}, \mathcal{O})$ in this group. Even more, TWO has a winning strategy in the game $\mathsf{G}_1(\mathcal{O}, \mathcal{O})$ in this group. More information about this example is given in Section 8.

Towards connecting covering properties with Ramseyan statements we now introduce a concept the significance of which will be clear soon: Following [**17**] we say that an open cover $\mathcal{U}$ of a topological space (X, τ) is an ω-cover if X is not a member of $\mathcal{U}$, but for each finite subset F of X there is a $U \in \mathcal{U}$ such that $F \subseteq U$. The symbol Ω denotes the collection of all ω-covers of the space X.

THEOREM 2.8. *Let (X, τ) be a topological space and let Y be a subset of X. The following are equivalent:*

(1) *(X, τ) satisfies $\mathsf{S}_1(\mathcal{O}, \mathcal{O}_Y)$.*
(2) *(X, τ) satisfies $\mathsf{S}_1(\Omega, \mathcal{O}_Y)$.*

PROOF. (2)$\Rightarrow$(1) requires a proof. Thus, let a sequence $(\mathcal{U}_n : n \in \mathbb{N})$ of elements of $\mathcal{O}$ be given.

Next, write $\mathbb{N} = \bigcup\{S_n : n \in \mathbb{N}\}$ where each S_n is infinite, and for $m \neq n$ we have $S_m \cap S_n = \emptyset$. Then, for each n define

$$\mathcal{V}_n = \{\bigcup\{U_x : x \in F\} : F \subset S_n \text{ nonempty, finite and } U_x \in \mathcal{U}_x\}$$

Without loss of generality, each $\mathcal{V}_n$ is an ω cover. Applying $\mathsf{S}_1(\Omega, \mathcal{O}_Y)$ to the sequence $(\mathcal{V}_n : n \in \mathbb{N})$, choose for each n a $V_n \in \mathcal{V}_n$ such that $\{V_n : n \in \mathbb{N}\}$ is an open cover of Y.

Now for each n, choose a finite subset F_n of S_n such that $V_n = \cup\{U_x : x \in F_n\}$, where each corresponding U_x is a member of the original open cover $\mathcal{U}_x$ of X. Then, for each $m \in F = \bigcup\{F_n : n \in \mathbb{N}\}$ we have a $U_m \in \mathcal{U}_m$ such that the set $\{U_m : m \in F\}$ is a cover of Y. By choosing $U_m \in \mathcal{U}_m$ arbitrary when $m \in \mathbb{N} \setminus F$, we find a sequence $(U_n : n \in \mathbb{N})$ that witnesses $\mathsf{S}_1(\mathcal{O}, \mathcal{O}_Y)$ for the given sequence $(\mathcal{U}_n : n \in \mathbb{N})$. ☐

At this point it is worth observing that when the topology arises from a uniformity Ψ for which X is not totally bounded, we obtain a specific subset of the collection of ω-covers as follows: For $U \in \Psi$ we define $\Omega(U)$ to be the set $\{\cup_{x \in F} U(x) : F \subset X \text{ finite}\}$. When X is not a member of $\Omega(U)$, then $\Omega(U)$ is indeed an ω-cover of X. The symbol Ω_Ψ denotes the open ω-covers of X arising in this way from uniformity Ψ. In light of the statement of Theorem 2.8 one might speculate that for a subspace Y of a uniform space (X, Ψ) the statements $\mathsf{S}_1(\mathcal{O}_\Psi, \mathcal{O}_Y)$ and $\mathsf{S}_1(\Omega_\Psi, \mathcal{O}_Y)$ are equivalent. This, however, is not so.

EXAMPLE 2.9. The real line with its standard metric topology satisfies the statement $\mathsf{S}_1(\Omega_{nbd}, \mathcal{O})$, but does not satisfy the statement $\mathsf{S}_1(\mathcal{O}_{nbd}, \mathcal{O})$. For let a sequence $(\epsilon_n : n \in \mathbb{N})$ of positive real numbers be given. Then for each n the set $\mathcal{U}_n = \{\cup\{(x - \epsilon_n, x + \epsilon_n) : x \in F, F \subset \mathbb{R} \text{ finite}\}$ is a member of Ω_{nbd} for the real line. For each n, by the compactness of the closed interval $[-n, n]$, choose a finite subset F_n of $\mathbb{R}$ such that $[-n, n] \subset S_n = \bigcup\{(x - \epsilon_n, x + \epsilon_n) : x \in F_n\}$. Note that $S_n \in \mathcal{U}_n$, and the set $\{S_n : n \in \mathbb{N}\}$ is an open cover of the real line. On the other hand, the real line does not have the property $\mathsf{S}_1(\mathcal{O}_{nbd}, \mathcal{O})$.

More information about the differences between $\mathsf{S}_1(\Omega_\Psi, \mathcal{O}_Y)$ and $\mathsf{S}_1(\mathcal{O}_\Psi, \mathcal{O}_Y)$ can be gleaned from [**2**].

3. A Theorem of F.P. Ramsey

We now arrive at the introduction of the other main concept of the paper, partition relations. Partition relations, as well as corresponding symbolic representations, were originally introduced by Erdös and Rado [**12**] as various Ramseyan statements. The following theorem, entering the mathematics literature ten years after the Borel covering property, is the original stimulus for the development of Ramsey theory as a subfield of mathematics:

THEOREM 3.1 (Ramsey [**27**]). *For all positive integers m and k, and for each function $f : [\mathbb{N}]^m \longrightarrow \{1,\ 2,\ \cdots,\ k\}$ there are an infinite subset M of $\mathbb{N}$, and an element i of $\{1,\ 2,\ \cdots,\ k\}$ such that f is constant, of value i, on the set $[M]^m$.*

The statement of Theorem 3.1 uses notation requiring introduction: For an arbitrary set S and for a positive integer m the symbol $[S]^m$ denotes the set of all m-element subsets of the set S. The following notation will also be needed: When λ is an infinite cardinal number and S is a set, the symbol $[S]^{<\lambda}$ denotes the set consisting of subsets of cardinality less than λ of the set S, while $[S]^\lambda$ denotes the set whose elements are the subsets A of S for which $|A| = \lambda$.

In terms of the notation that was introduced by Erdös and collaborators, Theorem 3.1 can be stated as follows:

$$\text{For all positive integers m and k, } \aleph_0 \longrightarrow (\aleph_0)^m_k.$$

More generally, for cardinal numbers κ and λ, the symbolic statement

$$\text{For positive integers m and k, } \kappa \longrightarrow (\lambda)^m_k.$$

denotes that for any set X of cardinality κ, and for any function

$$f : [X]^m \to \{1,\ \cdots,\ k\}$$

there is an $i \in \{1,\ \cdots,\ k\}$ and a subset Y of X such that $|Y| = \lambda$, and f is constant of value i on the set $[Y]^m$. In this paper we adopt the following more general notation:

DEFINITION 3.2. Let k and m be positive integers. Let $\mathcal{A}$, $\mathcal{B}_1, \cdots \mathcal{B}_k$ be families of sets. The symbol

$$\mathcal{A} \longrightarrow (\mathcal{B}_1, \cdots, \mathcal{B}_k)^m$$

denotes that for each $A \in \mathcal{A}$, and for each function $f : [A]^m \to \{1, \cdots, k\}$, there is an $i \in \{1, \cdots, k\}$ and a $B_i \subset A$ such that $B_i \in \mathcal{B}_i$, and f is constant of value i on $[B_i]^m$.

When for $1 \leq i \leq k$ it is the case that $\mathcal{B}_i = \mathcal{B}$, then $\mathcal{A} \longrightarrow (\mathcal{B}_1, \cdots, \mathcal{B}_k)^m$ is denoted $\mathcal{A} \longrightarrow (\mathcal{B})^m_k$. If in Definition 3.2 we take the special instances $\mathcal{A} = [\mathbb{N}]^{\aleph_0}$ and $\mathcal{B} = \mathcal{A}$, then the symbol $\mathcal{A} \longrightarrow (\mathcal{B})^m_k$ denotes the instance for m and k of Ramsey's theorem. If, instead, we take $\mathcal{A}$ to be Ω, the set of ω-covers of a topological space, then the following holds:

LEMMA 3.3. *Let (X, τ) be a topological space which has infinite ω-covers. Then for each positive integer k the partition relation*

$$\Omega \longrightarrow (\Omega)^1_k$$

holds.

PROOF. Suppose that on the contrary $\mathcal{U}$ is an ω-cover of X, and that there is a partition, say

$$\mathcal{U} = \mathcal{U}_1 \bigcup \cdots \bigcup \mathcal{U}_k$$

of $\mathcal{U}$ into $k > 1$ pieces such that none of the $\mathcal{U}_i$ is an ω-cover of X. For each $1 \le i \le k$ choose a finite set $F_i \subset X$ for which there is no $U \in \mathcal{U}_i$ such that $F_i \subset U$. Then the set $F = \bigcup\{F_i : 1 \le i \le k\}$ is a finite subset of X, and yet no element of $\mathcal{U}$ covers F, contradicting the fact that $\mathcal{U}$ is an ω-cover of X. $\square$

Later on in the paper we will need the following consequence of Lemma 3.3:

COROLLARY 3.4. *Assume that for each sequence $(\mathcal{U}_n : n \in \mathbb{N})$ of ω-covers of X there is a positive integer k and a sequence $(\mathcal{F}_n : n \in \mathbb{N})$ such that for each n it is the case that $\mathcal{F}_n \subset \mathcal{U}_n$ and $|\mathcal{F}| \le k$, and $\bigcup\{\mathcal{F}_n : n \in \mathbb{N}\}$ is an ω-cover of X. Then $\mathsf{S}_1(\Omega, \Omega)$ holds.*

PROOF. For let a sequence $(\mathcal{U}_n : n \in \mathbb{N})$ of ω-covers of X be given, and fix a positive integer k as in the hypothesis of the Corollary. For each n fix a $\le k$-element set $\mathcal{F}_n$ of $\mathcal{U}_n$ such that $\mathcal{F} = \bigcup\{\mathcal{F}_n : n \in \mathbb{N}\}$ is an ω-cover of X. For each n enumerate $\mathcal{F}_n$ bijectively as $\{F_1^n, \cdots, F_{j_n}^n\}$ where $j_n \le k$. For $i \le k$ set $\mathcal{G}_i = \{F_i^n : n \in \mathbb{N} \text{ and } i \le j_n\}$. Then the sets $\mathcal{G}_1, \cdots, \mathcal{G}_k$ are pairwise disjoint, and $\mathcal{G} = \bigcup\{\mathcal{G}_i : i \le k\}$ is an ω-cover. By Lemma 3.3, for an $i \le k$ the set $\mathcal{G}_i$ is an ω-cover. But $\mathcal{G}_i$ results from at most one selection from each $\mathcal{U}_n$, confirming $\mathsf{S}_1(\Omega, \Omega)$ for the given sequence of ω-covers of X. Since the given sequence was an arbitrary sequence of ω-covers of X, the corollary follows. $\square$

Recall that a topological space is said to be Lindelöf if each open cover of the space has a countable subset that is a cover of the space. The following classical result is another theoretical component towards establishing the connection between Ramsey theory and variations of the Borel covering property.

THEOREM 3.5 (Arkhangelskii, Pytkeev). *If (X, τ) is a topological space for which each finite power is Lindelöf, then each ω-cover of X has a countable subset that is an ω-cover.*

A proof of Therorem 3.5 can be found in Theorem II.1.1 of [**1**]. If a space is σ-compact, it is Lindelöf in all finite powers, and thus each of its ω-covers has, by the Arkhangel'skii-Pytkeev Theorem, a countable subset that is an ω-cover of X. We now arrive at our first result, Theorem 3.6, that establishes a Ramseyan equivalence of the Borel covering property. This result was proven in [**36**], Theorem 9, for the special context of σ-compact metric spaces. Its generalization to Theorem 3.6 was briefly explained previously in Theorem 3 of [**40**]. For the reader's convenience we provide a proof of Theorem 3.6 here.

THEOREM 3.6. *Let (X, Ψ) be a σ-compact uniform space and let Y be a subset of X. Then the following statements are equivalent:*

(1) *The selection principle $\mathsf{S}_1(\mathcal{O}, \mathcal{O}_Y)$ holds.*
(2) *For each positive integer k the partition relation $\Omega \to (\mathcal{O}_Y)^2_k$ holds.*

PROOF. $\underline{(1) \Rightarrow (2)}$: Let (X, τ) be a σ-compact uniform space for which the selection principle $\mathsf{S}_1(\mathcal{O}, \mathcal{O}_Y)$ holds. Let $\mathcal{U}$ be a given ω-cover of X. By Theorem

3.5 we may assume that $\mathcal{U}$ is countable. Fix an enumeration of $\mathcal{U}$, say $(U_n : n \in \mathbb{N})$. Also, let a coloring

$$f : [\mathcal{U}]^2 \longrightarrow \{1, \cdots, k\}$$

be given.

Now recursively construct two sequences $(\mathcal{U}_n : n \in \mathbb{N})$ and $(i_n : n \in \mathbb{N})$ such that

(1) $\mathcal{U}_1 := \{U_m : m > 1 \text{ and } f(\{U_1, U_m\}) = i_1\}$ is an ω-cover of X and
(2) For each n, $\mathcal{U}_{n+l} := \{U_m \in \mathcal{U}_n : m > n+1 \text{ and } f(\{U_{n+I}, U_m\}) = i_{n+l}\}$ is an ω-cover of Y.

To obtain $\mathcal{U}_1$ and i_1, observe that $\mathcal{V} = \mathcal{U} \setminus \{U_1\}$ is an ω-cover of X. Setting $\mathcal{V}_j = \{U \in \mathcal{V} : f(\{U_1, U\} = j\}$, we obtain a partition

$$\mathcal{V} = \mathcal{V}_1 \cup \cdots \cup \mathcal{V}_k$$

of the ω-cover $\mathcal{V}$ into finitely many parts. By Lemma 3.3 at least one of the parts is an ω-cover. Select i_1 so that $\mathcal{V}_{i_1}$ is an ω-cover, and put $\mathcal{U}_1 = \mathcal{V}_{i_1}$. Assuming that the ω-cover $\mathcal{U}_n$ and i_n have been determined, proceed in the same way to obtain the ω-cover $\mathcal{U}_{n+1}$ and i_{n+1} from $\mathcal{U}_n$, using the function f. Observe that for each n we have $\mathcal{U}_{n+1} \subset \mathcal{U}_n$.

Next, define for $j \in \{1, \cdots, k\}$ the set $\mathcal{W}_j = \{U_n : i_n = j\}$. Then for each n we have the partition

$$\mathcal{U}_n = (\mathcal{U}_n \cap \mathcal{W}_1) \cup \cdots \cup (\mathcal{U}_n \cap \mathcal{W}_k).$$

Applying Lemma 3.3 to each of these partitions we fix for each n a $j_n \in \{1, \cdots, k\}$ for which $\mathcal{U}_n \cap \mathcal{W}_{j_n}$ is an ω-cover. Since for each n we have $\mathcal{U}_{n+1} \subset \mathcal{U}_n$, we may assume that the selected j_n's are all of the same value, say j. Thus, for each n, $\mathcal{U}_n \cap \mathcal{W}_j$ is an ω-cover of X.

With the sequence of $\mathcal{U}_n \cap \mathcal{W}_j$ selected, define the following strategy, F, for ONE in the game $\mathsf{G}_1(\mathcal{O}, \mathcal{O}_Y)$ played on X. ONE's first move is $F(\emptyset) = \mathcal{U}_1 \cap \mathcal{W}_j$. When TWO plays $U_{n_1} \in F(\emptyset)$, ONE responds with $F(U_{n_1}) = \mathcal{U}_{n_1} \cap \mathcal{W}_j$. When TWO responds with a $U_{n_2} \in F(U_{n_1})$, ONE responds with $F(U_{n_1}, U_{n_2}) = \mathcal{U}_{n_2} \cap \mathcal{W}_j$, and so on. By hypothesis $\mathsf{S}_1(\mathcal{O}, \mathcal{O}_Y)$ holds. Since X is σ-compact, Theorem 2.5 implies that ONE does not have a winning strategy in the game $\mathsf{G}_1(\mathcal{O}, \mathcal{O}_Y)$. Thus, the strategy F just defined for player ONE is not a winning strategy in the game $\mathsf{G}_1(\mathcal{O}, \mathcal{O}_Y)$. Choose an F-play

$$F(\emptyset), U_{n_1}, F(U_{n_1}), U_{n_2}, \cdots, U_{n_m}, F(U_{n_1}, \cdots, U_{n_m}), U_{n_{m+1}}, \cdots$$

that is lost by ONE. Then the set $\mathcal{H} = \{U_{n_j} : j \in \mathbb{N}\}$ is an element of $\mathcal{O}_Y$ (that is, a cover of Y), and for all $u < v$ we have $f(\{U_{n_u}, U_{n_v}\}) = j$. But then $\mathcal{H} \subseteq \mathcal{U}$ witnesses that $\Omega \to (\mathcal{O}_Y)^2_k$ holds for the function f. Since the ω-cover $\mathcal{U}$ of X, and the function $f : [\mathcal{U}]^2 \to \{1, \cdots, k\}$ were arbitrary, this completes the proof of the implication (1) $\Rightarrow$ (2).

(2) $\Rightarrow$ (1) Since, by Theorem 2.8, for any topological space and a subspace Y of it the statements $\mathsf{S}_1(\mathcal{O}, \mathcal{O}_Y)$ and $\mathsf{S}_1(\Omega, \mathcal{O}_Y)$ are equivalent, it suffices to show that if the space satisfies the statement $\Omega \longrightarrow (\mathcal{O}_Y)^2_k$ for each positive integer k, then indeed that space satisfies the statement $\mathsf{S}_1(\Omega, \mathcal{O}_Y)$.

Thus, let a sequence $(\mathcal{U}_n : n \in \mathbb{N})$ of ω-covers of the space X be given. By the σ-compactness of X and Theorem 3.5 we may assume that each $\mathcal{U}_n$ is countable. Fix for each n a repetitionfree enumeration of $\mathcal{U}_n$, say $(U^n_m : m \in \mathbb{N})$. Then define

$\mathcal{U} = \{U_m^1 \cap U_n^m : m, n \in \mathbb{N}\}$. Then $\mathcal{U}$ is an ω-cover of X, for let a finite subset F of X be given. Since $\mathcal{U}_1$ is an ω-cover of X, choose $U_m^1 \in \mathcal{U}_1$ with $F \subseteq U_m^1$. Then, since $\mathcal{U}_m$ is an ω-cover of X, choose $U_n^m \in \mathcal{U}_m$ with $F \subseteq U_n^m$. Then F is a subset of the element $U_m^1 \cap U_n^m$ of $\mathcal{U}$.

Next define a function $f : [\mathcal{U}]^2 \to \{0, 1\}$ as follows:

$$f(\{U_m^1 \cap U_n^m, U_k^1 \cap U_\ell^k\}) = \begin{cases} 0 & \text{if } m = k \\ 1 & \text{otherwise} \end{cases}$$

By the hypothesis that $\Omega \longrightarrow (\mathcal{O}_Y)_2^2$ holds, select a subset $\mathcal{S}$ of $\mathcal{U}$ and an $i \in \{0, 1\}$ such that $\mathcal{S}$ is a cover of Y, and for any two $A, B \in \mathcal{S}$, we have $f(\{A, B\}) = i$.

If $i = 0$, then there is a fixed k such that each element of $\mathcal{V}$ is of the form $U_k^1 \cap U_m^k$. It follows that in this case $Y \subseteq U_k^1$. But then to obtain an element of $\mathcal{O}_Y$ that witnesses $\mathsf{S}_1(\Omega, \mathcal{O}_Y)$ for the sequence $(\mathcal{U}_n : n \in \mathbb{N})$, select U_k^1 from $\mathcal{U}_1$, and arbitrary elements from $\mathcal{U}_m$ when $m > 1$.

On the other hand, if $i = 1$, then $\mathcal{V}$ is of the form $\{U_{m_k}^1 \cap U_{n_k}^{m_k} : k \in \mathbb{N}\}$ where $m_k \neq m_\ell$ whenever $k \neq \ell$. In this case choose elements $V_j \in \mathcal{U}_j$ so that $V_1 = U_{m_1}^1$, $V_{m_k} = U_{n_k}^{m_k}$ when $j = m_k$, and for all other values of j, $V_j \in \mathcal{U}_j$ are arbitrarily chosen.

In either case we obtain an element of $\mathcal{O}_Y$. □

With the basic result that in σ-compact uniformizable spaces Borel's covering property is a Ramseyan property now established, the rest of the paper is dedicated to a deeper analysis of the Ramsey-theoretic aspects of covering properties analogous to Borel's covering property.

4. More Partition Relations for the Borel Covering Property.

Let a topological space (X, τ) and subspace Y be given. We now explore strengthening the partition relation $\Omega \longrightarrow (\mathcal{O}_Y)_k^2$: Two immediate targets are the exponent 2, and the type of open cover of Y that can be obtained in the partition relation. To this end let Ω_Y denote the set of ω-covers of the subspace Y, using sets that are open in X. Our first quest is identifying the conditions under which the stronger partition relation $\Omega \to (\Omega_Y)_k^2$ holds. This section encapsulates some of the results obtained in [**36**].

For the finite powers X^n and Y^n with the product topologies, use the following special notation for clarity.

$\mathcal{O}(n)$	The collection of open covers of X^n
$\Omega(n)$	The collection of open ω-covers of X^n
$\mathcal{O}_{Y^n}$	The collection of covers of Y^n by sets open in X^n
Ω_{Y^n}	The collection of ω-covers of Y^n by sets open in X^n

If (X, Ψ) is a σ-compact uniformizable space then by Theorem 2.5 the following equivalence holds: For a positive integer n, X^n satisfies $\mathsf{S}_1(\mathcal{O}(n), \mathcal{O}_{Y^n})$ if, and only if, the partition relation $\Omega(n) \longrightarrow (\mathcal{O}_{Y^n})_k^2$ holds for each positive integer k.

The following sequence of lemmas will be used in the main argument.

LEMMA 4.1. *Let X be a topological space and let $\mathcal{U}$ be an ω-cover for the finite power X^n. Then there is an ω-cover $\mathcal{V}$ of X such that for each $V \in \mathcal{V}$ there is a $U \in \mathcal{U}$ with $V^n \subseteq U$, and $\{V^n : V \in \mathcal{V}\}$ is an ω-cover for X^n.*

A proof of Lemma 4.1 can be found in Lemma 3.3 of [**20**]. Lemma 4.2, which is also Lemma 11 of [**36**], will be left as an exercise.

LEMMA 4.2. *For each n let Y_n be a subset of the σ-compact uniform space (X_n, Ψ_n), and let Ω_n be the set of ω-covers of X_n, and $\mathcal{O}_n$ the set of covers of Y_n by sets open in X_n. Let $\mathcal{O}_{\sum_n X_n}$ be the set of open covers of $\sum_n X_n$ and let $\mathcal{O}_{\sum Y_n}$ be the set of covers of $\sum_n Y_n$ by sets open in $\sum_n X_n$. If each X_n satisfies $\mathsf{S}_1(\Omega_n, \mathcal{O}_n)$, then the uniform space $\sum_n X_n$ satisfies $\mathsf{S}_1(\mathcal{O}_{\sum_n X_n}, \mathcal{O}_{\sum_n Y_n})$*

For the convenience of the reader we give the proof of the following lemma.

LEMMA 4.3. *Let $((X_n, \tau_n) : n \in \mathbb{N})$ be a sequence of topological spaces and for each n, let Y_n be a subspace of X_n. If for each n player ONE does not have a winning strategy in the game $\mathsf{G}_1(\mathcal{O}_{X_n}, \mathcal{O}_{Y_n})$, then ONE has no winning strategy in the game $\mathsf{G}_1(\mathcal{O}_{\Sigma X_n}, \mathcal{O}_{\Sigma Y_n})$.*

PROOF. Observe that open subsets of $\Sigma_n X_n$ are sets of the form $\Sigma_n U_n$ where for each n the set U_n is an open subset of the set X_n. Let F be a strategy of player ONE for the game $\mathsf{G}_1(\mathcal{O}_{\Sigma X_n}, \mathcal{O}_{\Sigma Y_n})$. Choose a well-ordering $\prec$ of the topology of $X = \Sigma_n X_n$. Also, choose a partition of $\mathbb{N}$ into infinitely many pairwise disjoint infinite sets, say $\mathbb{N} = \cup\{S_n : n \in \mathbb{N}\}$. Then for each natural number k fix m_k so that $k \in S_{m_k}$. Further, if O is an open subset of ΣX_m and k is a positive integer, then use O_k to denote the term U_k in the representation $O = \Sigma_m U_m$.

Define responses of player TWO to the strategy F as follows:

In the first inning, ONE plays $F(\emptyset)$, an open cover of ΣX_n. Fix m_1 so that $1 \in S_{m_1}$. Define a strategy F_{m_1} for ONE of the game $\mathsf{G}_1(\mathcal{O}_{X_{m_1}}, \mathcal{O}_{Y_{m_1}})$ as follows:

$$F_{m_1}(\emptyset) = \{U_{m_1} : \Sigma U_m \in F(\emptyset)\}$$

For a response T of TWO to F_{m_1}, choose the $\prec$-first $U = \Sigma U_m \in F(\emptyset)$ for which $U_{m_1} = T$, and put $T_1 = U$.

In the next inning ONE plays $F(T_1)$, an open cover of ΣX_n. Take m_2 so that $2 \in S_{m_2}$. Put $O_{m_2} = \{U_{m_2} : \Sigma U_m \in F(T_1)\}$. Then define the strategy F_{m_2} for ONE of the game $\mathsf{G}_1(\mathcal{O}_{X_{m_2}}, \mathcal{O}_{Y_{m_2}})$ as follows: If $m_2 = m_1$, then $F_{m_2}(T) = \{U_{m_2} : \Sigma U_m \in F(T_1)\}$. Else, $F_{m_2}(\emptyset) = \{U_{m_2} : \Sigma U_m \in F(T_1)\}$. For a response T of TWO in the game $\mathsf{G}_1(\mathcal{O}_{X_{m_2}}, \mathcal{O}_{Y_{m_2}})$, pick the $\prec$-first $U \in F(T_1)$ with $U_{m_2} = T$, and set $T_2 = U$.

In general, suppose that $k-1$ innings of the game have been played, producing the data

$$F(\emptyset), T_1, F(T_1, T_2), T_3, \cdots, F(T_1, \cdots, T_{k-2}), T_{k-1}$$

and numbers $m_1, \cdots, m_{k-1}$ such that for $1 \le i \le k-1$ we have $i \in S_{m_i}$.

Now when ONE plays $F(T_1, \cdots, T_{k-1})$, the k-th inning is underway. Then choose the longest possible sequence $i_1 < \cdots < i_j$ so that

- $i_j = k$
- $i_1 < \cdots < i_j$ and
- $m_{i_1} = \cdots = m_{i_j}$.

Define the strategy F_{m_k} for ONE of the game $\mathsf{G}_1(\mathcal{O}_{X_{m_k}}, \mathcal{O}_{Y_{m_k}})$ as follows:

If $j = 1$: Then we proceed as follows:

$$F_{m_k}(\emptyset) = \{U_{m_k} : \Sigma U_m \in F(T_1, \cdots, T_{k-1})\}$$

For a response T of TWO to F_{m_k}, choose the $\prec$-first $U = \Sigma U_m \in F(T_1, \cdots, T_{k-1})$ for which $U_{m_k} = T$, and put $T_k = U$.

<u>If $j > 1$:</u> Then we proceed as follows: The strategy F_{m_k} is the same as $F_{m_{i_1}}$, and we define

$$F_{m_{i_1}}((T_{i_1})_{m_k}, \cdots, (T_{i_{j-1}})_{m_k}) = \{U_{m_k} : U = \Sigma U_n \in F(T_1, \cdots, T_{k-1})\}$$

Now for a response T of TWO to the move $F_{m_{i_1}}((T_{i_1})_{m_k}, \cdots, (T_{i_{j-1}})_{m_k})$, choose the $\prec$-first $T_k \in F(T_1, \cdots, T_{k-1})$ so that $(T_k)_{m_k} = T$ and let this T_k be TWO's response to $F(T_1, \cdots, T_{k-1})$.

This process recursively defines for each $k \in \mathbb{N}$ the strategy F_k for player ONE of the game $\mathsf{G}_1(\mathcal{O}_{X_k}, \mathcal{O}_{Y_k})$. Each strategy F_k is by hypothesis not a winning strategy for ONE of the game $\mathsf{G}_1(\mathcal{O}_{X_k}, \mathcal{O}_{Y_k})$. Thus, for each k there is a play lost by ONE. For each k choose a play

$$F_k(\emptyset), T_1^k,\ F_k(T_1^k),\ T_2^k,\ \cdots,$$

lost by ONE in the game $\mathsf{G}_1(\mathcal{O}_{X_k}, \mathcal{O}_{Y_k})$.

Now consider the following F-play of the game $\mathsf{G}_1(\mathcal{O}_{\Sigma X_m},\ \mathcal{O}_{\Sigma Y_m})$ obtained as follows (recall the definition of the sequence $(m_k : k \in \mathbb{N})$):

Choose the $\prec$-first $T_1 \in F(\emptyset)$ with $T_{m_1} = F_{m_1}(\emptyset)$. Then choose the $\prec$-first $T_2 \in F(T_1)$ so that $(T_2)_{m_2} \in F_j(A)$ where $A = \emptyset$ if $m_2 \neq m_1$, and $A = T_{m_1}$ otherwise. Observe that when $m_1 = m_2$, then $(T_2)_{m_2} = T_2^{m_1}$, and when $m_1 \neq m_2$, then $(T_2)_{m_2} = T_1^{m_2}$.

With $T_1, \cdots, T_k$ selected, choose the $\prec$-first $T_{k+1} \in F(T_1, \cdots, T_k)$ so that

$$(T_{k+1})_{m_{k+1}} = T_{i_j}^{m_{k+1}} \in F_{m_{k+1}}((T_{i_1})_{m_{k+1}}, \cdots, (T_{i_j})_{m_{k+1}})$$

where i_j is such that $m_{i_1} = \cdots = m_{i_j}$.

It is left to the reader to verify that the F-play

$$F(\emptyset), T_1,\ F(T_1),\ T_2,\ \cdots, T_k,\ F(T_1, \cdots, T_k),\ \cdots$$

of the game $\mathsf{G}_1(\mathcal{O}_{\Sigma X_m}, \mathcal{O}_{\Sigma Y_m})$ that comes about in this way is lost by ONE. □

THEOREM 4.4. *Let Y be a subspace of a σ-compact uniform space (X, Ψ). The following are equivalent:*

(1) *For each n, X^n satisfies $\mathsf{S}_1(\Omega(n), \mathcal{O}_{Y^n})$;*
(2) *X satisfies $\mathsf{S}_1(\Omega, \Omega_Y)$*

PROOF. (1) $\Rightarrow$ (2) : Let $(\mathcal{U}_n : n \in \mathbb{N})$ be a sequence of ω-covers for the space X. Then for each n define

$$\mathcal{V}_n = \{U^k : U \in \mathcal{U}_n \text{ and } k \in \mathbb{N}\}.$$

Then each $\mathcal{V}_n$ is an open cover of the σ-compact uniformizable space $\sum_m X^m$. By hypothesis (1) each X^m has the property $\mathsf{S}_1(\Omega(m), \mathcal{O}_{Y^m})$, and thus by Theorem 2.8 each X^m has the property $\mathsf{S}_1(\mathcal{O}(m), \mathcal{O}_{Y^m})$. But then by Lemma 4.2, $\sum_m X^m$ satisfies the property $\mathsf{S}_1(\mathcal{O}_{\sum_m X^m}, \mathcal{O}_{\sum_m Y^m})$. Thus, select for each m a set $V_m \in \mathcal{V}_m$ so that $\{V_m : m \in \mathbb{N}\}$ is an open cover of $\sum_m Y^m$. For each m choose a positive integer k_m and a $U_m \in \mathcal{U}_m$ so that $V_m = U_m^{k_m}$. We claim that $\{U_m : m \in \mathbb{N}\}$ is an ω-cover for Y. For let a finite subset $\{x_1, \cdots, x_p\}$ of Y be given. Then the point $(x_1, \cdots, x_p)$ is an element of Y^p and thus of $\sum_m Y^m$. Choose an m such that $(x_1, \cdots, x_p) \in V_m$. Then $k_m = p$ and $(x_1, \cdots, x_p) \in U_m^p$, meaning that $\{x_1, \cdots, x_p\} \subset U_m$. This completes the proof of (1) $\Rightarrow$ (2).

$(2) \Rightarrow (1)$: Fix a positive integer n, and let $(\mathcal{U}_m : m \in \mathbb{N})$ be a sequence of ω-covers of X^n. By Lemma 4.1 choose for each m an ω-cover $\mathcal{V}_m$ of X such that $\{V^n : V \in \mathcal{V}_m\}$ refines $\mathcal{U}_m$ (and is an ω-cover of X^n). Applying the hypothesis $\mathsf{S}_1(\Omega, \Omega_Y)$ to the sequence $(\mathcal{V}_m : m \in \mathbb{N})$ of ω covers of X, choose for each m a $V_m \in \mathcal{V}_m$ so that $\{V_m : m \in \mathbb{N}\}$ is an ω-cover of Y. But then $\{V_m^n : m \in \mathbb{N}\}$ is an ω cover of Y^n by sets open in X^n. For each m choose a $V_m \in \mathcal{V}_m$ so that $U_m^n \subseteq V_m$. Then the set $\{V_m : m \in \mathbb{N}\}$ is an ω-cover of Y^n, and witnesses $\mathsf{S}_1(\Omega(n), \Omega_{Y^n})$ for the given sequence $(\mathcal{V}_m : m \in \mathbb{N})$ of ω-covers of X^n. □

The proof of Theorem 4.4 is a rewrite for uniform spaces of the proof of $(2) \Leftrightarrow (3)$ of Theorem 12 of [**36**], originally given for metric spaces. Next we expand the equivalences to include the appropriate game theoretic version.

THEOREM 4.5. *Let Y be a subspace of a σ-compact uniform space (X, Ψ). The following are equivalent:*

(1) *X satisfies $\mathsf{S}_1(\Omega, \Omega_Y)$*
(2) *On X ONE has no winning strategy in the game $\mathsf{G}_1(\Omega, \Omega_Y)$.*

PROOF. $(1) \Rightarrow (2)$: Assume that the space X has the property $\mathsf{S}_1(\Omega, \Omega_Y)$. Fix a strategy F for ONE in the game $\mathsf{G}_1(\Omega, \Omega_Y)$. We must show that F is not a winning strategy.

Define from the strategy F for ONE in the game $\mathsf{G}_1(\Omega, \Omega_Y)$ a corresponding strategy G for ONE in the game $\mathsf{G}_1(\mathcal{O}_{\sum_n X^n}, \mathcal{O}_{\sum_n Y^n})$ as follows, as depicted in Figure 1: Define ONE's first move, $G(\emptyset)$ in the game $\mathsf{G}_1(\mathcal{O}_{\sum_n X^n}, \mathcal{O}_{\sum_n Y^n})$ as the following open cover of $\sum_n X^n$ obtained from $F(\emptyset)$:

$$G(\emptyset) = \{U^k : k \in \mathbb{N}, U \in F(\emptyset)\}.$$

For a response $T_1 \in G(\emptyset)$ by TWO, define $G(T_1)$ as follows: Fix $U_1 \in F(\emptyset)$ so that $T_1 = U_1^{k_1}$ for some positive integer k_1, and define

$$G(T_1) = \{U^k : k \in \mathbb{N}, U \in F(U_1)\},$$

and so on.

$\mathsf{G}_1(\Omega, \Omega_Y)$ on X

ONE	TWO
$F(\emptyset)$	
	U_1
$F(U_1)$	
	U_2
$F(U_1, U_2)$	
	U_3
$\vdots$	$\vdots$

$\mathsf{G}_1(\mathcal{O}_{\sum_n X^n}, \mathcal{O}_{\sum_n Y^n})$ on $\sum_n X^n$

ONE	TWO
$G(\emptyset) = \{U^k : U \in F(\emptyset), k \in \mathbb{N}\}$	$T_1 = U_1^{k_1}$
$G(T_1) = \{U^k : U \in F(U_1), k \in \mathbb{N}\}$	$T_2 = U_2^{k_2}$
$G(T_1, T_2) = \{U^k : U \in F(U_1, U_2), k \in \mathbb{N}\}$	$T_3 = U_3^{k_3}$
$\vdots$	$\vdots$

FIGURE 1. Defining the strategy G from the strategy F

Now note that by (2) $\Rightarrow$ (1) of Theorem 4.4 and by Theorem 2.8, for each n the space X^n satisfies $\mathsf{S}_1(\mathcal{O}(n), \mathcal{O}_{Y^n})$. Theorem 2.5 implies that ONE has no winning strategy in the game $\mathsf{G}_1(\mathcal{O}(n), \mathcal{O}_{Y^n})$. Lemma 4.3 implies that ONE has no winning strategy in the game $\mathsf{G}_1(\mathcal{O}_{\sum_n X^n}, \mathcal{O}_{\sum_n Y^n})$.

It follows that the strategy G just defined is not a winning strategy for ONE in the game $\mathsf{G}_1(\mathcal{O}_{\sum_n X^n}, \mathcal{O}_{\sum_n Y^n})$. Thus, fix a G-play lost by ONE, say

$$G(\emptyset), T_1, G(T_1), T_2, G(T_1, T_2), \cdots, T_n, G(T_1, T_2, \cdots, T_n), \cdots.$$

Since this play is lost by ONE, the set $\{T_n : n \in \mathbb{N}\}$ is a cover of $\sum_n Y^n$.

Note from the definition of G that there is a corresponding sequence $((U_m, k_m) : m \in \mathbb{N})$ such that for each m

(1) $T_m = U_m^{k_m}$
(2) $U_1 \in F(\emptyset)$ and for each m, $U_{m+1} \in F(U_1, \cdots, U_m)$.

Thus,

$$F(\emptyset), U_1, F(U_1), U_2, F(U_1, U_2), \cdots, U_m, F(U_1, \cdots, U_m), \cdots$$

is a corresponding F-play of the game $\mathsf{G}_1(\Omega, \Omega_Y)$ on X. As in the proof of (1) $\Rightarrow$ (2) of Theorem 4.4, it follows that $\{U_m : m \in \mathbb{N}\}$ is an ω-cover of Y, and thus the corresponding F-play is lost by ONE. This completes the proof of (1) $\Rightarrow$ (2).

The implication (2) $\Rightarrow$ (1) follows from earlier remarks. $\square$

The proof of Theorem 4.5 is a rewrite for uniform spaces of the proof of (3) $\Leftrightarrow$ (4) of Theorem 12 of [**36**], originally given for metric spaces. In the case when $X = Y$, the following result is known:

THEOREM 4.6. *For a topological space (X, τ) the following are equivalent:*

(1) *For each natural number n the space X^n has property $\mathsf{S}_1(\mathcal{O}, \mathcal{O})$.*
(2) *The space X has the property $\mathsf{S}_1(\Omega, \Omega)$*
(3) *ONE has no winning strategy in the game $\mathsf{G}_1(\Omega, \Omega)$.*

Observe that in Theorem 4.6 the hypothesis that X is σ-compact is dropped, but $Y = X$ is assumed. The equivalence of (1) and (2) in Theorem 4.6 was discovered by M. Sakai [**30**], while the equivalence of (2) and (3) was proven in Theorem 2 of [**32**].

Next we expand the equivalences in Theorem 4.5 to include the appropriate Ramsey theoretic version. The corresponding work for Theorem 4.6 is done in Section 7.

THEOREM 4.7. *Let Y be a subspace of a σ-compact uniform space (X, Ψ). The following are equivalent:*

(1) *X satisfies $\mathsf{S}_1(\Omega, \Omega_Y)$*
(2) *For each k, X satisfies $\Omega \longrightarrow (\Omega_Y)^2_k$.*

PROOF. (1) $\Rightarrow$ (2) : Assume that for the subspace Y of X it is true that $\mathsf{S}_1(\Omega, \Omega_Y)$ holds. Let $\mathcal{U}$ be a given ω-cover of X. Since X is σ-compact the Arkhangel'skii-Pytkeev theorem implies that $\mathcal{U}$ has a countable subset that is an ω-cover of X. We may as well assume that $\mathcal{U}$ is countable, and enumerate it without repetitions as $(U_n : n \in \mathbb{N})$. We now use an argument analogous to that in the proof of Theorem 3.6.

Let k be a positive integer and let a function $f : [\mathcal{U}]^2 \to \{1, \cdots, k\}$ be given. Recursively construct two sequences $(\mathcal{U}_n : n \in \mathbb{N})$ and $(i_n : n \in \mathbb{N})$ such that

(1) $\mathcal{U}_1 := \{U_m : m > 1 \text{ and } f(\{U_1, U_m\}) = i_1\}$ is an ω-cover of X and
(2) For each n, $\mathcal{U}_{n+l} := \{U_m \in \mathcal{U}_n : m > n+1 \text{ and } f(\{U_{n+I}, U_m\}) = i_{n+l}\}$ is an ω-cover of Y.

For example, to obtain $\mathcal{U}_1$ and i_1, note that $\mathcal{V} = \mathcal{U} \setminus \{U_1\}$ is an ω-cover of X. Setting $\mathcal{V}_j = \{U \in \mathcal{V} : f(\{U_1, U\} = j\}$ defines a partition

$$\mathcal{V} = \mathcal{V}_1 \cup \cdots \cup \mathcal{V}_k$$

of the ω-cover $\mathcal{V}$ into finitely many parts. By Lemma 3.3 at least one of these parts is an ω-cover. Select a i_1 for which $\mathcal{V}_{i_1}$ is an ω-cover, and put $\mathcal{U}_1 = \mathcal{V}_{i_1}$. Assume that the ω-cover $\mathcal{U}_n$ and i_n have been determined, and proceed in the same way to obtain the ω-cover $\mathcal{U}_{n+1}$ and i_{n+1} from $\mathcal{U}_n$, using the function f. Observe that for each n we have $\mathcal{U}_{n+1} \subset \mathcal{U}_n$.

Next we define for $j \in \{1, \cdots, k\}$ the set $\mathcal{W}_j = \{U_n : i_n = j\}$. Then for each n we have the partition

$$\mathcal{U}_n = (\mathcal{U}_n \cap \mathcal{W}_1) \cup \cdots \cup (\mathcal{U}_n \cap \mathcal{W}_k).$$

Applying Lemma 3.3 to each of these partitions we fix for each n a $j_n \in \{1, \cdots, k\}$ for which $\mathcal{U}_n \cap \mathcal{W}_{j_n}$ is an ω-cover. Since for each n we have $\mathcal{U}_{n+1} \subset \mathcal{U}_n$, we may assume that the selected j_n's are all of the same value, say j. Thus, for each n, $\mathcal{U}_n \cap \mathcal{W}_j$ is an ω-cover of X.

With the sequence of $\mathcal{U}_n \cap \mathcal{W}_j$ selected, define the following strategy, F, for ONE in the game $\mathsf{G}_1(\Omega, \Omega_Y)$ played on X. ONE's first move is $F(\emptyset) = \mathcal{U}_1 \cap \mathcal{W}_j$. When TWO plays $U_{n_1} \in F(\emptyset)$, ONE responds with $F(U_{n_1}) = \mathcal{U}_{n_1} \cap \mathcal{W}_j$. When TWO responds with a $U_{n_2} \in F(U_{n_1})$, ONE responds with $F(U_{n_1}, U_{n_2}) = \mathcal{U}_{n_2} \cap \mathcal{W}_j$, and so on. By hypothesis $\mathsf{S}_1(\Omega, \Omega_Y)$ holds. Since X is σ-compact, Theorem 4.5 implies that ONE does not have a winning strategy in the game $\mathsf{G}_1(\Omega, \Omega_Y)$. Thus, the strategy F just defined for player ONE is not a winning strategy in the game $\mathsf{G}_1(\Omega, \Omega_Y)$. Choose an F-play

$$F(\emptyset), U_{n_1}, F(U_{n_1}), U_{n_2}, \cdots, U_{n_m}, F(U_{n_1}, \cdots, U_{n_m}), U_{n_{m+1}}, \cdots$$

that is lost by ONE. Then the set $\mathcal{H} = \{U_{n_j} : j \in \mathbb{N}\}$ is an element of Ω_Y (that is, an ω cover of Y), and for all $u < v$ we have $f(\{U_{n_u}, U_{n_v}\}) = j$. But then $\mathcal{H} \subseteq \mathcal{U}$ witnesses that $\Omega \to (\Omega_Y)^2_k$ holds for the function f. Since the ω-cover $\mathcal{U}$ of X, and the function $f : [\mathcal{U}]^2 \to \{1, \cdots, k\}$ were arbitrary, this completes the proof of the implication (1) $\Rightarrow$ (2).

(2) $\Rightarrow$ (1) Let a sequence $(\mathcal{U}_n : n \in \mathbb{N})$ of ω-covers of the space X be given. By the σ-compactness of X and Theorem 3.5 we may assume that each $\mathcal{U}_n$ is countable. Fix for each n a repetitionfree enumeration of $\mathcal{U}_n$, say $(U^n_m : m \in \mathbb{N})$. Then define $\mathcal{U} = \{U^1_m \cap U^m_n : m, n \in \mathbb{N}\}$. Then $\mathcal{U}$ is an ω-cover of X.

Next define a function $f : [\mathcal{U}]^2 \to \{0, 1\}$ as follows:

$$f(\{U^1_m \cap U^m_n, U^1_k \cap U^k_\ell\}) = \begin{cases} 0 & \text{if } m = k \\ 1 & \text{otherwise} \end{cases}$$

By the hypothesis that $\Omega \longrightarrow (\Omega_Y)^2_2$ holds, select a subset $\mathcal{S}$ of $\mathcal{U}$ and an $i \in \{0, 1\}$ such that $\mathcal{S}$ is an ω cover of Y, and for any two $A, B \in \mathcal{S}$, we have $f(\{A, B\}) = i$.

If $i = 0$, then there is a fixed k such that each element of $\mathcal{V}$ is of the form $U^1_k \cap U^k_m$. It follows that in this case $Y \subseteq U^1_k$, contradicting the fact that $\mathcal{S}$ is an ω-cover of Y.

Therefore, $i = 1$ and $\mathcal{V}$ is of the form $\{U^1_{m_k} \cap U^{m_k}_{n_k} : k \in \mathbb{N}\}$ where $m_k \neq m_\ell$ wheneve $k \neq \ell$. In this case choose elements $V_j \in \mathcal{U}_j$ so that $V_1 = U^1_{m_1}$, $V_{m_k} = U^{m_k}_{n_k}$ when $j = m_k$, and for all other values of j, $V_j \in \mathcal{U}_j$ are arbitrarily chosen. The set $\{V_j : j \in \mathbb{N}\}$ is an element of Ω_Y. □

The proof of Theorem 4.7 is a rewrite for uniform spaces of the proof of (4) ⇔ (5) of Theorem 12 of [**36**], originally given for metric spaces. We expect that there are also improvements to be made in the exponents appearing in the partition relation $\Omega \longrightarrow (\Omega_Y)^2_k$, but currently have no theoretical evidence for this.

PROBLEM 1. Let (X, Ψ) be a σ-compact uniformizable space. Is it true that if $\Omega \longrightarrow (\Omega_Y)^2_2$, then for all finite positive integers m and k, $\Omega \longrightarrow (\Omega_Y)^m_k$.

We shall later see that the answer to Problem 1 is "yes" for the special situation when $Y = X$.

5. Partition relations and Rothberger's covering property $\mathsf{S}_1(\mathcal{O}, \mathcal{O})$

In [**28**] Rothberger pointed out that if a set X of real numbers, endowed with the relative topology inherited from $\mathbb{R}$, satisfies the selection principle $\mathsf{S}_1(\mathcal{O}, \mathcal{O})$, then X has the Borel covering property. As it is independent of ZFC whether the Borel covering property and the Rothberger covering property coincide in separable metric spaces, the converse implication is not provable. In [**29**] Rothberger proved that the Continuum Hypothesis implies the existence of a set of real numbers that has Borel's covering property, but does not have Rothberger's covering property. Formally, the Rothberger covering property is stronger than the Borel covering property.

The Rothberger covering property can be characterized in terms of Ramseyan properties that are generally stronger than the Ramseyan properties characterizing the Borel covering property. We now describe the known relation between Ramsey theoretic properties and the Rothberger covering property. The preservation of a covering property under finite powers is significantly reflected by the Ramseyan covering properties of the underlying space. Correspondingly the treatment below is divided into two parts according to the behavior of the covering property under finite powers.

From this point on we work with a topological space (X, τ), and assume that the subspace Y considered before is equal to X. Our first goal is the following basic theorem:

THEOREM 5.1. *Let (X, τ) be a regular topological space. The following statements are equivalent:*

(1) *X has the covering property $\mathsf{S}_1(\mathcal{O}, \mathcal{O})$.*
(2) *ONE has no winning strategy in the game $\mathsf{G}_1(\mathcal{O}, \mathcal{O})$ played on X.*
(3) *For each positive integer k, $\Omega \longrightarrow (\mathcal{O})^2_k$ holds for X.*

PROOF. The proof of the fact that (1) implies (2) is Theorem 2.2, due to Pawlikowski, [**25**]. Note that this implication is stronger than the corresponding one proven in Theorem 3.6 since in that theorem we assumed that the space X is σ-compact. That (2) implies (3), and (3) implies (1) follows the argument in the proof of Theorem 3.6. □

We now consider the corresponding equivalence for the case where we consider whether for a fixed finite power of X, that power has the Rothberger covering property. For convenience we introduce the following notation: Let (X,τ) be a topological space and let n be a positive integer. Let $\mathcal{O}_n$ denote the collection of open covers $\mathcal{U}$ of X with the property that whenever $F \subset X$ has at most n elements, then there is a set $U \in \mathcal{U}$ for which $F \subseteq U$. Note that when (X,τ) is a topological space such that for a positive integer n the Tychonoff product space X^n satisfies the property $\mathsf{S}_1(\mathcal{O},\mathcal{O})$, then all smaller powers of X also have this property. The reason is that the property $\mathsf{S}_1(\mathcal{O},\mathcal{O})$ is preserved by homeomorphisms, a closed subset of a space inherits the property $\mathsf{S}_1(\mathcal{O},\mathcal{O})$ from the ambient space, and when $m < n$ are positive integers, then the space X^m embeds as a closed subspace of X^n.

In the proof of Theorem 5.3 we make use of the following lemma, the proof of which is left to the reader:

LEMMA 5.2. *If $\mathcal{U}$ is an (open) ω-cover for the space X, then $\{U^n : U \in \mathcal{U}\}$ is a ω-cover for the space X^n.*

THEOREM 5.3. *For a topological space (X,τ) and positive integer n, the following are equivalent:*

(1) *For each positive integer k, $\Omega \longrightarrow (\mathcal{O}_n)^2_k$ holds.*
(2) *The n-th power of X with the Tychonoff product topology satisfies the property $\mathsf{S}_1(\mathcal{O},\mathcal{O})$.*

PROOF. (2) $\Rightarrow$ (1) : Fix a positive integer n and an ω-cover $\mathcal{U}$ of X. Let $f : [\mathcal{U}]^2 \longrightarrow \{1,\cdots,k\}$ be given. By Lemma 5.2 the set $\mathcal{V} = \{U^n : U \in \mathcal{U}\}$ is an ω-cover of X^n. Since X^n has the property $\mathsf{S}_1(\mathcal{O},\mathcal{O})$, Theorem 5.3 implies that for each positive integer k, $\Omega \longrightarrow (\mathcal{O})^2_k$ holds for X^n.

Define, from the given f, a function $g : [\mathcal{V}]^2 \longrightarrow \{1,\cdots,k\}$ so that

$$g(\{U^n, V^n\}) = f(\{U,V\}).$$

Apply Theorem 5.1, and fix an $i \in \{1,\cdots,k\}$, and a subset $\mathcal{W}$ of $\mathcal{V}$ such that g is constant on $[\mathcal{W}]^2$, and $\mathcal{W}$ is a cover of X^n. Then consider the set $\mathcal{A} = \{U \in \mathcal{U} : U^n \in \mathcal{W}\}$. For a subset F of X such that $|F| = n$, say $F = \{y_1,\cdots,y_n\}$ consider the element $(y_1,\cdots,y_n)$ of X^n. For some $U \in \mathcal{A}$ we have $(x_1,\cdots,x_n) \in U^n$, from which it follows that $F \subseteq U$. Then $\mathcal{A}$ witnesses that (1) holds for X.

(1) $\Rightarrow$ (2) : It suffices to show that the space X^n satisfies the property $\mathsf{S}_1(\Omega,\mathcal{O})$. Thus, for each m let $\mathcal{U}_m$ be an ω-cover of X^n. For each m, applying Lemma 5.2, choose an ω-cover $\mathcal{V}_m$ of X such that for each $V \in \mathcal{V}_m$ there is a $U \in \mathcal{U}_m$ with $V^n \subseteq U$. Now applying hypothesis (1) to the sequence $(\mathcal{V}_n : n \in \mathbb{N})$ of ω-covers of X, choose for each m a $V_m \in \mathcal{V}_m$ such that there is for each subset $F \subset X$ with $|F| \leq n$, there is an m with $F \subseteq V_m$. Next, for each m choose a $U_m \in \mathcal{U}_m$ such that $V^n_m \subseteq U_m$. Then $\{U_m : m \in \mathbb{N}\}$ is an open cover of X^n. For let $(x_1,\cdots,x_n) \in X^n$ be given. Then $F = \{x_1,\cdots,x_n\} \subset X$ has at most n elements. Pick m with $F \subseteq V_m$. Then $(x_1,\cdots,x_n) \in V^n_m \subseteq U_m$. This completes the proof of (1) $\Rightarrow$ (2). □

6. A menu of Ramseyan statements.

We now introduce several new notational aids to describe the intimate connection between versions of Borel's covering property and variations on Ramsey's theorem. Each notational aid is introduced in a subsection. In the next section we will state a theorem, this paper's main exhibit of the intimate connections between Ramsey theory and a strengthened version of Borel's covering property.

The Ellentuck theorem and $\mathsf{E}(\mathcal{A},\mathcal{B})$. We describe the Ellentuck theorem in a slightly more abstract setting than the usual: For A an abstract countably infinite set fix a bijective enumeration $(a_n : n \in \mathbb{N})$ of A. Define for s and T nonempty subsets of A:

$$s < T \text{ if: } a_n \in s \text{ and } a_m \in T \Rightarrow n < m.$$

With the relation $s < T$ defined, define the Ellentuck topology on $[A]^{\aleph_0}$ as follows: For $s \in [A]^{<\aleph_0}$ and for $B \in [A]^{\aleph_0}$ use $s < B$ to denote that $s = \emptyset$ or $\max(s) < \min(B)$. For $s < B$ define $[s, B] = \{s \cup C \in [A]^{\aleph_0} : s < C \subseteq B\}$. The family $\{[s, B] : s \subset A \text{ finite and } s < B \in [A]^{\aleph_0}\}$ forms a basis for a topology on $[A]^{\aleph_0}$. This is the *Ellentuck topology* on $[A]^{\aleph_0}$ and was introduced in [**10**] for the special case when $A = \mathbb{N}$, the set of positive integers. If a subset $\mathcal{X}$ of $[A]^{\aleph_0}$ has the topology inherited from $[A]^{\aleph_0}$ endowed with the Ellentuck topology, we speak of "$\mathcal{X}$ with the Ellentuck topology". For $B \subseteq A$ and for finite set $s \subseteq A$ the symbol $B|s$ denotes $\{a_n \in B : s < \{a_n\}\}$.

Recall that a subset N of a topological space is *nowhere dense* if there is for each nonempty open set U of the space a nonempty open subset $V \subset U$ such that $N \cap V = \emptyset$. When N is the union of countably many nowhere dense sets, it is said to be *meager*. A subset of the form $(U \setminus M)\bigcup(M \setminus U)$ for some open set U and some meager set M of a topological space is said to have the *Baire property*. In this context the main result in [**10**] can now be stated as follows:

THEOREM 6.1 (Ellentuck). *Let A be a countably infinite set with a fixed enumeration defining the relation $<$ among subsets of A. For a set $R \subset [A]^{\aleph_0}$ the following are equivalent:*

(1) *R has the Baire property in the Ellentuck topology.*
(2) *For each finite set $s \subset A$ and for each infinite set $S \subset A$ with $s < S$ there is an infinite set $T \subset S$ such that either $[s, T] \subset R$, or else $[s, T] \cap R = \emptyset$.*

Next we refine the statement of Ellentuck's theorem by placing more constraints on the parameters in the theorem. This refinement motivates the first in the list of upcoming notational aids for this section. For families $\mathcal{A}$ and $\mathcal{B}$ we now define a sequence of statements:

$\mathsf{E}(\mathcal{A},\mathcal{B})$: For each countably infinite $A \in \mathcal{A}$ and for each set $R \subset [A]^{\aleph_0} \cap \mathcal{B}$ the implication (1)$\Rightarrow$(2) holds, where:

(1) R has the Baire property in the Ellentuck topology on $[A]^{\aleph_0} \cap \mathcal{B}$.
(2) For each $S \subset A$ with $S \in \mathcal{A}$ and each finite subset s of A, there is an infinite $B \subset S|s$ with $B \in \mathcal{B}$ such that $[s, B] \cap \mathcal{B} \subseteq R$ or $[s, B] \cap \mathcal{B} \cap R = \emptyset$.

Thus, for a given countably infinite set S, $\mathsf{E}([S]^{\aleph_0}, [S]^{\aleph_0})$ is Ellentuck's theorem. We shall prove that for a topological space (X, τ) the statement $\mathsf{S}_1(\Omega, \Omega)$ is equivalent to the statement $\mathsf{E}(\Omega, \Omega)$.

The Galvin-Prikry Theorem and $\mathsf{GP}(\mathcal{A},\mathcal{B})$. Galvin and Prikry proved a precursor of Theorem 6.1: If R is a Borel set in the topology inherited from 2^A via representing sets by their characteristic functions, then R has property (2) in

Theorem 6.1. Silver and Mathias subsequently gave metamathematical proofs that analytic sets (in the 2^A-topology) have this property. Theorem 6.1 at once yields all these prior results. The original papers [**10**] and [**16**] give a nice overview of these facts, and more. The proof of (1) $\Rightarrow$ (2) of Theorem 6.1 is nontrivial but uses only the techniques of Galvin and Prikry [**16**].

One might wonder to what extent the Galvin-Prikry Theorem, which is a consequence of Ellentuck's Theorem, adapts to the more constrained context in the definition of the notation $\mathsf{E}(\mathcal{A},\mathcal{B})$. Towards considering this idea, we give a corresponding abstract formulation of the Galvin-Prikry theorem, and introduce the second notational aid:

> $\mathsf{GP}(\mathcal{A},\mathcal{B})$: For each countably infinite $A \in \mathcal{A}$ and each $R \subset [A]^{\aleph_0} \cap \mathcal{B}$ the implication (1) $\Rightarrow$ (2) holds:
> (1) R is open in the 2^A topology on $[A]^{\aleph_0} \cap \mathcal{B}$.
> (2) For each $S \in [A]^{\aleph_0} \cap \mathcal{A}$ there is a set $B \in [S]^{\aleph_0} \cap \mathcal{B}$ such that either $([B]^{\aleph_0} \cap \mathcal{B}) \subseteq R$, or else $[B]^{\aleph_0} \cap \mathcal{B} \cap R = \emptyset$.

Thus, $\mathsf{GP}([\mathbb{N}]^{\aleph_0}, [\mathbb{N}]^{\aleph_0})$ is part of the Galvin-Prikry theorem.

Galvin's generalization of Ramsey's theorem and $\mathsf{FG}(\mathcal{A},\mathcal{B})$. An earlier generalization of Ramsey's Theorem, due to Fred Galvin, can also be formulated in the more general context as follows:

DEFINITION 6.2. A subset $\mathcal{S}$ of $[A]^{<\aleph_0}$ is:
(1) dense if for each $B \in [A]^{\aleph_0} \cap \mathcal{A}$, $\mathcal{S} \cap [B]^{<\aleph_0} \neq \emptyset$.
(2) thin if no element of $\mathcal{S}$ is an initial segment of another element of $\mathcal{S}$.

The following is an abstract formulation of Galvin's generalization of Ramsey's Theorem, announced in [**14**] and in [**16**] derived from Theorem 1 there, and is the third notational aid introduced in this section:

> $\mathsf{FG}(\mathcal{A},\mathcal{B})$: For each countably infinite $A \in \mathcal{A}$ and for each dense set $\mathcal{S} \subset [A]^{<\aleph_0}$ there is a $B \in [A]^{\aleph_0} \cap \mathcal{B}$ such that each $C \in [B]^{\aleph_0} \cap \mathcal{B}$ has an initial segment in $\mathcal{S}$.

In this notation Galvin's generalization of Ramsey's theorem reads that $\mathsf{FG}([\mathbb{N}]^{\aleph_0}, [\mathbb{N}]^{\aleph_0})$.

The Nash-Williams theorem and $\mathsf{NW}(\mathcal{A},\mathcal{B})$. Nash-Williams also discovered a generalization of Ramsey's theorem. The following is an abstract formulation of Nash-Williams' theorem, and introduces the fourth notational aid:

> $\mathsf{NW}(\mathcal{A},\mathcal{B})$: For each countably infinite $A \in \mathcal{A}$ and for each thin family $\mathcal{T} \subset [A]^{<\aleph_0}$ and for each n, and each partition $\mathcal{T} = \mathcal{T}_1 \cup \mathcal{T}_2 \cup \cdots \cup \mathcal{T}_n$ there is a $B \in [A]^{\aleph_0} \cap \mathcal{B}$ and an $i \in \{1, \cdots, n\}$ such that $[B]^{<\aleph_0} \cap \mathcal{T} \subseteq \mathcal{T}_i$.

In this notation Nash-Williams' theorem reads that $\mathsf{NW}([\mathbb{N}]^{\aleph_0}, [\mathbb{N}]^{\aleph_0})$.

Square bracket partition relations. Square bracket partition relations for cardinal numbers were introduced by Erdös, Hajnal and Rado in Section 18 of the paper [**11**]. This partition relation has been explored extensively for other combinatorial structures also, including ultrafilters. More remarks about the work on ultrafilters will be given later in the paper. To formulate the square bracket partition relation in sufficient generality for our anticipated application, we introduce the following, the fifth of the notational aids for this section.

DEFINITION 6.3. The symbol $\mathcal{A} \longrightarrow [\mathcal{B}]^m_{k/\leq \ell}$ denotes the statement that: For each $A \in \mathcal{A}$ and for each function $f : [A]^m \longrightarrow \{1, \cdots, k\}$ there is a $B \subseteq A$ such that $B \in \mathcal{B}$ such that $|\{f(\{x_1, \cdots, x_m\}) : \{x_1 \cdots, x_m\} \in [B]^m\}| \leq \ell$.

In the special case when $\ell = k - 1$, we use the notation $\mathcal{A} \longrightarrow [\mathcal{B}]^m_k$ instead of $\mathcal{A} \longrightarrow [\mathcal{B}]^m_{k/\leq k-1}$. In the special case when $\mathcal{A} = \mathcal{B} = \Omega$, the set of ω-covers of a topological space, the following two results will be used.

LEMMA 6.4. *Let X be a topological space in which each ω-cover has a countable subset that is an ω-cover. If $\Omega \rightarrow [\Omega]^2_3$, then for every positive integer k, $\Omega \rightarrow [\Omega]^2_{k/\leq 2}$.*

Lemma 6.4 is the previously proven Theorem 1 of [**35**], where a proof can be found. There are several other partition relations, for example polarized partition relations, hybrids of the square bracket partition relation and the polarized partition relation, that we have not represented in this section. We will make some remarks in a separate section about some of the well-known Ramseyan theorems not featured in the main result. The reason for our limited choice is related to the length of this paper and time limitations rather than to the importance of the concepts and ideas.

7. The Rothberger covering property in all finite powers

We now focus on a specific strengthening of Borel's covering property, namely the Rothberger covering property in all finite powers. Much of the material in this section is proven in prior papers, including [**39**]. For the convenience of the reader we provide a proof of the implications depicted in Figure 2. In light of theorems appearing earlier in this paper, one can prove the following for the selection principle $\mathsf{S}_1(\Omega, \Omega)$ the following:

THEOREM 7.1. *Let (X, τ) be a topological space. The following statements are equivalent:*

(1) *(X, τ) has the property $\mathsf{S}_1(\Omega, \Omega)$*
(2) *$\Omega \longrightarrow (\Omega)^2_2$ holds for (X, τ)*

Instead of giving a direct proof of the equivalence of the statements in Theorem 7.1, we follow a more roundabout way to illustrate the connection with several significant strengthenings of the classical Ramsey Theorem. Here is the statement of the target theorem:

THEOREM 7.2. *For a topological space (X, τ) in which each ω cover has a countable subset that is an ω-cover, the following statements are equivalent:*

(1) $\mathsf{S}_1(\Omega, \Omega)$.
(2) $\mathsf{E}(\Omega, \Omega)$.
(3) $\mathsf{GP}(\Omega, \Omega)$.
(4) $\mathsf{FG}(\Omega, \Omega)$.
(5) $\mathsf{NW}(\Omega, \Omega)$.
(6) *For all n and k, $\Omega \rightarrow (\Omega)^n_k$.*
(7) *For all n and k, $\Omega \rightarrow [\Omega]^n_{k/\leq 2}$.*
(8) $\Omega \rightarrow [\Omega]^2_3$.
(9) $\Omega \rightarrow (\Omega, 4)^3$.
(10) $\Omega \rightarrow (\Omega)^2_2$.

(11) $\Omega \to (\Omega, \mathcal{O})^2$.

The theorem's proof is organized into several subsections, each treating a specific implication, as depicted in Figure 2, of Theorem 7.2.

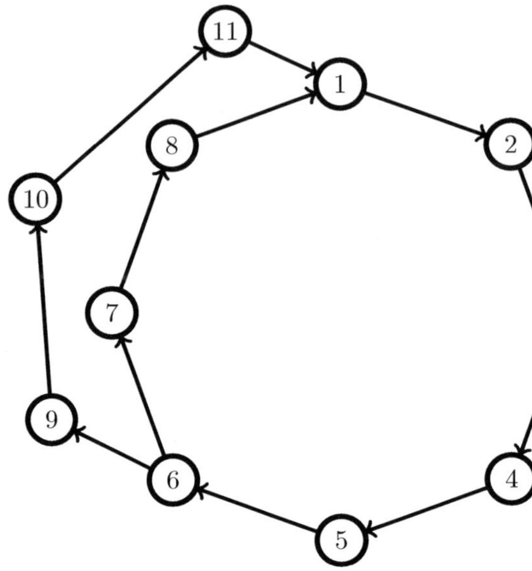

FIGURE 2. The directed edges between two labeled vertices denote the statements between which direct implications are proved in the paper. It is possible to arrange the proof so that the corresponding directed graph denoting the proved implications has a single closed path.

7.1. A proof of the implication $\mathsf{S}_1(\Omega,\Omega) \Rightarrow \mathsf{E}(\Omega,\Omega)$: Assume that X has the property $\mathsf{S}_1(\Omega,\Omega)$. Fix a countable ω-cover A of X and fix a set $R \subset [A]^{\aleph_0} \cap \Omega$. Also fix a bijective enumeration $(a_n : n \in \mathbb{N})$ of the set A. The relation $<$ among subsets of A is interpreted in terms of this enumeration. Recall that sets of the form $[s, C] = \{D : s < C \text{ and } s \subset D \subseteq s \cup C\}$ constitute a basis for the Ellentuck topology on $[A]^{\aleph_0}$.

We now introduce terminology from the paper [**16**]:

DEFINITION 7.3. For a finite set $s \subset A$ and for $B \in [A|s]^{\aleph_0} \cap \Omega$:

(1) B accepts s if $[s, B] \cap \Omega \subseteq R$.
(2) B rejects s if no $C \in [B]^{\aleph_0} \cap \Omega$ accepts s.

Lemma 7.4 will be used without special reference:

LEMMA 7.4. *Let a finite set $s \subset A$ and a set $B \in [A|s]^{\aleph_0} \cap \Omega$ be given:*

(1) *B accepts s if, and only if, each $C \in [B]^{\aleph_0} \cap \Omega$ accepts s.*
(2) *B rejects s if, and only if, each $C \in [B]^{\aleph_0} \cap \Omega$ rejects s.*

LEMMA 7.5. *For each finite set $s \subset A$, there is a $B \in [A|s]^{\aleph_0} \cap \Omega$ such that B accepts s or B rejects s.*

PROOF. If $A|s$ does not reject s, choose a $B \in [A|s]^{\aleph_0} \cap \Omega$ accepting s. □

LEMMA 7.6. *Let $t \subset A$ be a finite set. Let $B \in [A]^{\aleph_0} \cap \Omega$ be such that for each finite set $s \subset (t \cup B)$, $B|s$ accepts s or $B|s$ rejects s. If $B|t$ rejects t then $C = \{u \in B : B|(t \cup \{u\})$ rejects $t \cup \{u\}\}$ is a member of Ω.*

PROOF. Suppose not. Then $D = t \cup (B \setminus C) \in \Omega$, and for each $u \in D|t$, $B|(t \cup \{u\})$ accepts $t \cup \{u\}$. Thus for each $u \in D|t$, $D|(t \cup \{u\})$ accepts $t \cup \{u\}$. This means that $[t, D|t] = \cup_{u \in D}[t \cup \{u\}, D|(t \cup \{u\})] \subseteq R$, and so $D|t$ accepts t. This contradicts Lemma 7.4 (2) since $D \in [B]^{\aleph_0} \cap \Omega$ and $B|t$ rejects t. □

ω-covers accepting or rejecting all finite subsets. Recall that

THEOREM 7.7. *For a topological space Y the following are equivalent:*

(1) *Y has property $\mathsf{S}_1(\Omega, \Omega)$.*
(2) *ONE has no winning strategy in $\mathsf{G}_1(\Omega, \Omega)$.*

THEOREM 7.8. *If Y has property $\mathsf{S}_1(\Omega, \Omega)$, then for each finite set $t \subset A$ and for each $B \in [A|t]^{\aleph_0} \cap \Omega$ there is a $C \in [B]^{\aleph_0} \cap \Omega$ such that for each finite set $s \subset t \cup C$, $C|s$ accepts s or $C|s$ rejects s.*

PROOF. Let t and $B \in [A|t]^{\aleph_0} \cap \Omega_Y$ be given. Define a strategy σ for ONE of $\mathsf{G}_1(\Omega, \Omega)$ as follows:

Enumerate the set of all subsets of t as $\{t_1, \cdots, t_n\}$. Using Lemma 7.5 recursively choose $B_1 \supset B_2 \supset \cdots \supset B_n$ in $[B]^{\aleph_0} \cap \Omega$ such that for each i, B_i accepts t_i or B_i rejects t_i. Then define:

$$\sigma(\emptyset) = B_n.$$

If TWO now chooses $T_1 \in \sigma(\emptyset)$ then use Lemma 7.5 in the same way to choose

$$\sigma(T_1) \in [\sigma(\emptyset)|\{T_1\}]^{\aleph_0} \cap \Omega$$

such that for each set $F \subset t \cup \{T_1\}$, $\sigma(T_1)$ accepts F, or rejects F.

When TWO responds with $T_2 \in F(T_1)$, enumerate the subsets of $t \cup \{T_1, T_2\}$ as $(t_1, \cdots, t_n)$ say, and choose by Lemma 7.5 sets $B_1, \cdots B_n \in [\sigma(T_1)|\{T_2\}]^{\aleph_0} \cap \Omega$ such that B_j accepts t_j or B_j rejects t_j for $1 \leq j \leq n$ and $B_j \subset B_{j-1}$. Finally put

$$\sigma(T_1, T_2) = B_n.$$

Note that for each finite subset F of $t \cup \{T_1, T_2\}$, $\sigma(T_1, T_2)$ accepts F or rejects it.

It is clear how player ONE's strategy is defined. By Theorem 7.7 σ is not a winning strategy for ONE. Consider a σ-play lost by ONE, say

$$\sigma(\emptyset),\ T_1,\ \sigma(T_1),\ T_2,\ \sigma(T_1, T_2),\ \cdots,\ T_n,\ \sigma(T_1, \cdots, T_n),\ \cdots$$

Then $C = t \cup \{T_n : n \in \mathbb{N}\} \subset B$ is an element of Ω.

We claim that for each finite subset s of $t \cup C$, $C|s$ accepts s or $C|s$ rejects s. For consider such a s. If $s \subseteq t$, then as $C \subset F(\emptyset)$ and $F(\emptyset)$ accepts or rejects s, also C does. If $s \not\subseteq t$, then put $n = \max\{m : T_m \in s\}$. Then s is a subset of $t \cup \{T_1, \cdots, T_n\}$, so that s is accepted or rejected by $\sigma(T_1, \cdots, T_n)$. But $C|s \subseteq \sigma(T_1, \cdots, T_n)$, and so $C|s$ accepts or rejects s. □

Completely Ramsey sets. The subset R of $[A]^{\aleph_0} \cap \Omega$ is said to be *completely Ramsey* if there is for each finite set $s \subset A$ and for each $B \in [A|s]^{\aleph_0} \cap \Omega$ a set $C \in [B]^{\aleph_0} \cap \Omega$ such that

(1) either $([s, C] \cap \Omega) \subseteq R$,
(2) or else $([s, C] \cap \Omega) \cap R = \emptyset$.

LEMMA 7.9. *If R and S are completely Ramsey subsets of $[A]^{\aleph_0} \cap \Omega$, then so is $R \bigcup S$.*

PROOF. Let a finite set $s \subset A$ and $B \in [A|s]^{\aleph_0} \cap \Omega$ be given. Since R is completely Ramsey, choose $C \in [B]^{\aleph_0} \cap \Omega$ such that $([s, C] \cap \Omega \subset R$, or $([s, C] \cap \Omega) \cap R = \emptyset$. If the former hold we are done. In the latter case, since S is completely Ramsey, choose $D \in [C]^{\aleph_0} \cap \Omega$ such that $([s, D] \cap \Omega) \subseteq S$, or $([s, D] \cap \Omega) \cap S = \emptyset$. In either case the proof is complete. □

The following Lemma is obviously true.

LEMMA 7.10. *If R is completely Ramsey, then so is $([A]^{\aleph_0} \cap \Omega) \setminus R$.*

COROLLARY 7.11. *If R and S are completely Ramsey subsets of $[A]^{\aleph_0} \cap \Omega$, then so is $R \bigcap S$.*

PROOF. Lemmas 7.9 and 7.10, and De Morgan's laws. □

Open sets in the Ellentuck topology. We are still subject to the hypothesis that X satisfies $\mathsf{S}_1(\Omega, \Omega)$.

LEMMA 7.12. *For each finite set $t \subset A$ and for each $B \in [A|t]^{\aleph_0} \cap \Omega$ such that for each finite subset F of $t \cup B$, $B|F$ accepts, or rejects F the following holds: For each finite set $s \subset t \cup B$ such that $B|s$ rejects s, there is a $C \in [B|s]^{\aleph_0} \cap \Omega$ such that for each finite set $F \subset C$, $C|F$ rejects $s \cup F$.*

PROOF. Fix B and s as in the hypotheses. Define a strategy σ for ONE in $\mathsf{G}_1(\Omega, \Omega)$ as follows: By Lemma 7.6

$$\sigma(\emptyset) = \{U \in B : s < \{U\} \text{ and } B|\{U\} \text{ rejects } s \cup \{U\}\} \in \Omega_X.$$

Notice that $\sigma(\emptyset)$ accepts or rejects each of its finite subsets, it rejects s, and for each $U \in \sigma(\emptyset)$, $\sigma(\emptyset)|\{U\}$ rejects $s \cup \{U\}$.

If TWO now chooses $T_1 \in \sigma(\emptyset)$, then by Lemma 7.6

$$\sigma(T_1) = \{U \in \sigma(\emptyset) \setminus \{T_1\} : \sigma(\emptyset)|F \text{ rejects } s \cup F \text{ for each finite } F \subset \{T_1, U\}\}$$

is in Ω. As before, $\sigma(T_1)$ accepts or rejects each of its finite subsets, and for any $U \in \sigma(T_1)$, for each finite subset F of $\{U, T_1\}$, $\sigma(T_1)|F$ rejects $s \cup F$.

If next TWO chooses $T_2 \in \sigma(T_1)$, then by Lemma 7.6

$$\sigma(T_1, T_2) = \{U \in \sigma(T_1) \setminus \{T_2\} : \sigma(T_1)|F \text{ rejects } s \cup F \text{ for any finite } F \subset \{T_1, T_2, U\}\}$$

is an element of Ω.

Continuing in this way we define a strategy σ for ONE in $\mathsf{G}_1(\Omega, \Omega)$. Since X satisfies $\mathsf{S}_1(\Omega, \Omega)$, σ is not a winning strategy for ONE. Consider a σ-play lost by ONE, say:

$$\sigma(\emptyset),\ T_1,\ \sigma(T_1),\ T_2,\ \sigma(T_1, T_2),\ T_3,\ \sigma(T_1, T_2, T_3),\ \cdots$$

Put $C = \{T_n : n \in \mathbb{N}\}$. Then $C \in [B|s]^{\aleph_0} \cap \Omega$. We claim that for each finite set $F \subset C$, $C|F$ rejects $s \cup F$.

For choose a finite set $F \subset C$. Then $F \cap s = \emptyset$. Fix $n = \max\{m : T_m \in F\}$. Then $C|F \subset \sigma(T_1, \cdots, T_n)$, and the latter rejects $s \cup F$ for all finite subsets F of $\{T_1, \cdots, T_n\}$. Thus $C|F$ rejects $s \cup F$. □

THEOREM 7.13. *If X has property $\mathsf{S}_1(\Omega, \Omega)$, then every open subset of $[A]^{\aleph_0} \cap \Omega$ is completely Ramsey.*

PROOF. Let $R \subset [A]^{\aleph_0} \cap \Omega$ be open in this subspace. Consider a finite set $s \subset A$ and a $B \in [A|s]^{\aleph_0} \cap \Omega$. Since $(X, d) \models \mathsf{S}_1(\Omega, \Omega)$, choose by Theorem 7.8 a $C \in [B]^{\aleph_0} \cap \Omega$ such that for each finite set $F \subset (s \cup C)$, $C|F$ accepts or rejects F.

If C accepts s then we have $[s, C] \cap \Omega \subseteq R$, and we are done. Thus, assume that C does not accept s. Then C rejects s, and we choose by Lemma 7.12 a $D \in [C|s]^{\aleph_0} \cap \Omega$ such that for each finite subset F of D, $D|F$ rejects $s \cup F$.

We claim that $([s, D] \cap \Omega) \cap R = \emptyset$. For suppose not. Choose $E \in [s, D] \cap \Omega \cap R$. Since R is open, choose an Ellentuck neighborhood of E contained in R, say $[t, K] \cap \Omega$. Then we have $s \subset E \subset s \cup D$ and $t \subset E \subset t \cup K$. But then $s \cup t \subset E \subset t \cup K$ and $[s \cup t, K|s] \subset R$, whence also $[s \cup t, E|(s \cup t)] \subset R$. But then $E|(s \cup t)$ accepts $s \cup t$ where t is a finite subset of $s \cup D$, and $E|(s \cup t) \subset D|t$, and $D|t$ rejects $s \cup t$, a contradiction. □

Meager subsets in the Ellentuck topology. If the subset R of $[A]^{\aleph_0} \cap \Omega$ is nowhere dense in the topology, then for each $B \in [A]^{\aleph_0} \cap \Omega$ and for each finite set $s \subset A$, $B|s$ rejects s. We now examine the meager subsets of $[A]^{\aleph_0} \cap \Omega$.

LEMMA 7.14. *If R is nowhere dense, then there is for each $B \in [A]^{\aleph_0} \cap \Omega$ and each finite set $t \subset A$ a set $C \in [B|t]^{\aleph_0} \cap \Omega$ such that for each finite set $s \subset t \cup C$, $C|s$ rejects s.*

PROOF. Since R is nowhere dense, no ω-cover contained in A can accept a finite set. Thus each ω-cover contained in A rejects each finite subset of A. □

LEMMA 7.15. *Assume $\mathsf{S}_1(\Omega, \Omega)$. If R is a closed nowhere dense subset of $[A]^{\aleph_0} \cap \Omega$ then there is for each finite subset $s \subset A$ and for each $B \in [A|s]^{\aleph_0} \cap \Omega$ a $C \in [B]^{\aleph_0} \cap \Omega$ such that $[s, C] \cap R = \emptyset$.*

PROOF. First, note that closed nowhere dense subsets are complements of open dense sets. By Theorem 7.13, each open set is completely Ramsey. By Lemma 7.10 each closed, nowhere dense set is completely Ramsey. By Lemma 7.14 the rest of the statement follows. □

By taking closures, the preceding lemma implies:

COROLLARY 7.16. *Assume $\mathsf{S}_1(\Omega, \Omega)$. If R is a nowhere dense subset of $[A]^{\aleph_0} \cap \Omega$ then there is for each finite subset $s \subset A$ and for each $B \in [A|s]^{\aleph_0} \cap \Omega$ a $C \in [B]^{\aleph_0} \cap \Omega$ such that $[s, C] \cap R = \emptyset$.*

And now we prove:

THEOREM 7.17. *Assume* $\mathsf{S}_1(\Omega,\Omega)$. *For a subset* N *of* $[A]^{\aleph_0}\cap\Omega$ *the following are equivalent:*

(1) N *is nowhere dense.*
(2) N *is meager.*

PROOF. We must show that (2)$\Rightarrow$(1). Thus, assume that N is meager and write $N=\bigcup_{n\in\mathbb{N}}N_n$, where for each n we have $N_n\subseteq N_{n+1}$, and N_n is nowhere dense in $[A]^{\aleph_0}\cap\Omega$. Consider any basic open set $[s,B]$ of $[A]^{\aleph_0}\cap\Omega$. Define a strategy σ for ONE in the game $\mathsf{G}_1(\Omega,\Omega)$ as follows:

Since N_1 is nowhere dense, choose by Corollary 7.16 an $O_1\in[B]^{\aleph_0}\cap\Omega$ with $[s,O_1]\cap N_1=\emptyset$. Define $\sigma(\emptyset)=O_1$.

When TWO chooses $T_1\in\sigma(\emptyset)$ choose by Corollary 7.16 an $O_2\in[\sigma(\emptyset)|\{T_1\}]^{\aleph_0}\cap\Omega_X$ with $[s,O_2]\cap N_2=\emptyset$, and define $\sigma(T_1)=O_2$.

Now when TWO chooses $T_2\in\sigma(T_1)$, find by Corollary 7.16 an $O_3\in[\sigma(T_1)|\{T_2\}]^{\aleph_0}\cap\Omega_X$ with $[s,O_3]\cap N_3=\emptyset$, and define $\sigma(T_1,T_2)=O_3$.

It is clear how to define ONE's strategy σ. By Theorem 7.7 F is not a winning strategy for ONE. Consider a play

$$\sigma(\emptyset),\ T_1,\ \sigma(T_1),\ T_2,\ \cdots,\ \sigma(T_1,\cdots,T_n),\ T_{n+1},\ \cdots$$

lost by ONE. Put $C=\{T_n:n\in\mathbb{N}\}$. Then $C\in[B]^{\aleph_0}\cap\Omega$. Observe that by the definition of σ we have for each k and each finite set $F\subset\{T_1,\cdots,T_k\}$ that $[s\cup F,\sigma(T_1,\cdots,T_k)]\cap N_k=\emptyset$.

Claim: $[s,C]\cap N=\emptyset$.
For suppose that instead $[s,C]\cap N\neq\emptyset$. Choose $V\in[s,C]\cap N$, and then choose m so that $V\in N_m$. Choose the least $k>m$ with $T_k\in V|s$. This is possible because s is finite. Observe also that $s\subseteq V\subseteq s\cup C=s\cup\{T_j:j\in\mathbb{N}\}$. Put $F=V\cap\{T_1,\cdots,T_k\}$. Thus we have that $[s\cup F,V|F]\cap N_k\neq\emptyset$, which contradicts the fact that $V|F\subset\sigma(T_1,\cdots,T_k)$, and $[s\cup F,\sigma(T_1,\cdots,T_k)]\cap N_k=\emptyset$. This completes the proof of the claim. □

Using Lemmas 7.9 and 7.10 and Corollary 7.11 we have:

THEOREM 7.18. *Suppose* X *satisfies* $\mathsf{S}_1(\Omega,\Omega)$. *Then for each* $A\in\Omega$, *every subset of* $[A]^{\aleph_0}\cap\Omega$ *which has the Baire property is completely Ramsey.*

7.2. A proof of the implication $\mathsf{E}(\Omega,\Omega)\Rightarrow\mathsf{GP}(\Omega,\Omega)$**:** Note that a set open in the $2^{\mathbb{N}}$ topology is also open in the Ellentuck topology. The implication (2) $\Rightarrow$ (3) of Theorem 7.2 follows from this remark.

7.3. A proof of the implication $\mathsf{GP}(\Omega,\Omega)\Rightarrow\mathsf{FG}(\Omega,\Omega)$**:**

LEMMA 7.19. *Assume* $\mathsf{GP}(\Omega,\Omega)$. *Then* $\mathsf{FG}(\Omega,\Omega)$ *holds.*

PROOF. Let $\mathcal{S}\subset[A]^{<\aleph_0}$ be dense and define $\mathcal{I}$ to be the set $\{D\in[A]^{\aleph_0}\cap\Omega_X:$ D has an initial segment in $\mathcal{S}\}$. Then we have:

$$\mathcal{I}=\cup\{[s,D|s]:s\in\mathcal{S},\ D\in[A]^{\aleph_0}\cap\Omega_X\text{ and s an initial segment of }D\}$$

is a $2^{\mathbb{N}}$-open subset of $[A]^{\aleph_0}\cap\Omega$. Choose a $B\in[A]^{\aleph_0}\cap\Omega$ such that $[B]^{\aleph_0}\cap\Omega\subset\mathcal{I}$, or $[B]^{\aleph_0}\cap\Omega\cap\mathcal{I}=\emptyset$. But the second alternative implies the contradiction that $[B]^{<\aleph_0}\cap\mathcal{S}=\emptyset$. It follows that the first alternative holds. □

7.4. A proof of the implication $\mathsf{FG}(\Omega,\Omega) \Rightarrow \mathsf{NW}(\Omega,\Omega)$:

THEOREM 7.20. *Assume* $\mathsf{FG}(\Omega,\Omega)$. *Then* $\mathsf{NW}(\Omega,\Omega)$ *holds.*

PROOF. Fix a thin family $\mathcal{T} \subset [A]^{<\aleph_0}$ and positive integer n, and a partition $\mathcal{T} = \mathcal{T}_1 \cup \mathcal{T}_2 \cup \cdots \cup \mathcal{T}_n$. We may assume $n = 2$. If $\mathcal{T}_1$ is not dense, we can choose $B \in [A]^{\aleph_0} \cap \Omega$ such that $[B]^{<\aleph_0} \cap \mathcal{T} \subseteq \mathcal{T}_2$. Thus, assume $\mathcal{T}_1$ is dense. Choose, by the hypothesis, a $B \in [A]^{\aleph_0} \cap \Omega$ such that for each $C \in [B]^{\aleph_0} \cap \Omega$, some initial segment of C is in $\mathcal{T}_1$.

Consider any $s \in \mathcal{T} \cap [B]^{<\aleph_0}$, and put $D = s \cup (B|s)$. Then s is an initial segment of D, and $D \in [B]^{\aleph_0} \cap \Omega$, and so some initial segment of D, say t, is in $\mathcal{T}_1$. Since both t and s are initial segments of D and are both in $\mathcal{T}$, and since $\mathcal{T}$ is thin, we have $s = t$, and so $s \in \mathcal{T}_1$. Consequently we have $[B]^{<\aleph_0} \cap \mathcal{T} \subseteq \mathcal{T}_1$. □

7.5. A proof of the implication $\mathsf{NW}(\Omega,\Omega) \Rightarrow (\forall n)(\forall k)(\Omega \longrightarrow (\Omega)^n_k)$.

THEOREM 7.21. *Assume that* $\mathsf{NW}(\Omega,\Omega)$ *holds. Then: For each n and k we have* $\Omega \to (\Omega)^n_k$.

PROOF. Let $A \in \Omega$ be countable. Let positive integers n and k be given. Put $\mathcal{T} = [A]^n$. Then $\mathcal{T}$ is thin. Apply the hypothesis. □

7.6. A proof of the implication $(\forall n)(\forall k)(\Omega \longrightarrow (\Omega)^n_k \Rightarrow (\forall n)(\forall k)(\Omega \longrightarrow [\Omega]^n_{k/\leq 2})$. Examination of the definitions reveals that this implication is evidently true.

7.7. A proof of the implication $\Omega \longrightarrow [\Omega]^2_3 \Rightarrow \mathsf{S}_1(\Omega,\Omega)$.

PROOF. Assume that $\Omega \longrightarrow [\Omega]^2_3$ holds. By Lemma 6.4 $\Omega \longrightarrow [\Omega]^2_{5/\leq 2}$ holds. Let $(\mathcal{U}_n : n \in \mathbb{N})$ be a sequence of ω-covers of X. By hypothesis, each ω-cover of X has a countable subset which is still an ω-cover of X. Thus we may assume that each of the $\mathcal{U}_n$s is countable. For each n, enumerate $\mathcal{U}_n$ bijectively as $(U^n_m : m \in \mathbb{N})$. Define a new ω-cover $\mathcal{V}$ by $\{U^1_k \cap U^k_n \cap U^n_\ell : k,n,\ell \in \mathbb{N}\} \setminus \{\emptyset\}$.

For each element V of $\mathcal{V}$ choose a representation $V = U^1_k \cap U^k_n \cap U^n_\ell$. The term U^k_n will be called the middle term of the representation, while the term U^n_ℓ will be called the end-term of the representation. Define a partition $f : [\mathcal{V}]^2 \longrightarrow \{0,1,2,3,4\}$ as follows: (in the displayed formula we are assuming that the two arguments of f are listed so that (k_1,n_1) lexicographically precedes, or is equal to (k_2,n_2)).

$$f(\{U^1_{k_1} \cap U^{k_1}_{n_1} \cap U^{n_1}_{\ell_1}, U^1_{k_2} \cap U^{k_2}_{n_2} \cap U^{n_2}_{\ell_2}\}) = \begin{cases} 0 & \text{if} \quad k_1 = k_2 \text{ and } n_1 = n_2, \\ 1 & \text{if} \quad k_1 = k_2 \text{ and } n_1 < n_2, \\ 2 & \text{if} \quad k_1 < k_2 \text{ and } n_1 < n_2, \\ 3 & \text{if} \quad k_1 < k_2 \text{ and } n_1 > n_2, \\ 4 & \text{if} \quad k_1 < k_2 \text{ and } n_1 = n_2 \end{cases}$$

Choose an ω-cover $\mathcal{W} \subseteq \mathcal{V}$ on which f is at most two-valued. List $\mathcal{W}$ as $(U^1_{k_1} \cap U^{k_1}_{n_1} \cap U^{n_1}_{\ell_1},\ U^1_{k_2} \cap U^{k_2}_{n_2} \cap U^{n_2}_{\ell_2},\ \cdots)$ according to the lexicographic order of the triples (k_i,n_i,ℓ_i) which occur in the chosen representations of elements of $\mathcal{W}$. There are four main cases to be considered.

Case 1: f does not have the values 0 or 1 on $[\mathcal{W}]^2$. In this case we have $k_1 < k_2 < \cdots < k_n < \cdots$ and so if we set $V_{k_i} = U^{k_i}_{n_i}$, and for $m \notin \{k_i : i = 1,2,3,\cdots\}$ we choose $V_m \in \mathcal{U}_m$ arbitrarily, then the sequence $(V_m : m = 1,2,3,\cdots)$ constitutes an ω-cover of X and for each m we have $V_m \in \mathcal{U}_m$.

Case 2: $\{0, 1\} \cap f[[\mathcal{W}]^2] = \{1\}$.

Subcase 2.1: $f[[W]^2] \subseteq \{1, 2\}$. In this case, we see that $n_i \neq n_j$ whenever $i \neq j$. But then $\mathcal{W}$ refines the sequence $(U_{\ell_i}^{n_i} : i = 1, 2, 3, \cdots)$, whence the latter constitutes an ω-cover of X. This sequence can then be augmented to one of the form $(U_n : n < \omega)$ where for each n, $U_n \in \mathcal{U}_n$, and still $\{U_n : n \in \mathbb{N}\}$ is an ω-cover of X.

Subcase 2.2: $f[[W]^2] \subseteq \{1, 3\}$. This case doesn't occur. To see this, first observe that 3 cannot be attained with only finitely many different k's, since then $\mathcal{W}$ would be a refinement of a finite subset of $\mathcal{U}_1$, and thus not an ω-cover. But for Subcase 2.2, value 3 also cannot be attained with infinitely many k's. To see this, list the subscripts of the U_k^1's occurring in the representations of elements of $\mathcal{W}$ monotonically, say

$$k_1 \leq k_2 \leq \cdots \leq k_j < \cdots .$$

If 3 occurs with infinitely many k's, then it happens infinitely often that $k_j < k_{j+1}$. Define $i_1 = 1$ and $i_{n+1} > i_n$, to be minimal such that $k_{i_{n+1}} > k_{i_n}$. Look at the subset $\{U_{k_{i_j}}^1 \cap U_{n_{i_j}}^{k_{i_j}} \cap U_{\ell_{i_j}}^{n_{i_j}} : 1, 2, 3, \cdots\}$ of $\mathcal{W}$.. It is homogeneous of color 3, and thus we find $n_{i_1} > n_{i_2} > \cdots > n_{i_j} > \cdots$, an infinite descending sequence of natural numbers, a contradiction.

Subcase 2.3: $f[[W]^2] \subseteq \{1, 4\}$. If the value 4 is taken with only finitely many k's, then $\mathcal{W}$ would be a refinement of a finite subset of $\mathcal{U}_1$, and thus not an ω-cover of X. If the value 1 is taken with only finitely many k's, then all but finitely many of the middle terms $U_{n_i}^{k_i}$ could be assigned to distinct ones of the covers $\mathcal{U}_n$, and would constitute an ω-cover, in which case we would be done. Thus we must treat the case where each of 1 and 4 is attained with infinitely many k's. Then

$$(k_1, n_1), (k_2, n_2), \cdots, (k_r, n_r), \cdots$$

forms (in the lexicographic order) a strictly increasing sequence such that for each i, either $k_i < k_{i+l}$ and $n_i = n_{i+1}$ (in which case $f(\{U_{k_i}^1 \cap U_{n_i}^{k_i} \cap U_{\ell_i}^{n_i}, U_{k_{i+1}}^1 \cap U_{n_{i+1}}^{k_{i+1}} \cap U_{\ell_{i+1}}^{n_{i+1}},\}) = 4$), or else $k_i = k_{i+1}$ and $n_i < n_{i+1}$ (in which case $f(\{U_{k_i}^1 \cap U_{n_i}^{k_i} \cap U_{\ell_i}^{n_i}, U_{k_{i+1}}^1 \cap U_{n_{i+1}}^{k_{i+1}} \cap U_{\ell_{i+1}}^{n_{i+1}},\}) = 1$).

Consider the two sequences

$$(U_{n_i}^{k_i} : i = 1, 2, 3, \cdots \text{ and } k_{i-1} < k_i \text{ or } k_i < k_{i+1})$$

which consists of certain middle terms of the three-set intersections composing the elements of $\mathcal{W}$, and

$$(U_{\ell_i}^{n_i} : i = 1, 2, 3, \cdots \text{ and } n_{i-1} < n_i \text{ or } n_i < n_{i+1})$$

which consists of certain end terms of the three-set intersections composing the elements of $\mathcal{W}$. Since $\mathcal{W}$ refines the totality of sets belonging to these two sequences, these two sequences constitute an ω-cover of X. For each m the set $\{i : k_i = m \text{ or } n_i = m\}$ has at most four elements.

Thus this ω-cover can be partitioned into four-element sets each of which could be assigned to distinct terms of the original sequence $(\mathcal{U}_n : n < \omega)$. Being an ω-cover, we can find a new ω-cover by selecting one term per each of these four-element sets (i.e., $\Omega \longrightarrow (\Omega)_4^1$ holds). In this way we find a selector for the original sequence of ω-covers in such a way that the selector is also an ω-cover.

Case 3. $\{0,1\} \cap f[[\mathcal{W}]^2] = \{0\}$.

Subcase3 .1. $f[[\mathcal{W}]^2] \subseteq \{0,2\}$.

If the value 2 occurs at only a finite number of distinct k's, then $\mathcal{W}$ is a refinement of a finite subset of $\mathcal{U}_1$, and thus not an ω-cover of X. So, the value 2 is achieved at infinitely many k's. Each time it is achieved and only then, both k_i and n_i increase in value. Let $i_l < i_2 < \cdots$ be such that

- $i_l = 1$,
- for each j, if $i_j \leq t < i_{j+1}$, then $k_{i_j} = k_t$ and $n_{i_j} = n_t$,
- for each j, $k_{i_j} < k_{i_{j+1}}$, and
- $\{(k_{I_j}, n_{i_j}) : j = 1,2,3,\cdots\} = \{(k_j, n_j.) : j = 1,2,3,\cdots\}$

For each j put $V_{k_{i_j}} = U_{n_{i_j}}^{k_{i_j}}$ and for n not in $\{k_{i_j} : j = 1,2,3,\cdots\}$, choose V_n from $\mathcal{U}_n$ arbitrarily. Then for each n, $V_n \in \mathcal{U}_n$, and $\mathcal{W}$ is a refinement of $\{V_n : n = 1,2,3,\cdots\}$, and thus the latter is an ω-cover of X.

Subcase3 .2. $f[[\mathcal{W}]]^2] \subseteq \{0,3\}$. The value 3 can be attained at only a finite number of distinct k's, lest we have an infinite descending sequence of natural numbers. But then $\mathcal{W}$ is a refinement of a finite subset of $\mathcal{U}_1$, and so not an ω-cover of X. It follows that this case doesn't occur.

Subcase 3.3. $f[[\mathcal{W}]^2] \subseteq \{0,4\}$. The value 4 must be attained at an infinite number of distinct k's, else $\mathcal{W}$ would be a refinement of a finite subset of $\mathcal{U}_1$, hence not an ω-cover of X. But then an argument as in Subcase 3.1 shows that there is a sequence $(U_n : n \in \mathbb{N})$ such that for each n $U_n \in \mathcal{U}_n$, and $\{U_n : n \in \mathbb{N}\}$ is an ω-cover of X.

Case 4. $f[[\mathcal{W}]^2] \subseteq \{0,1\}$. Then there is a fixed k such that each element of $\mathcal{W}$ is a subset of U_k^1. Since $X \neq U_k^1$ it follows that $\mathcal{W}$ doesn't even cover X. Consequently, this case doesn't occur. □

7.8. A proof of the implication $\Omega \longrightarrow (\Omega, 4)_2^3 \Rightarrow (\Omega \longrightarrow (\Omega)_2^2$. The proof of this implication is modeled after the proof of (iii)⇒(i) of Theorem 2.1 of [**3**]. Let $\mathcal{U}$ be a countable ω-cover of X, enumerated bijectively as $(U_n : n \in \mathbb{N})$. Assume that the partition relation $\Omega \longrightarrow (\Omega, 4)_2^3$ holds, and let a function $f : [\mathcal{U}]^2 \longrightarrow \{1,\ 2\}$ be given.

Define a new function $g : [\mathcal{U}]^3 \longrightarrow \{1,2\}$ as follows: Let $\{U_{n_1},\ U_{n_2},\ U_{n_3}\}$ be an element of $[\mathcal{U}]^3$, written so that $n_1 < n_2 < n_3$. Define

$$g(\{U_{n_1},\ U_{n_2},\ U_{n_3}\}) = \begin{cases} 2 & \text{if } f(\{U_{n_1}, U_{n_2}\}) = 1 \text{ and } f(\{U_{n_2}, U_{n_3}\}) = 2 \\ 1 & \text{otherwise} \end{cases}$$

Now apply the partition relation $\Omega \longrightarrow (\Omega, 4)_2^3$ to the function g.

Claim 1: There is no four-element subset of $\mathcal{U}$ on whose triples g has value 2.

For consider a 4-tuple $\mathcal{F} = \{U_{n_1},\ U_{n_2},\ U_{n_3},\ U_{n_4}\}$ with $n_1 < n_2 < n_3 < n_4$ and suppose that g has value 2 at each three element subset of the set $\mathcal{F}$. Then, considering the subset $\{U_{n_1},\ U_{n_2},\ U_{n_3}\}$ we have $f(\{U_{n_1},\ U_{n_2}\}) = 1$ and $f(\{U_{n_2},\ U_{n_3}\}) = 2$, and considering $\{U_{n_2},\ U_{n_3},\ U_{n_4}\}$, we have $f(\{U_{n_2},\ U_{n_3}\}) = 1$ and $f(\{U_{n_3},\ U_{n_4}\}) = 2$, a contradiction. This completes the proof of Claim 1.

With Claim 1 established, we find a subset $\mathcal{V}$ of $\mathcal{U}$ such that $\mathcal{V}$ is an ω-cover, and f is constant of value 1 on $[\mathcal{V}]^3$. If f is constant of value 2 on $[\mathcal{V}]^2$, there is nothing more to prove. Thus, assume that there is a two-element subset, say

$\{U_{n_1},\ U_{n_2}\}$ with $n_1 < n_2$, of $\mathcal{V}$ for which $f(\{U_{n_1},\ U_{n_2}\}) = 1$. Then the set $\mathcal{W} = \{U_n \in \mathcal{V} : n > n_2\}$ is still an ω-cover of X.

Claim 2: f has value 1 on the set $[\mathcal{W}]^2$. For if not, choose a two-element set $\{U_k,\ U_m\} \subset \mathcal{W}$ with $k < m$ and $f(\{U_k,\ U_m\}) = 2$. By the construction of $\mathcal{W}$ we have $n_1 < n_2 < k < m$. Also, we have $f(\{U_{n_1}, U_{n_2}\}) = 1$ and $f(\{U_k,\ U_m\}) = 2$. If $f(\{U_{n_2}, U_k\}) = 1$, then by definition of the function g we have $g(\{U_{n_2},\ U_m,\ U_k\}) = 2$, contradicting the fact that g has value 1 on $[\mathcal{V}]^3$. Thus, we must have $f(\{U_{n_2}, U_k\}) = 2$. But in this case we have $f(\{U_{n_1},\ U_{n_2}\}) = 1$, so that by definition $g(\{U_{n_1},\ U_{n_2},\ U_k\}) = 2$ also contradicting the fact that g has value 1 on $[\mathcal{V}]^3$. These contradictions follow from assuming that f is not constant of value 1 on the set $[\mathcal{W}]^2$. This completes the proof of Claim 2, and of the implication $\Omega \longrightarrow (\Omega, 4)^3_2 \Rightarrow (\Omega \longrightarrow (\Omega)^2_2$.

7.9. A proof of the implication $\Omega \longrightarrow (\Omega)^2_2 \Rightarrow (\Omega \longrightarrow (\Omega, \mathcal{O})^2$. Since Ω is a subset of $\mathcal{O}$, this implication holds.

7.10. A proof of the implication $\Omega \longrightarrow (\Omega, \mathcal{O})^2_2 \Rightarrow \mathsf{S}_1(\Omega, \Omega)$. Let a sequence $(\mathcal{U}_n : n \in \mathbb{N})$ of ω-covers of the space X be given. We may assume that each $\mathcal{U}_n$ is countable, and enumerate it bijectively as $(U^n_m : m \in \mathbb{N})$. Then define the ω-cover $\mathcal{V} = \{U^1_m \cap U^m_k : 1 < m,\ k \in \mathbb{N}\} \setminus \{\emptyset\}$. Also, for each element of $\mathcal{V}$ fix a specific choice of k and m to represent the element as $U^1_k \cap U^k_m$.

Define a function $f : [\mathcal{V}]^2 \to \{0,\ 1\}$ so that

$$f(\{U^1_{k_1} \cap U^{k_1}_{m_1}, U^1_{k_2} \cap U^{k_2}_{m_2}\}) = \begin{cases} 0 & \text{if } k_1 \neq k_2 \\ 1 & \text{otherwise} \end{cases}$$

Applying the hypothesis that the partition relation $\Omega \longrightarrow (\Omega, \mathcal{O})^2_2$ holds, choose a subset $\mathcal{W}$ of $\mathcal{V}$ such that f is constant on $[\mathcal{W}]^2$, and $\mathcal{W}$ witnesses this partition relation for f. Assume that f does not have value 0 on $[\mathcal{W}]^2$, and thus that $\mathcal{W}$ is an open cover of X but not necessarily an ω-cover. Then f has value 1 on $[\mathcal{W}]^2$. Then for any $U^1_{k_1} \cap U^{k_1}_{m_1} \neq U^1_{k_2} \cap U^{k_2}_{m_2}$ in the set $\mathcal{W}$ we have $k_1 = k_2$, and thus all elements of $\mathcal{W}$ are subsets of one and the same U^1_k. But U^1_k being an element of an ω-cover of X is a proper subset of X, and thus $\mathcal{W}$ is not a cover of X, a contradiction. Consequently f has value 0 on $[\mathcal{W}]^2$, and $\mathcal{W}$ is an ω-cover of X. By the definition of the function f we also find that if $U^1_{k_1} \cap U^{k_1}_{m_1}$ and $U^1_{k_2} \cap U^{k_2}_{m_2}$ are two distinct elements of $\mathcal{W}$, then $k_1 \neq k_2$.

For each k such that $U^1_k \cap U^k_m$, with indices as in the originally chosen representation, appears in $\mathcal{W}$, pick $V_k = U^k_m \in \mathcal{U}_k$. For all other k choose $V_k \in \mathcal{U}_k$ arbitrarily. Then the set $\{V_k : k \in \mathbb{N}\}$ witnesses the statement $\mathsf{S}_1(\Omega, \Omega)$ for the given sequence $(\mathcal{U}_n : n \in \mathbb{N})$ of ω-covers of X.

The proof of Theorem 7.2 is complete. □

8. The Rothberger covering property for σ-compact spaces.

As shown in Theorem 2.5, in the case of a σ-compact uniform space (X, Ψ), for a subset Y of X the selection principles $\mathsf{S}_1(\mathcal{O}_\Psi, \mathcal{O}_Y)$ and $\mathsf{S}_1(\mathcal{O}, \mathcal{O}_Y)$ are equivalent. In the case when $Y = X$, the latter is the Rothberger covering property. In this section we focus on this scenario: The topological space (X, τ) is σ-compact and has the covering property $\mathsf{S}_1(\mathcal{O}, \mathcal{O})$. According to Theorem 2.5 for such a space, ONE has no winning strategy in the game $\mathsf{G}_1(\mathcal{O}, \mathcal{O})$. In fact, much stronger conclusions - see Theorem 8.5 - can be drawn.

LEMMA 8.1. *If* (X, τ) *is a compact* T_2*-space which has the property* $\mathsf{S}_1(\mathcal{O}, \mathcal{O})$, *then the space has an isolated point.*

PROOF. For assume the contrary. Let (X, τ) be a compact space which has no isolated points. We show that ONE has a winning strategy in the game $\mathsf{G}_1(\mathcal{O}, \mathcal{O})$, implying by Pawlikowski's Theorem that (X, τ) does not have the property $\mathsf{S}_1(\mathcal{O}, \mathcal{O})$.

Define a strategy F for ONE in the game $\mathsf{G}_1(\mathcal{O}, \mathcal{O})$ on X as follows: SInce X has no isolated points, it is infinite. Choose two distinct points x_0 and $_1$ from X. As X is compact and T_2, choose open sets U_0 and U_1 such that $x_0 \in U_0$ and $x_1 \in U_1$, and the closures of U_0 and U_1 are disjoint. ONE's first move is

$$F(\emptyset) = \{X \setminus \overline{U}_0,\ X \setminus \overline{U}_1\}.$$

When TWO chooses $T_1 \in F(\emptyset)$, say $T_1 = X \setminus \overline{U}_{i_1}$, then as X has no isolated points, the nonempty open set U_{i_1} has no isolated points. Thus choose two distinct points $x_{i_1,0}$ and $x_{i_1,1}$ in U_{i_1} and apply the fact that X is compact T_2 again to find two open sets $U_{i_1,0}$ and $U_{i_1,1}$ for which their closures are disjoint subsets of U_{i_1}, and $x_{i_1,0} \in U_{i_1,0}$ and $x_{i_1,1} \in U_{i_1,1}$. ONE's response is

$$F(T_1) = \{X \setminus \overline{U}_{i_1,0},\ X \setminus \overline{U}_{i_1,1}\}$$

When TWO chooses $T_2 \in F(T_1)$, say $T_2 = X \setminus \overline{U}_{i_1,i_2}$, then as X has no isolated points, the nonempty open set U_{i_1,i_2} has no isolated points. Thus choose two distinct points $x_{i_1,i_2,0}$ and $x_{i_1,i_2,1}$ in U_{i_1,i_2} and apply the fact that X is compact T_2 again to find two open sets $U_{i_1,i_2,0}$ and $U_{i_1,i_2,1}$ for which their closures are disjoint subsets of U_{i_1,i_2}, and $x_{i_1,i_2,0} \in U_{i_1,i_2,0}$ and $x_{i_1,i_2,1} \in U_{i_1,i_2,1}$. ONE's response is

$$F(T_1, T_2) = \{X \setminus \overline{U}_{i_1,i_2,0},\ X \setminus \overline{U}_{i_1,i_2,1}\}$$

When TWO chooses $T_3 \in F(T_1,\ T_2)$, say $T_3 = X \setminus \overline{U}_{i_1,i_2,i_3}$, then as X has no isolated points, the nonempty open set U_{i_1,i_2,i_3} has no isolated points. Thus choose two distinct points $x_{i_1,i_2,i_3,0}$ and $x_{i_1,i_2,i_3,1}$ in U_{i_1,i_2,i_3} and apply the fact that X is compact T_2 again to find two open sets $U_{i_1,i_2,i_3,0}$ and $U_{i_1,i_2,i_3,1}$ for which their closures are disjoint subsets of U_{i_1,i_2,i_3}, and $x_{i_1,i_2,i_3,0} \in U_{i_1,i_2,i_3,0}$ and $x_{i_1,i_2,i_3,1} \in U_{i_1,i_2,i_3,1}$. ONE's response is

$$F(T_1, T_2, T_3) = \{X \setminus \overline{U}_{i_1,i_2,i_3,0},\ X \setminus \overline{U}_{i_1,i_2,i_3,1}\}$$

and so on.

Consider an F-play

$$F(\emptyset),\ T_1,\ F(T_1),\ T_2,\ \cdots .\ F(T_1, \cdots, T_n),\ T_{n+1}, \cdots$$

of the game $\mathsf{G}_1(\mathcal{O}, \mathcal{O})$ on X. From the definition of F there is for each n a nonempty open subset $U_{i_1,\cdots,i_n}$ of X such that

(1) $T_n = X \setminus \overline{U}_{i_1,\cdots,i_n}$
(2) the set $\overline{U}_{i_1,\cdots,i_n}$ is a nonempty subset of $U_{i_1,\cdots,i_{n-1}}$,
(3) and $U_{i_1,\cdots,i_n}$ is an open set containing the point $x_{i_1,\cdots,i_n}$ of X.

Since (X, τ) is compact, the sequence $(\overline{U}_{i_1,\cdots,i_n} : n \in \mathbb{N})$ as a descending sequence of nonempty compact subsets of X has a nonempty intersection. Let the point y be a member of this intersection. Then for each n, y is not a member of T_n. It follows that ONE wins the F play, and as we considered an arbitrary F play, F is a winning strategy for ONE in the game $\mathsf{G}_1(\mathcal{O}, \mathcal{O})$ on X. By Pawlikowski's Theorem, (X, τ) does not have the selection property $\mathsf{S}_1(\mathcal{O}, \mathcal{O})$. $\square$

LEMMA 8.2. *If (X, τ) is a compact space with property $\mathsf{S}_1(\mathcal{O}, \mathcal{O})$, then each nonempty subset has an isolated point.*

PROOF. If not, suppose that (X, τ) is a compact space with property $\mathsf{S}_1(\mathcal{O}, \mathcal{O})$ and let U be a nonempty set with no isolated points. Then also $\overline{U}$ has no isolated points, but is a compact space with the property $\mathsf{S}_1(\mathcal{O}, \mathcal{O})$, contradicting Theorem 8.1. □

A topological space is scattered if each nonempty set has an isolated point. Thus, compact spaces with property $\mathsf{S}_1(\mathcal{O}, \mathcal{O})$ are scattered. For a topological space let $\mathcal{I}(X)$ denote the set of isolated points of X. Then we define the Cantor-Bendixson sequence of X as follows:

- $X^0 = X$
- For an ordinal α, $X^{\alpha+1} = X^\alpha \setminus \mathcal{I}(X^\alpha)$.
- For a limit ordinal α, $X^\alpha = \bigcap_{\beta<\alpha} X^\beta$

Observe that (X, τ) is a scattered space if there is an ordinal α for which $X^\alpha = \emptyset$. The least value of α for which $X^\alpha = \emptyset$ is said to be the scattered height of X and is denoted $ht(X)$. One can show that of (X, τ) is a compact scattered space, then $ht(X)$ is a successor ordinal.

LEMMA 8.3. *A compact scattered space has the property $\mathsf{S}_1(\mathcal{O}, \mathcal{O})$.*

PROOF. We prove the lemma by induction on the ordinal α for which $ht(X) = \alpha + 1$. When $\alpha = 0$, then X is a discrete finite set, and there is nothing to prove.

Assume that the lemma is proven for all compact scattered spaces (X, τ) with $ht(X) = \beta + 1$ where $\beta < \alpha$. Now consider a compact scattered space (X, τ) with $ht(X) = \alpha + 1$. Then X has a finite set of isolated points, being compact. Let $\{x_1, \cdots, x_n\}$ be the set of isolated points of X. Let $(\mathcal{U}_n : n \in \mathbb{N})$ be a given sequence of open covers of X. For $1 \le i \le n$ choose $U_i \in \mathcal{U}_i$ so that $x_i \in U_i$. If $X = \cup\{U_i : i \le n\}$, we are done. Else, the subspace $Y = X \setminus \cup\{U_i : i \le n\}$ is a compact scattered space and $ht(Y) < \alpha + 1$. Thus, by the induction hypothesis, Y has the property $\mathsf{S}_1(\mathcal{O}, \mathcal{O})$. Applying this property to the sequence $(\mathcal{U}_m : n < m \text{ and } m \in \mathbb{N})$, we find for each m with $n < m$ a set $U_m \in \mathcal{U}_m$ such that Y is covered by the sets U_i, $n < i$. But then the selection $(U_j : j \in \mathbb{N})$ witnesses for teh given sequence of covers that (X, τ) has the property $\mathsf{S}_1(\mathcal{O}, \mathcal{O})$. □

Lemmas 8.2 and 8.3 prove that a compact T_2-space is scattered if, and only if, it has the property $\mathsf{S}_1(\mathcal{O}, \mathcal{O})$.

LEMMA 8.4. *For a compact T_2 space (X, τ) the following statements are equivalent:*

(1) *(X, τ) has the property $\mathsf{S}_1(\mathcal{O}, \mathcal{O})$*
(2) *TWO has a winning strategy in the game $\mathsf{G}_1(\mathcal{O}, \mathcal{O})$*

PROOF. We must show that (1) implies (2). Thus, assume that (X, τ) has the property $\mathsf{S}_1(\mathcal{O}, \mathcal{O})$. By Lemma 8.2 the space (X, τ) is scattered, and has scattered height $ht(X) = \alpha + 1$ for some ordinal α. We prove the Lemma by induction on α.

When $\alpha = 0$ then X is a finite set, and TWO's winning strategy is to in each inning cover a previously uncovered point. Thus, assume that $\alpha > 0$ and that for each $\beta < \alpha$ it has been proven that for any compact T_2 space of scattered height $\beta + 1$, TWO has a winning strategy in the game $\mathsf{G}_1(\mathcal{O}, \mathcal{O})$. Now let (X, τ) be a compact scattered space of height $\alpha + 1$. Since X has isolated points and is compact, its set of isolated points is finite, say the set of isolated points of X is $\{x_1, \cdots, x_n\}$. TWO's winning strategy on X proceeds as follows. During inning i for $i \le n$, TWO chooses an element T_i of ONE's open cover $\mathcal{U}_i$ such that $x_i \in T_i$. Then after the first n innings passed, the space $Y = X \setminus \cup\{T_i : i \le n\}$ is compact, scattered, and of height $\beta + 1 < \alpha + 1$. By the induction hypothesis TWO has a winning strategy F_Y in the game $\mathsf{G}_1(\mathcal{O}, \mathcal{O})$ on Y. For the remainder of the game on X, TWO follows the winning strategy F_Y on Y. □

THEOREM 8.5. *Consider a σ-compact topological space (X, τ). The following statements are equivalent:*

(1) *(X, τ) has the property $\mathsf{S}_1(\mathcal{O}, \mathcal{O})$*
(2) *TWO has a winning strategy in the game $\mathsf{G}_1(\mathcal{O}, \mathcal{O})$*

PROOF. We must prove that (1) implies (2). Fix a well-ordering $\prec$ of the family of open subsets of X. Since X is σ-compact, write X as an increasing union of countably many subsets X_n, $n \in \mathbb{N}$, where each X_n is compact. By Lemma 8.4, for each n TWO has a winning strategy F_n in the game $\mathsf{G}_1(\mathcal{O}, \mathcal{O})$ played on X_n. We now define a strategy for TWO in the game on X, utilizing the strategies F_n. First, write $\mathbb{N}$ as a union of countably many pairwise disjoint infinite sets, S_n, $n \in \mathbb{N}$. For each n enumerate S_n as $(s_k^n : k \in \mathbb{N})$ so that for $k < \ell$, $s_k^n < s_\ell^n$. Now define TWO's strategy F for the game $\mathsf{G}_1(\mathcal{O}, \mathcal{O})$ on X as follows:

Given inning k, and the first k moves $O_1, \cdots, O_k$ of ONE, first find n so that $k \in S_n$, and then identify ℓ so that $k = s_\ell^n$. Then define $F(O_1, \cdots, O_k)$ as follows: For $1 \le i \le \ell$ define $\mathcal{V}_i$ to be the set $\{U \cap X_n : U \in O_{s_i^n}\}$, and then compute

$$T = F_n(O_{s_1^n}, \cdots, O_{s_\ell^n}),$$

TWO's response in the game $\mathsf{G}_1(\mathcal{O}_{X_n}, \mathcal{O}_{X_n})$ using the winning strategy F_n. Then choose the $\prec$-first $U \in O_k$ for which $U \cap X_n = T$. Declare $F(O_1, \cdots, O_k) = U$.

We claim that F as defined here is a winning strategy for TWO in the game $\mathsf{G}_1(\mathcal{O}, \mathcal{O})$ played on X. For consider any F-play, say

$$O_1, F(O_1), O_2, F(O_1, O_2), \cdots, O_k, F(O_1, \cdots O_k), \cdots$$

Consider a point $x \in X$, and choose the smallest n with $x \in X_n$. Recall that S_n is enumerated as $s_1^n < s_2^n < \cdots < s_\ell^n < \cdots$. Consider the subsequence

$$O_{s_1^n}, F(O_1, \cdots, O_{s_1^n}), O_{s_2^n}, F(O_1, \cdots, O_{s_2^n}), \cdots, O_{s_k^n}, F(O_1, \cdots, O_{s_k^n}), \cdots$$

of the given F-play.

In then notation above used in the definition of F we have $\mathcal{V}_i = \{U \cap X_n : U \in O_{s_i^n}\}$ is for each i an open cover of X_n, and $X_n \cap F(O_1, \cdots, O_{s_i^n}) = F_n(\mathcal{V}_1, \cdots, \mathcal{V}_i)$ for each$i$. Since

$$\mathcal{V}_1, F_n(\mathcal{V}_1), \mathcal{V}_2, F_n(\mathcal{V}_1, \mathcal{V}_2), \cdots$$

is an F_n play of the game $\mathsf{G}_1(\mathcal{O}_{X_n}, \mathcal{O}_{X_n})$ on X_n, and F_n is a winning strategy for TWO in this game, we find that for some k the point x is an element of $F(O_1, \cdots, O_{s_k^n})$. □

For σ-compact spaces the fact that TWO has a winning strategy in the game $\mathsf{G}_1(\mathcal{O}, \mathcal{O})$ is in in turn equivalent to a strong Ramseyan partition relation on the ω-covers of the space. Towards formulating this stronger property we introduce yet another class of open covers for topological spaces, first studied by Gerlits and Nagy in [**17**]. An open cover $\mathcal{U}$ of a topological space (X, τ) is said to be a γ-cover if it is infinite and for each $x \in X$, for all but finitely many $U \in \mathcal{U}$ it is the case that $x \in U$. The symbol Γ denotes the set of all γ-covers of X. In Theorem 1 of [**17**] the authors prove

THEOREM 8.6 (Gerlis and Nagy). *For a $\mathsf{T}_{3\frac{1}{2}}$-space (X, τ) the following are equivalent:*

(1) *TWO has a winning strategy in the game $\mathsf{G}_1(\mathcal{O}, \mathcal{O})$ on X*
(2) *TWO has a winning strategy in the game $\mathsf{G}_1(\Omega, \Gamma)$ on X*

It is evident that when TWO has a winning strategy in the game $\mathsf{G}_1(\Omega, \Gamma)$, then ONE does not have a winning strategy in that game, whence in turn $\mathsf{S}_1(\Omega, \Gamma)$ holds.

The main theorem of this section is as follows:

THEOREM 8.7. *For a σ-compact space (X, τ) the following are equivalent:*

(1) *(X, τ) has the property $\mathsf{S}_1(\mathcal{O}, \mathcal{O})$.*
(2) *TWO has a winning strategy in the game $\mathsf{G}_1(\mathcal{O}, \mathcal{O})$.*
(3) *(X, τ) has the property $\mathsf{S}_1(\Omega, \Gamma)$.*
(4) *For all m and k the partition relation $\Omega \to (\Gamma)^m_k$ holds.*

PROOF. The equivalence of (1) and (2) is proven in Theorem 8.5. That (2) implies (3) follows from Theorem 8.6. To see that (3) implies (4), let $\mathcal{U}$ be an ω-cover of X. We may assume that $\mathcal{U}$ is countable. Let a function $f : [\mathcal{U}]^m \to \{1, \cdots, k\}$ be given. By (3) let $\mathcal{V} \subset \mathcal{U}$ be such that $\mathcal{V}$ is a γ-cover of X. Note that any infinite subset of $\mathcal{V}$ is also a γ-cover of X. Now apply Ramsey's Theorem to the coloring f restricted to the set of m-tuples of $\mathcal{V}$. We find an infinite subset $\mathcal{W}$ of $\mathcal{V}$, and an $i \in \{1, \cdots, k\}$, such that on $[\mathcal{W}]^m$ the function f is constant and has value i. Finally, to see that (4) implies (1), note that since $\Gamma \subset \Omega$, (4) implies that for all m and k the partition relation $\Omega \to (\Omega)^m_k$ holds. But then by Theorem 7.2 the space (X, τ) has the property $\mathsf{S}_1(\Omega, \Omega)$, and thus by Theorem 4.6 has the property $\mathsf{S}_1(\mathcal{O}, \mathcal{O})$. □

To see that the hypothesis of being a compact space with the property $\mathsf{S}_1(\mathcal{O}, \mathcal{O})$ is not "trivial", note the following: Metrizable spaces that are compact and have property $\mathsf{S}_1(\mathcal{O}, \mathcal{O})$ are necessarily countable. But for each infinite cardinal number κ there is a T_2 topological space of cardinality κ that is σ-compact and has the property $\mathsf{S}_1(\mathcal{O}, \mathcal{O})$. The following examples of this phenomenon are based on a result of Corson (Proposition 4 of [**8**]), and were pointed out in [**40**]:

THEOREM 8.8 (Corson). *If $\{X_i : i \in I\}$ is a family of σ-compact topological groups where for each i the identity element of X_i is e_i, then the group*

$$G = \{f \in \prod_{i \in I} X_i : |\{j \in I : f(j) \neq e_j\}| < \aleph_0\}$$

is σ-compact.

PROPOSITION 8.9. *For each infinite cardinal number κ there is a $\mathsf{T}_{3\frac{1}{2}}$ σ-compact topological group of cardinality κ which has the property $\mathsf{S}_1(\mathcal{O}, \mathcal{O})$.*

PROOF. Let cardinal number κ be given. In Theorem 8.8 take I to be κ and for each $i \in I$ take X_i to be the additive group of integers, $\mathbb{Z}$. Now observe that the group G as in Corson's Theorem in fact has the property $\mathsf{S}_1(\mathcal{O}, \mathcal{O})$. To see this take for each n a neighborhood U_n of the identity element of G. We may assume that each U_n is a basic open set, and thus that there is a finite set $F_n \subseteq I$ such that U_n is of the form $\{f \in G : (\forall i \in F_n)(f(i) = 0)\}$.

The set $C = \bigcup_{n<\omega} F_n$ is a countable subset of I and thus the subgroup $G_C = \{f\lceil_C : f \in G\}$ of the group $\prod_{i\in C} X_i$ is a countable set. Thus, the property $\mathsf{S}_1(\mathcal{O}_{nbd}, \mathcal{O}_{G_C})$ evidently holds. For each n choose a $g_n \in G_C$ such that $G_C \subseteq \bigcup_{n<\omega} g_n * U_n\lceil_C$. For each n choose $f_n \in G$ with $f_n\lceil_C = g_n$. Then it follows that $G \subseteq \bigcup_{n<\omega} f_n * U_n$. Thus G satisfies the property $\mathsf{S}_1(\mathcal{O}_{nbd}, \mathcal{O})$. As G is a σ-compact Theorem 2.5 implies that G satisfies the property $\mathsf{S}_1(\mathcal{O}, \mathcal{O})$. □

Theorem 14 of [**40**] gave an *ad hoc* argument that TWO has a winning strategy in the game $\mathsf{G}_1(\mathcal{O}, \mathcal{O})$ in these specific examples. Theorem 8.5 generalizes that prior result.

It is further worth noting that the existence of a winning strategy for TWO in the game $\mathsf{G}_1(\mathcal{O}, \mathcal{O})$ on an uncountable $\mathsf{T}_{3\frac{1}{2}}$ space does not imply that the space is σ-compact. Thus is illustrated using the following example of Comfort and Ross ([**7**], Example 3.2). Consider the set $G := \{f \in {}^{\omega_1}2 : |\{\alpha : f(\alpha) \neq 0\}| < \aleph_0\}$ endowed with the operation $\oplus$ of coordinate-wise addition modulo 2. Then $(G, \oplus)$ is an Abelian group. Endow G with the G_δ topology. In [**7**] it is proven that the group $(G, \oplus)$ endowed with this G_δ topology is Lindelöf and thus T_4.

For convenience we review a few facts about so-called P-spaces - see for example Section 5 of [**18**]. A topological space is said to be a P-space if each G_δ set is open. Every subspace of a P-space is a P-space. Every countably infinite subspace of a T_2 P-space is closed and discrete. It follows that a compact P-space is finite, and thus a σ-compact P-space is countable. Thus, no uncountable P-space is a closed subspace of a σ-compact T_2-space. If a topological group $(G, *)$ is a Lindelöf P-space then it is has the property $\mathsf{S}_1(\mathcal{O}_{nbd}, \mathcal{O})$ in a strong sense: Let $(U_n : n < \omega)$ be a sequence of neighborhoods for the identity. Then $U = \bigcap_{n<\omega} U_n$ is a neighborhood for the identity. Since the group is Lindelöf, fix a sequence $(x_n : n < \omega)$ of elements of the group such that $x_n * U$, $n < \omega$ covers the group. Then the sequence $(x_n * U_n : n < \omega)$ witnesses that the uncountable group $(G, \oplus)$ has the property $\mathsf{S}_1(\mathcal{O}_{nbd}, \mathcal{O})$. Moreover this group is not contained as a closed subspace of any σ-compact group.

Theorem 2.3 of Comfort in [**6**] implies the following generalization of the example just given:

THEOREM 8.10 (Comfort). *Let $(G_i, *_i)$, $i \in I$, be a family of countable topological groups. Endow the product $\prod_{i\in I} G_i$ with the G_δ topology. Then the subgroup*

$$G := \{f \in \prod_{i\in I} G_i : |\{j \in I : f(j) \neq \mathsf{id}_j\}| < \aleph_0\}$$

is a Lindelöf P-group.

In particular, for each uncountable cardinal number κ there is a T_0 Lindelöf P group of cardinality κ. Now Galvin proved a result - see the Lemma in Section 2 of [**17**] - that implies that if a space is a Lindelöf P-space then has the property $\mathsf{S}_1(\mathcal{O}, \mathcal{O})$. Thus, Lindelöf P groups satisfy the stronger selection principle $\mathsf{S}_1(\mathcal{O}, \mathcal{O})$.

PROPOSITION 8.11. *For each uncountable cardinal κ there is a T_0 Lindelöf P-group of cardinality κ such that TWO has a winning strategy in $\mathsf{G}_1(\mathcal{O}, \mathcal{O})$.*

Thus for the examples based on Comfort's work cited above, none is homeomorphic to a closed subspace of a σ-compact space, yet TWO has a winning strategy in the game $\mathsf{G}_1(\mathcal{O},\mathcal{O})$. By arguments as in the proof of Theorem 8.7, this class of examples also satisfies the partition relation $\Omega \rightarrow (\Gamma)^m_k$ for all natural numbers m and k.

9. A question about the polarized partition relation

The polarized partition relation was studied by Rado in [**26**], and formally introduced by Erdös and Rado in Section 9 of [**12**]. Generalizing their notion we define the following: Let $\mathcal{A}_1$, $\mathcal{A}_2$ $\mathcal{B}_{i,1}$ and $\mathcal{B}_{i,2}$ be families of sets for $1 \le i \le n$.

DEFINITION 9.1. The symbol

$$\begin{pmatrix} \mathcal{A}_1 \\ \mathcal{A}_2 \end{pmatrix} \longrightarrow \begin{pmatrix} \mathcal{B}_{1,1} & \cdots & \mathcal{B}_{n,1} \\ \mathcal{B}_{1,2} & \cdots & \mathcal{B}_{n,2} \end{pmatrix}^{1,1}$$

denotes the statement that for each $A_1 \in \mathcal{A}_1$ and each $A_2 \in \mathcal{A}_2$, and for each function $f : A_1 \times A_2 \rightarrow \{1, \cdots, n\}$ there exists an i with $1 \le i \le n$, and sets $B_{i,1} \in \mathcal{B}_{i,1}$ and $B_{i,2} \in \mathcal{B}_{i,2}$ such that $B_{i,1} \subseteq A_1$ and $B_{i,2} \subseteq A_2$, and f has the value i on the set $B_{i,1} \times B_{i,2}$.

First observe that by Theorem 47 of [**12**], for each infinite cardinal number κ the negative relation

$$\begin{pmatrix} \kappa \\ \kappa \end{pmatrix} \not\rightarrow \begin{pmatrix} \kappa & 1 \\ 1 & \kappa \end{pmatrix}^{1,1}$$

holds. This in turn implies that for infinite topological spaces (X, τ) the relation

$$\begin{pmatrix} \Omega \\ \Omega \end{pmatrix} \not\rightarrow \begin{pmatrix} \Omega & 1 \\ 1 & \Omega \end{pmatrix}^{1,1}$$

holds. Thus, we relax the polarized partition relation to a square bracket version, analogous to the weakening of the ordinary partition relation.

DEFINITION 9.2. Let $\mathcal{A}_1$, $\mathcal{A}_2$, $\mathcal{B}_1$ and $\mathcal{B}_2$ be families of sets. The symbol

$$\begin{pmatrix} \mathcal{A}_1 \\ \mathcal{A}_2 \end{pmatrix} \longrightarrow \begin{bmatrix} \mathcal{B}_1 \\ \mathcal{B}_2 \end{bmatrix}^{1,1}_{k/\le \ell}$$

denotes the statement that for each $A_1 \in \mathcal{A}_1$ and $A_2 \in \mathcal{A}_2$, and for each function $f : A_1 \times A_2 \rightarrow \{1, \cdots, k\}$ there are sets $B_1 \subseteq A_1$ and $B_2 \subseteq A_2$ such that $B_1 \in \mathcal{B}_1$, $B_2 \in \mathcal{B}_2$, and $|\{f(b_1, b_2) : b_1 \in B_1 \text{ and } b_2 \in B_2\}| \le \ell$.

It is not clear if there is a version of the square bracket polarized partition relation $\begin{pmatrix} \Omega \\ \Omega \end{pmatrix} \longrightarrow \begin{bmatrix} \Omega \\ \Omega \end{bmatrix}^{1,1}_{k/\le \ell}$ that characterizes whether a space has the property $\mathsf{S}_1(\Omega, \Omega)$. We conjecture:

CONJECTURE 1. *If a topological space (X, τ) satisfies $\begin{pmatrix} \Omega \\ \Omega \end{pmatrix} \longrightarrow \begin{bmatrix} \Omega \\ \Omega \end{bmatrix}^{1,1}_{k/<3}$ for any positive integer k, then (X, τ) has the property $\mathsf{S}_1(\Omega, \Omega)$.*

We give a complete analysis in the proof of Theorem 9.4 to indicate to the reader the single case of the analysis which does not support the conjecture. We believe that this one "counter" case may be a shortcoming of the proof technique. The following concept, and corresponding facts about its relation to the conjecture are important. Space X satisfies property $\mathsf{Split}(\Omega, \Omega)$ if there is for each ω-cover $\mathcal{U}$

of X two ω-covers $\mathcal{B}_1$ and $\mathcal{B}_2$ such that $\mathcal{B}_1 \cap \mathcal{B}_2 = \emptyset$ and $\mathcal{B}_1 \cup \mathcal{B}_2 = \mathcal{U}$. This property was introduced in [19] and studied further in [15].

THEOREM 9.3. *Let X be a space which has the Lindelöf covering property in all finite powers. If for X the polarized partition relation* $\begin{pmatrix} \Omega \\ \Omega \end{pmatrix} \longrightarrow \begin{bmatrix} \Omega \\ \Omega \end{bmatrix}^{1,1}_{k/<3}$ *holds, then X has property* $\mathsf{Split}(\Omega, \Omega)$.

PROOF. Let $\mathcal{U}$ be an ω-cover of X . We may assume that it is countable. Enumerate it bijectively as $(U_m : m \in \mathbb{N})$. Then define $f : \mathcal{U} \times \mathcal{U} \to \{0, 1, 2\}$ so that

$$f(U_m, U_n) = \begin{cases} 0 & \text{i f } m = n \\ 1 & \text{if } m < n \\ 2 & \text{if } m > n \end{cases}$$

Apply the polarized relation to find ω-covers $\mathcal{V}$ and $\mathcal{W}$ contained in $\mathcal{U}$ such that on $\mathcal{V} \times \mathcal{W}$ f is two-valued. We show that $\mathcal{V}$ and $\mathcal{W}$ are disjoint by showing that 0 is not a value of f. Suppose on the contrary that 0 is a value of f on $\mathcal{V} \times \mathcal{W}$. Choose n such that $(U_n, U_n) \in \mathcal{V} \times \mathcal{W}$. Since $\mathcal{V}$ and $\mathcal{W}$ are ω-covers they are infinite. Thus there are $U_m \in \mathcal{V}$ with $m > n$ and $U_k \in \mathcal{W}$ with $k >$n. This implies that 1 and 2 are also values of f, so that f takes three distinct values on $\mathcal{V} \times \mathcal{W}$ instead of just two. □

THEOREM 9.4. *Let X be a space. Of the following statements, each implies the next.*

(1) $\mathsf{S}_1(\Omega, \Omega)$

(2) *For each positive integer k,* $\begin{pmatrix} \Omega \\ \Omega \end{pmatrix} \longrightarrow \begin{bmatrix} \Omega \\ \Omega \end{bmatrix}^{1,1}_{k/<3}$

(3) $\mathsf{S}_{fin}(\Omega, \Omega)$

PROOF. (1) $\Rightarrow$ (2): Let $\mathcal{U}$ and $\mathcal{V}$ be ω-covers of X. Since X has property $\mathsf{S}_1(\Omega, \Omega)$ we may assume that each of these is countable. Enumerate $\mathcal{U}$ bijectively as $(U_n : n \in \mathbb{N})$. Let k be a positive integer and let $f : \mathcal{U} \times \mathcal{V} \to \{1, ..., k\}$ be given.

Choose a chain of subsets of $\mathcal{V}$ as follows: Put $\mathcal{V}_0 = \mathcal{V}$ and for $0 \leq n < \omega$ choose $i_{n+1} \in \{1, ..., k\}$ such that

$$\mathcal{V}_{n+1} = \{V \in \mathcal{V}_n : f(U_{n+1}, V) = i_{n+1}\} \in \Omega$$

Since $\mathcal{U}$ is an ω-cover, fix an $i \in \{1, \cdots, k\}$ and $n_1 < n_2 < \cdots$ such that $\{U_{n_1}, U_{n_2}, \cdots\} \in \Omega$ and for each j, $i_{n_j} = i$. Apply property $\mathsf{S}_1(\Omega, \Omega)$ to the sequence $(\mathcal{V}_{n_1}, \mathcal{V}_{n2}, \cdots, \mathcal{V}_{n_k}, \cdots)$ of ω-covers of X and select for each m a $V_m \in \mathcal{V}_{n_m}$, such that $\{V_m : m \in \mathbb{N}\}$ is an ω-cover of X . We may assume that this enumeration is bijective. At this stage we have that if $j \leq m$, then $f(U_{n_j}, V_m) = i$.

Put $\mathcal{U}_{-1} = \{U_{n_l}, U_{n_2}, \cdots\}$ and choose for each m a $j_{m+l} \in \{1, \cdots, k\}$ such that $\mathcal{U}_m = \{U_{n_r} \in \mathcal{U}_m : f(U_{n_r}, V_{m+l}) = j_{m+l}\} \in \Omega$. Fix a j and $m_l < m_2 < \cdots$ such that for each r $j_{m_r} = j$, and $\{V_{m_1}, V_{m_2}, \cdots\}$ is an ω-cover of X . Next we apply $\mathsf{S}_1(\Omega, \Omega)$ to the sequence $(\mathcal{U}_{m_1}, \mathcal{U}_{m_2}, \cdots)$ of ω-covers of X, as follows: From $\mathcal{U}_{m_t}$ choose $U_{n_{h_t}}$, such that for each t, $m_t < h_t < h_{t+l}$ and $\{U_{n_{h_t}} : t \in \mathbb{N}\}$ is an ω-cover of X . One way of seeing that this can be done is as follows. In [21] it was shown that X has property $\mathsf{S}_1(\Omega, \Omega)$ if, and only if, ONE has no winning strategy in the following game, denoted $\mathsf{G}_1(\Omega, \Omega)$: In the n-th inning ONE chooses an ω-cover O_n, of X ; TWO responds by selecting $T_n \in O_n$. A play $(O_1, T_1, \cdots, O_n, T_n, \cdots)$ is won by TWO if $\{T_I, T_2, T_3, \cdots\}$ is an ω-cover of X ; otherwise, ONE wins.

Now consider the strategy which calls on ONE to play in the t-th inning the set of elements of $\mathcal{U}_{m_t}$ of the form U_{n_h}, where h exceeds $\max\{m_1, \cdots, m_t\}$ as well as the maximum of all subscripts of elements chosen thus far by TWO. Since this is not a winning strategy, there is a play according to it which is won by TWO. Such a play provides us with a sequence of $U_{n_{h_t}}$'s as above. But then on $\{U_{n_{h_l}}, U_{n_{h_2}}, \cdots\} \times \{V_{m_1}, V_{m_2}, \cdots\}$ the function f takes only the two values, i and j .

(2) $\Rightarrow$ (3): Let $(\mathcal{U}_n : n \in \mathbb{N})$ be a sequence of ω-covers of X. We may assume each $\mathcal{U}_n$ is countable and enumerate it as $(U^n_m : m \in \mathbb{N})$. Define $\mathcal{V}$ to be the non-empty intersections of the form $U^1_k \cap U^k_n$. Then $\mathcal{V}$ is an ω–cover of X . For each element of $\mathcal{V}$ choose a specific representation of the form $U^1_k \cap U^k_n$. Define a coloring $f : \mathcal{V} \times \mathcal{V} \to \{0, 1, 2, 3, 4, 5, 6, 7, 8\}$ such that

$$f((U^1_{k_1} \cap U^{k_1}_{n_1}, U^1_{k_2} \cap U^{k_2}_{n_2})) = \begin{cases} 0 & \text{if} \quad k_1 = k_2 \text{ and } n_1 = n_2 \\ 1 & \text{if} \quad k_1 = k_2 \text{ and } n_1 < n_2 \\ 2 & \text{if} \quad k_l = k_2 \text{ and } n_1 > n_2 \\ 3 & \text{if} \quad k_1 < k_2 \text{ and } n_1 = n_2 \\ 4 & \text{if} \quad k_1 < k_2 \text{ and } n_1 < n_2 \\ 5 & \text{if} \quad k_1 < k_2 \text{ and } n_1 > n_2 \\ 6 & \text{if} \quad k_l > k_2 \text{ and } n_1 = n_2 \\ 7 & \text{if} \quad k_1 > k_2 \text{ and } n_1 < n_2 \\ 8 & \text{if} \quad k_1 > k_2 \text{ and } n_1 > n_2. \end{cases}$$

Applying the partition relation to this f we find ω-covers $\mathcal{K}$ and $\mathcal{L}$, subsets of $\mathcal{V}$, such that f has no more than two values on $\mathcal{K} \times \mathcal{L}$. First, note that f cannot have all its values in $\{0, 1, 2\}$. For in this case there is a fixed k such that all elements of $\mathcal{K}$ and of $\mathcal{L}$ are subset of U^1_k, contradicting the fact that $\mathcal{K}$ and $\mathcal{L}$ are ω-covers.

By Theorem 9.3 choose ω-covers $\mathcal{A}$ and $\mathcal{B}$ such that $\mathcal{A} \cup \mathcal{B} \subset \mathcal{V}$ and $\mathcal{A} \cap \mathcal{B} = \emptyset$, and consider the function f restricted to $\mathcal{A} \times \mathcal{B}$. By the hypothesis choose ω-covers $\mathcal{K} \subset \mathcal{A}$ and $\mathcal{L} \subset \mathcal{B}$ such that f has an image set of at most two values on the set $\mathcal{K} \times \mathcal{L}$. First note that on $\mathcal{K} \times \mathcal{L}$ the function f does not have value 0 as $\mathcal{K}$ and $\mathcal{L}$ are disjoint. Secondly, by a prior remark, the range of f cannot be in $\{0, 1, 2\}$.

The rest of the argument consists of an extensive case analysis starting from a hypothesis like $range(f) \subseteq \{a, b\}$ where a and b are elements of $\{0, 1, \cdots, 8\}$. In all cases, except one, the conclusion is that either that case is not possible, or else that case produces a witness that $\mathsf{S}_1(\Omega, \Omega)$ holds for the given sequence $(\mathcal{U}_n : n \in \mathbb{N})$ of ω-covers of the space X. We leave the examination of these cases to the reader. The one exceptional case is the case when $range(f) = \{4, 8\}$:

Observe that in this case, for any $(U^1_{k_1} \cap U^{k_1}_{n_1},\ U^1_{k_2} \cap U^{k_2}_{n_2}) \in \mathcal{K} \times \mathcal{L}$ it is the case that either $k_1 < k_2$ and $n_1 < n_2$, or else $k_1 > k_2$ and $n_1 > n_2$. Thus, if we define $M = \{k : U^1_k \cap U^k_n \in \mathcal{K}\}$ and $N = \{k \in \mathbb{N} : U^1_{k_2} \cap U^{k_2}_{n_2} \in \mathcal{L}\}$, then M and N are disjoint. Moreover, both M and N are infinite. Enumerate N in increasing order as $(\ell_1,\ \ell_2,\ \cdots, \ell_m, \cdots)$. Moreover, for each k with a $U^1_k \cap U^k_n \in \mathcal{L}$, let $n(k)$ denote the least such n.

Now proceed to choose for each $\mathcal{U}_k$ a finite subset $\mathcal{F}_k$ as follows:

For $k \in M$ with. $k \leq \ell_1$ choose $U^k_n \in \mathcal{U}_k$ such that $n \leq n(\ell_1)$ and let $\mathcal{F}_k$ be the (finite) set of such chosen elements of $\mathcal{U}_k$. In general, for $k \in M$ such that

$\ell_j < k \leq \ell_{j+1}$ choose $U_n^k \in \mathcal{U}_k$ such that $n \leq n(\ell_{j+1})$, and let $\mathcal{F}_k$ be the (finite) set of such chosen elements of $\mathcal{U}_k$. For k not in M, let $\mathcal{F}_k$ consist of any single element of $\mathcal{U}_k$.

We claim that the sequence $(\mathcal{F}_k : k \in \mathbb{N})$ witnesses $\mathsf{S}_{fin}(\Omega, \Omega)$ for the given sequence of ω-covers of X. The reason is that for any $U_{k_1}^1 \cap U_{n_1}^{k_1} \in \mathcal{K}$, the set $U_{n_1}^{k_1}$ is an element of $\mathcal{F}_{k_1}$, and thus $\mathcal{K}$ is a refinement of $\bigcup\{\mathcal{F}_k : k \in \mathbb{N}\}$, and the latter is an ω-cover of X since $\mathcal{K}$ is. □

As suggested by the conjecture above, we expect that in fact $\mathsf{S}_1(\Omega, \Omega)$ is a consequence of the polarized square bracket relation, and most likely a different approach to a proof is needed.

10. Remarks

The results above are given for the family $\mathcal{A} = \Omega$ of ω-covers of certain topological spaces. A study of the proofs will reveal that several of these equivalences also hold for several other families $\mathcal{A}$. The main requirements on a family $\mathcal{A}$ are that each element of $\mathcal{A}$ has a countable subset in $\mathcal{A}$, that for each k $\mathcal{A} \to (\mathcal{A})_k^1$ holds, and that $\mathsf{S}_1(\mathcal{A}, \mathcal{A})$ is equivalent to ONE not having a winning strategy in $\mathsf{G}_1(\mathcal{A}, \mathcal{A})$, and that this is equivalent to the Ramseyan statement $\mathcal{A} \to (\mathcal{A})_2^2$.

As a non-topological example, consider the following: A collection $\mathcal{C}$ of subsets of a set S is said to be a *combinatorial ω-cover* of S if $S \notin \mathcal{C}$, but for each finite subset F of S there is a $C \in \mathcal{C}$ with $F \subseteq C$. For an infinite cardinal number κ let Ω_κ be the set of *countable* combinatorial ω-covers of κ. Let $\mathsf{cov}(\mathcal{M})$ be the least infinite cardinal number κ such that the real line is a union of κ first category sets. By the Baire Category Theorem $\mathsf{cov}(\mathcal{M})$ is uncountable. Using the techniques of this paper one can prove:

THEOREM 10.1. *For an infinite cardinal number κ the following are equivalent:*

(1) $\kappa < \mathsf{cov}(\mathcal{M})$.
(2) $\mathsf{S}_1(\Omega_\kappa, \Omega_\kappa)$.
(3) $\mathsf{E}(\Omega_\kappa, \Omega_\kappa)$.
(4) $\mathsf{GP}(\Omega_\kappa, \Omega_\kappa)$.
(5) $\mathsf{FG}(\Omega_\kappa, \Omega_\kappa)$.
(6) $\mathsf{NW}(\Omega_\kappa, \Omega_\kappa)$.
(7) *For all positive integers n and k, $\Omega_\kappa \to (\Omega_\kappa)_k^n$.*

Here are three more, among many, topological examples. For a topological space X and an element $x \in X$, define $\Omega_x = \{A \subset X \setminus \{x\} : x \in \overline{A}\}$. According to **[30]** X has strong countable fan tightness at x if the selection principle $\mathsf{S}_1(\Omega_x, \Omega_x)$ holds. Consider for a Tychonoff space X the subspace of the Tychonoff product $\Pi_{x \in X}\mathbb{R}$ consisting of the continuous functions from X to $\mathbb{R}$. The symbol $\mathsf{C}_p(X)$ denotes this subspace with the inherited topology. Since $\mathsf{C}_p(X)$ is homogeneous, the truth of $\mathsf{S}_1(\Omega_f, \Omega_f)$ at some point f implies the truth of $\mathsf{S}_1(\Omega_f, \Omega_f)$ at any point f. Thus we may confine attention to $\Omega_{\mathbf{o}}$, where $\mathbf{o}$ is the function which is zero on X. Using the techniques above one can prove:

THEOREM 10.2. *For a Tychonoff space X the following are equivalent for $\mathsf{C}_p(X)$:*

(1) $\mathsf{S}_1(\Omega_{\mathbf{o}}, \Omega_{\mathbf{o}})$.
(2) $\mathsf{E}(\Omega_{\mathbf{o}}, \Omega_{\mathbf{o}})$.

(3) $\mathsf{GP}(\Omega_{\mathbf{o}}, \Omega_{\mathbf{o}})$.
(4) $\mathsf{FG}(\Omega_{\mathbf{o}}, \Omega_{\mathbf{o}})$.
(5) $\mathsf{NW}(\Omega_{\mathbf{o}}, \Omega_{\mathbf{o}})$.
(6) *For all* n *and* k, $\Omega_{\mathbf{o}} \to (\Omega_{\mathbf{o}})^n_k$.

For a topological space X let $\mathcal{D}$ denote the collection whose members are of the form $\mathcal{U}$, a family of open subsets of X, such that no element of $\mathcal{U}$ is dense in X, but $\cup\mathcal{U}$ is dense in X. And let $\mathcal{D}_\Omega$ be the set of $\mathcal{U} \in \mathcal{D}$ such that for each finite family $\mathcal{F}$ of nonempty open subsets of X there is a $U \in \mathcal{U}$ with $U \cap F \neq \emptyset$ for each $F \in \mathcal{F}$. The families $\mathcal{D}$ and $\mathcal{D}_\Omega$ were considered in [**38**] where it was proved that for X a set of real numbers, and $\mathsf{PR}(X)$ the Pixley-Roy space over X, the following holds:

THEOREM 10.3. *If X is a set of real numbers, the following are equivalent for* $\mathsf{PR}(X)$*:*
(1) $\mathsf{S}_1(\mathcal{D}_\Omega, \mathcal{D}_\Omega)$.
(2) *ONE has no winning strategy in the game* $\mathsf{G}_1(\mathcal{D}_\Omega, \mathcal{D}_\Omega)$.
(3) *For each* n *and* k $\mathcal{D}_\Omega \to (\mathcal{D}_\Omega)^n_k$.
Each of these statements is equivalent to X *having* $\mathsf{S}_1(\Omega_X, \Omega_X)$.

Using the techniques above one can prove:

THEOREM 10.4. *For a set X of reals the following are equivalent for* $\mathsf{PR}(X)$*:*
(1) $\mathsf{S}_1(\mathcal{D}_\Omega, \mathcal{D}_\Omega)$.
(2) $\mathsf{E}(\mathcal{D}_\Omega, \mathcal{D}_\Omega)$.
(3) $\mathsf{GP}(\mathcal{D}_\Omega, \mathcal{D}_\Omega)$.
(4) $\mathsf{FG}(\mathcal{D}_\Omega, \mathcal{D}_\Omega)$.
(5) $\mathsf{NW}(\mathcal{D}_\Omega, \mathcal{D}_\Omega)$.
(6) *For all* n *and* k, $\mathcal{D}_\Omega \to (\mathcal{D}_\Omega)^n_k$.

For a non-compact topological space X call an open cover $\mathcal{U}$ a k-cover if there is for each compact $C \subset X$ a $U \in \mathcal{U}$ such that $C \subseteq U$, and if $X \notin \mathcal{U}$. Let $\mathcal{K}$ denote the collection of k-covers of such an X. If X is a separable metric space then each member of $\mathcal{K}$ has a countable subset which still is a member of $\mathcal{K}$. Using the techniques above one can prove:

THEOREM 10.5. *For separable metric spaces X the following are equivalent:*
(1) *ONE has no winning strategy in* $\mathsf{G}_1(\mathcal{K}, \mathcal{K})$.
(2) $\mathsf{S}_1(\mathcal{K}, \mathcal{K})$.
(3) $\mathsf{E}(\mathcal{K}, \mathcal{K})$.
(4) $\mathsf{GP}(\mathcal{K}, \mathcal{K})$.
(5) $\mathsf{FG}(\mathcal{K}, \mathcal{K})$.
(6) $\mathsf{NW}(\mathcal{K}, \mathcal{K})$.
(7) *For all* n *and* k, $\mathcal{K} \to (\mathcal{K})^n_k$.

The equivalence of (2) and (7) for n=2 and k=2 is Theorem 8 of [**9**]. The equivalence of (1) and (2) is a result of [**42**]. The remaining equivalences are then derived as was done above for Ω.

In the above examples we did not list all the equivalences proved in Theorem 7.2. Indeed, we did not check all of these. To illustrate that the remaining equivalences are worth checking, consider the following example: When $\mathcal{A}$ is a free ultrafilter on $\mathbb{N}$, the following theorem holds:

THEOREM 10.6. *For a non-trivial ultrafilter $\mathcal{U}$ on $\mathbb{N}$ the following statements are equivalent:*

(1) *Selection principle $\mathsf{S}_1(\mathcal{U},\mathcal{U})$ holds.*
(2) *ONE has no winning strategy in the game $\mathsf{G}_1(\mathcal{U},\mathcal{U})$.*
(3) *For all natural numbers m and k the partition relation $\mathcal{U} \to (\mathcal{U})^m_k$ holds.*

Ultrafilters $\mathcal{U}$ satisfying (1) of Theorem 10.6 have been called selective ultrafilters, while ones satisfying (3) have been called Ramsey ultrafilters. It is known that the existence of an ultrafilter as in Theorem 10.6 is independent of ZFC. The equivalence of statements (1) and (3) is due to Kunen, while the equivalence of (1) and (2) is due to Galvin and McKenzie. Note that these implications are analogous to the ones for Ω appearing in Theorem 7.2 among others. But not all equivalences in Theorem 7.2 also hold for ultrafilters $\mathcal{U}$ on $\mathbb{N}$. For example: In [**4**] Blass proved that under the Continuum Hypothesis there is an ultrafilter $\mathcal{U}$ on $\mathbb{N}$ such that $\mathcal{U} \to [\mathcal{U}]^2_3$, but $\mathcal{U}$ does not satisfy the selection principle $\mathsf{S}_1(\mathcal{U},\mathcal{U})$. Thus (8) $\Rightarrow$ (1) of Theorem 7.2 for Ω, does not hold for ultrafilters on $\mathbb{N}$.

11. Acknowledgement

The referee found numerous instances of lack of clarity. I thank the referee for pointing these out.

References

[1] A. V. Arkhangel′skiĭ, *Topological function spaces*, Mathematics and its Applications (Soviet Series), vol. 78, Kluwer Academic Publishers Group, Dordrecht, 1992. Translated from the Russian by R. A. M. Hoksbergen. MR1144519

[2] L. Babinkostova, Lj. D. R. Kočinac, and M. Scheepers, *Combinatorics of open covers. XI. Menger- and Rothberger-bounded groups*, Topology Appl. **154** (2007), no. 7, 1269–1280, DOI 10.1016/j.topol.2005.09.013. MR2310460

[3] J. E. Baumgartner and A. D. Taylor, *Partition theorems and ultrafilters*, Trans. Amer. Math. Soc. **241** (1978), 283–309, DOI 10.2307/1998845. MR491193

[4] A. Blass, *Ultrafilter mappings and their Dedekind cuts*, Trans. Amer. Math. Soc. **188** (1974), 327–340, DOI 10.2307/1996783. MR351822

[5] E. Borel, *Sur la classification des ensembles de mesure nulle* (French), Bull. Soc. Math. France **47** (1919), 97–125. MR1504785

[6] W. W. Comfort, *Compactness-like properties for generalized weak topological sums*, Pacific J. Math. **60** (1975), no. 1, 31–37. MR431088

[7] W. W. Comfort and K. A. Ross, *Pseudocompactness and uniform continuity in topological groups*, Pacific J. Math. **16** (1966), 483–496. MR207886

[8] H. H. Corson, *Normality in subsets of product spaces*, Amer. J. Math. **81** (1959), 785–796, DOI 10.2307/2372929. MR107222

[9] G. Di Maio, L. D. R. Kočinac, and E. Meccariello, *Applications of k-covers*, Acta Math. Sin. (Engl. Ser.) **22** (2006), no. 4, 1151–1160, DOI 10.1007/s10114-005-0753-8. MR2245246

[10] E. Ellentuck, *A new proof that analytic sets are Ramsey*, J. Symbolic Logic **39** (1974), 163–165, DOI 10.2307/2272356. MR349393

[11] P. Erdős, A. Hajnal, and R. Rado, *Partition relations for cardinal numbers*, Acta Math. Acad. Sci. Hungar. **16** (1965), 93–196, DOI 10.1007/BF01886396. MR202613

[12] P. Erdös and R. Rado, *A partition calculus in set theory*, Bull. Amer. Math. Soc. **62** (1956), 427–489, DOI 10.1090/S0002-9904-1956-10036-0. MR81864

[13] F. Galvin, J. Mycielski, and R. M. Solovay, *Strong measure zero and infinite games*, Arch. Math. Logic **56** (2017), no. 7-8, 725–732, DOI 10.1007/s00153-017-0541-z. MR3696064

[14] F. Galvin, *A generalization of Ramsey's Theorem*, Notices of the American Mathematical Society 15 (1968), p. 548. Abstract 68T-368.

[15] F. Galvin, *Indeterminacy of point-open games* (English, with Russian summary), Bull. Acad. Polon. Sci. Sér. Sci. Math. Astronom. Phys. **26** (1978), no. 5, 445–449. MR493925

[16] F. Galvin and K. Prikry, *Borel sets and Ramsey's theorem*, J. Symbolic Logic **38** (1973), 193–198, DOI 10.2307/2272055. MR337630

[17] J. Gerlits and Zs. Nagy, *Some properties of $C(X)$. I*, Topology Appl. **14** (1982), no. 2, 151–161, DOI 10.1016/0166-8641(82)90065-7. MR667661

[18] L. Gillman and M. Henriksen, *Concerning rings of continuous functions*, Trans. Amer. Math. Soc. **77** (1954), 340–362, DOI 10.2307/1990875. MR63646

[19] K. Gödel, *The Consistency of the Continuum Hypothesis*, Annals of Mathematics Studies, no. 3, Princeton University Press, Princeton, N. J., 1940. MR0002514

[20] W. Just, A. W. Miller, M. Scheepers, and P. J. Szeptycki, *The combinatorics of open covers. II*, Topology Appl. **73** (1996), no. 3, 241–266, DOI 10.1016/S0166-8641(96)00075-2. MR1419798

[21] J. L. Kelley, *General topology*, D. Van Nostrand Company, Inc., Toronto-New York-London, 1955. MR0070144

[22] R. Laver, *On the consistency of Borel's conjecture*, Acta Math. **137** (1976), no. 3-4, 151–169, DOI 10.1007/BF02392416. MR422027

[23] N.N. Lusin, *Sur un problème de M. Baire*, Comptes Rendus de l'Académie des Sciences Paris 158 (1914), 1258?1261

[24] L. Nachbin, *Sur les espaces uniformes ordonnés* (French), C. R. Acad. Sci. Paris **226** (1948), 774–775. MR24120

[25] J. Pawlikowski, *Undetermined sets of point-open games*, Fund. Math. **144** (1994), no. 3, 279–285. MR1279482

[26] R. Rado, *Direct decomposition of partitions*, J. London Math. Soc. **29** (1954), 71–83, DOI 10.1112/jlms/s1-29.1.71. MR65616

[27] F. P. Ramsey, *On a Problem of Formal Logic*, Proc. London Math. Soc. (2) **30** (1929), no. 4, 264–286, DOI 10.1112/plms/s2-30.1.264. MR1576401

[28] F. Rothberger, *Eine Verschärfung der Eigenschaft* C, **Fundamenta Mathematicae** 30 (1938), 50 - 55.

[29] F. Rothberger, *Sur les familles indénombrables de suites de nombres naturels et les problèmes concernant la propriété C* (French), Proc. Cambridge Philos. Soc. **37** (1941), 109–126. MR4281

[30] M. Sakai, *Property C'' and function spaces*, Proc. Amer. Math. Soc. **104** (1988), no. 3, 917–919, DOI 10.2307/2046816. MR964873

[31] M. Scheepers, *Combinatorics of open covers. I. Ramsey theory*, Topology Appl. **69** (1996), no. 1, 31–62, DOI 10.1016/0166-8641(95)00067-4. MR1378387

[32] M. Scheepers, *Combinatorics of open covers. III. Games,* $\mathsf{C}_\mathrm{p}(X)$, Fund. Math. **152** (1997), no. 3, 231–254. MR1444715

[33] M. Scheepers, *The least cardinal for which the Baire category theorem fails*, Proc. Amer. Math. Soc. **125** (1997), no. 2, 579–585, DOI 10.1090/S0002-9939-97-03597-1. MR1350961

[34] M. Scheepers, *Rothberger's property and partition relations*, J. Symbolic Logic **62** (1997), no. 3, 976–980, DOI 10.2307/2275582. MR1472133

[35] M. Scheepers, *Open covers and the square bracket partition relation*, Proc. Amer. Math. Soc. **125** (1997), no. 9, 2719–2724, DOI 10.1090/S0002-9939-97-03898-7. MR1396995

[36] M. Scheepers, *Finite powers of strong measure zero sets*, J. Symbolic Logic **64** (1999), no. 3, 1295–1306, DOI 10.2307/2586631. MR1779763

[37] M. Scheepers, *Open covers and partition relations*, Proc. Amer. Math. Soc. **127** (1999), no. 2, 577–581, DOI 10.1090/S0002-9939-99-04513-X. MR1458262

[38] M. Scheepers, *Combinatorics of open covers. V. Pixley-Roy spaces of sets of reals, and ω-covers*, Topology Appl. **102** (2000), no. 1, 13–31, DOI 10.1016/S0166-8641(98)00141-2. MR1739261

[39] M. Scheepers, *Rothberger's property in all finite powers*, Topology Appl. **156** (2008), no. 1, 93–103, DOI 10.1016/j.topol.2007.10.011. MR2463828

[40] M. Scheepers, *Rothberger bounded groups and Ramsey theory*, Topology Appl. **158** (2011), no. 13, 1575–1583, DOI 10.1016/j.topol.2011.05.025. MR2812467

[41] W. Sierpiński, *Sur un ensemble non dénombrable, donc toute image continue est de mesure nulle*, **Fundamenta Mathematicae** 11 (1928), 301-304.

[42] N. Samet and B. Tsaban, *Ramsey theory of open covers: Lecture 3*, preprint.

[43] B. Tsaban, *Algebra, selections, and additive Ramsey theory*, Fund. Math. **240** (2018), no. 1, 81–104, DOI 10.4064/fm342-1-2017. MR3720919

[44] A. Weil, *Sur les Espaces à Structure Uniforme et sur la Topologie Générale*, Herrman & Cie 1937
[45] S. Willard, *General topology*, Addison-Wesley Publishing Co., Reading, Mass.-London-Don Mills, Ont., 1970. MR0264581

Department of Mathematics, Boise State University, Boise, Idaho 83725
Email address: mscheepe@boisestate.edu

Contemporary Mathematics
Volume **755**, 2020
https://doi.org/10.1090/conm/755/15185

On the algebraic union of strongly measure zero sets and their relatives with sets of real numbers

Tomasz Weiss

ABSTRACT. In this article we do the following:
(1) We present classical results on the algebraic union of strongly measure zero and strongly meager sets (denoted by $\mathcal{SN}$ respectively, by $\mathcal{SM}$) with sets of real numbers, which are small in usual senses of smallness.
(2) We characterize natural subclasses of $\mathcal{SN}$ and $\mathcal{SM}$ sets, and we investigate their algebraic unions with sets of real numbers. We indicate also a possible measure counterpart to the family of perfectly meager sets.
(3) We determine Ramseyan properties of sets that appeared earlier in Section 1 and 2.

1. Introduction

In this paper, we are concerned with subsets of the Cantor space 2^ω equipped with the standard metric, topology and measure and with the coordinatewise modulo 2 addition denoted by $+$. However, most of the results presented in Section 1, 2 and in the first part of Section 3 remain true in other σ-compact metric spaces like R with the usual addition and in $[0,1)$ interval with the modulo 1 addition. We apply the following notation and terminology. For $s \in 2^{<\omega}$ $[s] = \{x \in 2^\omega : x \restriction \text{length}(s) = s\}$ is called a basic open (or clopen) set in 2^ω. If A is a set, then $[A]^{<\omega}$ denotes the collection of all finite subsets of A, and $[A]^\omega$ is the set all countable subsets of A. $\omega^{\omega\uparrow}$ is the family of all strictly increasing functions that belong to ω^ω. $\mathbb{O}$ (respectively, $\mathbb{1}$) are reserved for a constant function, which equals 0 (respectively, 1) on 2^ω, and Q is the set of all sequences in 2^ω, which are eventually equal to zero. We write $\exists_n^\infty$ for $\forall_m \exists_{n\geq m}$ and $\forall_n^\infty$ for $\exists_m \forall_{n\geq m}$. In the article, $\mathcal{M}$ stands for the σ-ideal of meager subsets of 2^ω, $\mathcal{N}$ is the σ-ideal of measure zero sets in 2^ω, and by $\mathcal{E}$ we denote the σ-ideal generated by F_σ measure zero subsets of 2^ω. We identify $\mathcal{M} \cap \mathcal{N}$ with the σ-ideal of sets that are meager and of measure zero. For $X, Y \subseteq 2^\omega$

$$X + Y = \{x + y : x \in X,\ y \in Y\}.$$

We begin with the classical definition of strongly measure zero sets, which is due to Émile Borel (see [**Bor**]).

2010 *Mathematics Subject Classification.* Primary 03E05, 03E75, 28A05.

Key words and phrases. strongly null sets, strongly meager sets, Luzin sets, Sierpiński sets, algebraic union, Cantor space.

DEFINITION 1.1 (Borel). An $X \subseteq 2^\omega$ is said to be *strongly measure zero*, or *strongly null* if for any sequence $\{\varepsilon_n\}_{n\in\omega}$ of real numbers with $\varepsilon_n > 0$ for $n \in \omega$ one can find a sequence of sets $\{X_n\}_{n\in\omega}$ such that each $\operatorname{diam}(X_n) \leq \varepsilon_n$, and $X \subseteq \bigcup_{n\in\omega} X_n$.

By $\mathcal{SN}$ we denote the class of all strongly measure zero sets in 2^ω.

Behind many of our considerations is the following important theorem (see [**GMS**]).

THEOREM 1.2 (Galvin, Mycielski, Solovay). *A set $X \subseteq 2^\omega$ is strongly measure zero if and only if for every meager set $A \subseteq 2^\omega$ $X + A \neq 2^\omega$.*

PROOF (SKETCH). For the easy direction, which is due to K. Prikry, let $\{\varepsilon_n\}_{n\in\omega}$ be a given sequence with each $\varepsilon_n > 0$. Suppose also that a dense set G is equal to a union of basic open sets $\{I_n\}_{n\in\omega}$ each having diameter ε_n. If for some $x_0 \in 2^\omega$ $X \subseteq G + x_0$, then $X \subseteq \bigcup_{n\in\omega}(I_n + x_0)$.

The other implication is more difficult, and can be proven in the following way. Suppose that F is a closed nowhere dense set, and I is a basic open set with $\operatorname{diam}(I) > 0$. By the continuity of the $+$ operation and by nowhere density of F there is for every $x \in 2^\omega$ a basic open set U_x with $x \in U_x$ such that $U_x + F$ misses some basic open set contained in I. Thus we can apply a standard compactness argument to show that there are $\varepsilon > 0$, $\delta > 0$ such that for any basic open set J with $\operatorname{diam}(J) \leq \varepsilon$ $J + F$ leaves uncovered some basic open set $I_0 \subseteq I$ with $\operatorname{diam}(I_0) \geq \delta$. Using this fact infinitely many times for a fixed set $\bigcup_{n\in\omega} F_n$ with F_n closed nowhere dense, and $F_n \subseteq F_{n+1}$ for $n \in \omega$ we obtain two sequences $\{\varepsilon_n\}_{n\in\omega}$, $\{\delta_n\}_{n\in\omega}$ with each $\varepsilon_n > 0$ and $\delta_n > 0$ such that: if

$$X \subseteq \bigcap_{n\geq m} \bigcup_{m\in\omega} J_n = \{x \in 2^\omega : \exists_n^\infty \; x \in J_n\},$$

where $\operatorname{diam}(J_n) \leq \varepsilon_n$ for $n \in \omega$, and $A \subseteq \bigcup_{n\in\omega} F_n$, then $X + A$ is disjoint with $\{x_0\}$, where $x_0 \in \bigcap_{n\in\omega} I_n$, and $\{I_n\}_{n\in\omega}$ is a descending sequence of basic open sets satisfying $\operatorname{diam}(I_n) \geq \delta_n$, $(J_n + F_n) \cap I_n = \emptyset$ for $n \in \omega$. $\square$

REMARK. In [**K**] and [**F**], the authors went even further to show that Theorem 1.2 holds in σ-compact spaces. A separate subject of Galvin–Mycielski–Solovay theorem and of some stronger notions than $\mathcal{SN}$ in an arbitrary Polish group is investigated thoroughly in recent articles by M. Hrušák and O. Zindulka (see [**HZ**] and [**Z1**]).

THEOREM 1.3 (folklore). If X is a strongly measure zero set, and F is an F_σ measure zero set, then $X + F$ is measure zero.

PROOF. Since F is F_σ and of measure zero there are (see [**BJ**, Lemma 2.6.3]) $f \in \omega^{\omega\uparrow}$ and a sequence $\{A_n\}_{n\in\omega}$, with each $A_n \subseteq 2^{[f(n),f(n+1))}$, and

$$\sum_{n\in\omega} \frac{|A_n|}{2^{f(n+1)-f(n)}} < +\infty$$

such that

$$F \subseteq \{x \in 2^\omega : \forall_n^\infty \; x \restriction [f(n), f(n+1)) \in A_n\}.$$

As $X \in \mathcal{SN}$, there exists $g \in 2^\omega$, so that

$$X \subseteq \{x \in 2^\omega : \exists_n^\infty \; x \restriction [f(n), f(n+1))\} = g \restriction [f(n), f(n+1))\}.$$

Then

$$X + F \subseteq \{x \in 2^\omega : \exists_n^\infty \; x \restriction [f(n), f(n+1)) \in g \restriction [f(n), f(n+1)) + A_n\},$$

and the latter set is clearly null. □

The reverse fact, due to J. Pawlikowski, follows immediately from Galvin–Mycielski–Solovay theorem and the crucial lemma below, which is much harder to prove.

LEMMA 1.4 (Pawlikowski). *For every comeager set $G \subseteq 2^\omega$ there exists a closed null set $F \subseteq 2^\omega$ such that for every $X \subseteq 2^\omega$ with $F + X$ being null there is $t \in 2^\omega$, so that $X + t \subseteq G$.*

PROOF. See [**P**, Lemma 1]. □

Thus we obtain the following characterization of $\mathcal{SN}$ sets.

THEOREM 1.5. *X is a strongly measure zero set if and only if for every closed measure zero set F $X + F$ is null.*

PROOF. See [**P**, Theorem, page 172]. □

Next we discuss a dual notion to $\mathcal{SN}$ sets.

DEFINITION 1.6. We shall say that $X \subseteq 2^\omega$ is *strongly meager* if for every measure zero set $A \subseteq 2^\omega$ $X + A \neq 2^\omega$.

$\mathcal{SM}$ stands for the class of all strongly meager subsets of 2^ω.

Contrary to $\mathcal{SN}$ sets, which form a σ-ideal, strongly meager sets behave badly, and it is consistent with ZFC that the class $\mathcal{SM}$ does not even form an ideal (see [**BS**, Theorem 1.3]).

Also, by uniform continuity of a continuous function $f : 2^\omega \to 2^\omega$ we have that for $X \in \mathcal{SN}$ the image $f(X)$ is strongly null as well. This fact does not hold for members of $\mathcal{SM}$, namely under CH one can construct a continuous function $f : 2^\omega \to 2^\omega$ and $X \in \mathcal{SM}$ such that $f(X) = 2^\omega$ (see Theorem 2 in [**BNW**]).

REMARK. Notice that if $X \in \mathcal{SN}$, then for every $A \in \mathcal{M}$ the set $\{t : t \notin X + A\}$ is not meager. To prove this suppose that the opposite holds for some $X \in \mathcal{SN}$ and $A \in \mathcal{M}$, that is $\{t : t \notin X + A\} = B \in \mathcal{M}$. Then $X + (A \cup B) = (X + A) \cup (X + B) = 2^\omega$, as we can assume without loss of generality that $\mathbb{O} \in X$. Thus we obtain a contradiction with the fact that $X \in \mathcal{SN}$. A similar argument shows that if $X \in \mathcal{SM}$, then for each $A \in \mathcal{N}$ the set $\{t : t \notin X + A\}$ is not null.

THEOREM 1.7 (Recław, unpublished). *If $X \subseteq 2^\omega$ is strongly meager, then for every F_σ null set F $X + F$ is meager.*

PROOF. Let F be an F_σ measure zero set. Then by the preceding considerations there are $f \in \omega^{\omega\uparrow}$ and a sequence $\{A_n\}_{n\in\omega}$ with $A_n \subseteq 2^{[f(n), f(n+1))}$,

$$\frac{|A_n|}{2^{f(n+1)-f(n)}} \leq \frac{1}{2^n} \quad \text{for } n \in \omega$$

such that

$$F \subseteq \{x \in 2^\omega : \forall_n^\infty \ x \restriction [f(n), f(n+1)) \in A_n\}.$$

Since X is strongly meager

$$t \notin X + \{x \in 2^\omega : \exists_n^\infty \ x \restriction [f(n), f(n+1)) \in A_n\} \qquad \text{for some } t \in 2^\omega.$$

As

$$\{x \in 2^\omega : \forall_n^\infty \ x \restriction [f(n), f(n+1)) \in A_n\} + \{x \in 2^\omega : \exists_n^\infty \ x \restriction [f(n), f(n+1)) \equiv \mathbb{O}\} \\ = \{x \in 2^\omega : \exists_n^\infty \ x \restriction [f(n), f(n+1)) \in A_n\},$$

we obtain that $X + F$ is disjoint with

$$t + \{x : \exists_n^\infty \ x \restriction [f(n), f(n+1)) \equiv \mathbb{O}\}.$$

Thus $X + F$ is meager. □

Next we consider sets that have "small" cardinality on every set from $\mathcal{M}$ (respectively, $\mathcal{N}$). Assume that an uncountable cardinal κ is not bigger than $\mathfrak{c}$.

DEFINITION 1.8. A set $X \subseteq 2^\omega$ is called a *κ-Luzin set* if $|X| \geq \kappa$ and for every $A \in \mathcal{M}$ $|X \cap A| < \kappa$. An $X \subseteq 2^\omega$ is said to be a *κ-Sierpiński set* if $|X| \geq \kappa$ and for each $B \in \mathcal{N}$ $|X \cap B| < \kappa$.

To construct κ-Luzin or κ-Sierpiński set we use transfinite induction and some special axioms of set theory like CH (the continuum hypothesis) or certain assumptions on the cardinality of $\mathrm{cov}(\mathcal{M})$, $\mathrm{cov}(\mathcal{N})$ and $\mathrm{cof}(\mathcal{M})$, $\mathrm{cof}(\mathcal{N})$. A classical argument for the construction of $\aleph_1$-Luzin and $\aleph_1$-Sierpiński set can be found in [**Bu**, Theorem 8.26] and in [**M**, Theorem 2.1 and 2.2].

LEMMA 1.9. *If there exists a κ-Luzin set, then every set of cardinality smaller than $cf(\kappa)$ is strongly measure zero, and dually, the existence of a κ-Sierpiński set implies that every set of cardinality smaller than $cf(\kappa)$ is strongly meager.*

PROOF. See [**Bu**, Lemma 8.34]. □

The $\aleph_1$ version of the following theorem is due to W. Sierpiński (see [**Sier1**]).

THEOREM 1.10. *For a regular cardinal κ every κ-Luzin set is strongly measure zero.*

PROOF. Suppose that X is κ-Luzin, and $\{\varepsilon_n\}_{n\in\omega}$ is a given sequence of positive real numbers. We apply an argument, which is quite standard in theory of $\mathcal{SN}$ sets. Let $\{q_n\}_{n\in\omega}$ be a bijective enumeration of Q. Consider an open dense set $A = \bigcup_{n\in\omega} A_n$, where each A_n is a basic open set with $\mathrm{diam}(A_n) \leq \varepsilon_{2n}$ and containing q_n. By Lemma 1.9, $X \setminus A$, as a strongly measure zero set, can be covered by a sequence of sets with diameters not bigger than ε_{2n+1} for $n \in \omega$. Thus $X = (X \setminus A) \cup (X \cap A)$ belongs to $\mathcal{SN}$. □

The proof of the measure analogue for $\kappa = \aleph_1$ is highly notrivial.

THEOREM 1.11 (Pawlikowski). *Every $\aleph_1$-Sierpiński set is strongly meager.*

PROOF. See [**BJ**, Theorem 8.5.13]. □

Since $\aleph_1$-Luzin set is not meager, and $\aleph_1$-Sierpiński set can not be null, we conclude that it is consistent with ZFC that $\mathcal{SN}$ and $\mathcal{SM}$ families contain sets, which are quite big.

2

In this section we introduce other dramatis personae that are included in $\mathcal{SN}$ and $\mathcal{SM}$ families.

DEFINITION 2.1. A set $X \subseteq 2^\omega$ is called *meager additive* if for every meager set $A \subseteq 2^\omega$ $X + A$ is meager. An $X \subseteq 2^\omega$ is *null additive* if for each null set $A \subseteq 2^\omega$ $X + A$ is null.

We denote the class of all meager additive sets in by $\mathcal{M}^*$ and the family of all null additive sets by $\mathcal{N}^*$.

Analogously, we say that $X \subseteq 2^\omega$ is a member of $\mathcal{E}^*$ if for every $A \in \mathcal{E}$ $X + A$ belongs to $\mathcal{E}$.

DEFINITION 2.2. By $\mathfrak{b}$ we denote the minimum cardinality of a set $X \subseteq \omega^\omega$ such that for every $f \in \omega^\omega$ there is $g \in X$ satisfying $\exists_n^\infty\ f(n) < g(n)$. $\mathfrak{d}$ stands for the minimum cardinality of a set $X \subseteq \omega^\omega$ such that for each $f \in \omega^\omega$ there exists $g \in X$ with $\forall_n^\infty\ f(n) < g(n)$.

In [**B**], the author proves (using a method of S. Todorčević) the following statement.

THEOREM 2.3. *Assume that $\mathfrak{b} = \aleph_1$. Then there exists $X \subseteq 2^\omega$, $|X| = \aleph_1$, and $X \in \mathcal{M}^*$. Suppose that $\mathfrak{d} = \aleph_1$. Then there is $X \subseteq 2^\omega$, $|X| = \aleph_1$, and $X \in \mathcal{N}^*$.*

PROOF. See Theorem 2 in [**B**]. □

The following combinatorial property characterizes meager additive sets in 2^ω.

THEOREM 2.4 (Bartoszyński, Judah, Shelah). *For $X \subseteq 2^\omega$ the following are equivalent.*

(1) $X \in \mathcal{M}^*$,
(2) *for every $f \in \omega^{\omega\uparrow}$ there exist $g \in \omega^{\omega\uparrow}$ and $y \in 2^\omega$ such that*

$$\forall x \in X\ \forall_n^\infty\ \exists k\,(g(n) \le f(k) < f(k+1) < g(n+1),$$
$$\text{and } x \restriction [f(k), f(k+1))) = y \restriction [f(k), f(k+1)).$$

PROOF. See [**BJ**, Theorem 2.7.17]. □

A recent work by O. Zindulka (see [**Z1**]) led to the following interesting result.

THEOREM 2.5 (Zindulka). *$X \in \mathcal{E}^*$ if and only if $X \in \mathcal{M}^*$.*

PROOF (SKETCH). The implication $\Leftarrow$ is quite easy, and can be derived from the following fact. Suppose that

$$A = \{x \in 2^\omega :\ \exists_{n, m_1 \le n < m_2} x \restriction [f(n), f(n+1)) = y \restriction [f(n), f(n+1))\},$$

where $m_1, m_2,\ n \in \omega$, $f \in \omega^{\omega\uparrow}$, $y \in 2^{[f(m_1), f(m_2))}$, and

$$B = \{x \in 2^\omega :\ \forall_{n, m_1 \le n < m_2} x \restriction [f(n), f(n+1)) \in A_n\}$$

with $A_n \subseteq 2^{[f(n), f(n+1))}$. Then

$$A + B \subseteq \{x \in 2^\omega :\ \exists_{n, m_1 \le n < m_2} x \restriction [f(n), f(n+1)) \in y \restriction [f(n), f(n+1)) + A_n\}.$$

Thus if $f \in \omega^{\omega\uparrow}$ and $\{A_n\}_{n\in\omega}$ are as in the characterization of a set F in $\mathcal{E}$ (see Theorem 1.7 above), then

$$\mu(A + B) \le \frac{1}{2^{m_1}} + \frac{1}{2^{m_1+1}} + \ldots + \frac{1}{2^{m_2-1}}.$$

This implies that if X satisfies the above characterization of a meager additive set, and $Y \in \mathcal{E}$, then $X + Y$ is an F_σ measure zero set. The other implication is a consequence of the following neat observation of O. Zindulka (see Proposition 5.7 in [**Z1**]). Assume that $X + F \in \mathcal{N}$, where $F \in \mathcal{E}$. Then there exists X', an F_σ set containing X such that $X' + F \in \mathcal{N}$. So let $X \in \mathcal{E}^*$ and $A \in \mathcal{M}$. By Lemma 1.4 of Pawlikowski for a given comeager $G = 2^\omega \setminus (A + Q)$ there is $F \in \mathcal{E}$, so that the condition $X + F \in \mathcal{N}$ implies $X' + A \neq t + Q$. Since $X' + A$ is an F_σ set, we have that $X + A$ is disjoint with a comeager set. □

We recall subsets of 2^ω, which are defined in terms of topological selections, and we connect them with subfamilies of $\mathcal{SN}$ and $\mathcal{SM}$ sets. More information on these sets can be found in [**JMSSz**].

DEFINITION 2.6. A set $X \subseteq 2^\omega$ has *the Hurewicz property* if there is for each sequence $\{\mathcal{U}_n\}_{n\in\omega}$ of open covers of a X a sequence $\{\mathcal{V}_n\}_{n\in\omega}$ such that every $\mathcal{V}_n \in [\mathcal{U}_n]^{<\omega}$ and each $x \in X$ is in all but finitely many of the sets $\bigcup \mathcal{V}_n$. $X \subseteq 2^\omega$ has *the Rothberger property* if for each sequence $\{\mathcal{U}_n\}_{n\in\omega}$ of open covers of X there is a sequence $\{U_n\}_{n\in\omega}$ with each $U_n \in \mathcal{U}_n$, so that $X \subseteq \bigcup_{n\in\omega} U_n$.

THEOREM 2.7 (folklore). *Every $\aleph_1$-Luzin set has the Rothberger property.*

PROOF. Let X be an $\aleph_1$-Luzin set, and suppose that $D = \{d_n\}_{n\in\omega}$ is a dense set in X. Then for any open set G containing D we have that $X \setminus G$ is countable. Suppose that $\{\mathcal{U}_n\}_{n\in\omega}$ is a sequence of open covers of X. Choose $V_n \in \mathcal{U}_{2n}$ for $n \in \omega$ such that $d_n \in V_n$. Next pick a sequence $\{\widetilde{V}_n\}_{n\in\omega}$ with $\widetilde{V}_n \in \mathcal{U}_{2n+1}$ for $n \in \omega$ to cover $X' = X \setminus \bigcup_{n\in\omega} V_n$. □

THEOREM 2.8 (Fremlin, Miller). *Every $\aleph_1$-Sierpiński set has the Hurewicz property.*

PROOF. We adopt the argument from the proof of Theorem 7 in [**MF**]. Let X be an $\aleph_1$-Sierpiński set. Consider $S = X + Q$. Then $\mu^*(S) = 1$, where μ^* denotes the outer measure. For $n \in \omega$ let $\mathcal{U}_n$ be an open cover of S. Choose $U_n \in [\mathcal{U}_{2n}]^{<\omega}$ for $n \in \omega$ such that $\mu(\bigcup U_n) \geq 1 - \frac{1}{2^n}$. Then by the fact that the second set in the below formula is of measure zero

$$S \cap \left(2^\omega \setminus \bigcup_{m\in\omega} \bigcap_{n\geq m} \bigcup U_n \right) = S'$$

is a countable set. Now it is easy to pick $V_n \in [\mathcal{U}_{2n+1}]^{<\omega}$ for $n \in \omega$, so that

$$S' \subseteq \bigcup_{m\in\omega} \bigcap_{n\geq m} \bigcup V_n. \qquad \square$$

The following notion, due to Gerlits and Nagy, is less common (see [**GN**]).

DEFINITION 2.9. A set $X \subseteq 2^\omega$ is *add($\mathcal{M}$)-small* if there is for every sequence $\{\mathcal{U}_n\}_{n\in\omega}$ of open covers of X a sequence $\{U_n\}_{n\in\omega}$ and $f \in \omega^{\omega\uparrow}$ such that

(1) for every $n \in \omega$ $U_n \in \mathcal{U}_n$,
(2) each $x \in X$ is in all but finitely many sets of the form

$$\bigcup_{f(n)\leq i<f(n+1)} U_i.$$

THEOREM 2.10 (Nowik, Scheepers, Weiss). *For $X \subseteq 2^\omega$ the following three conditions are equivalent.*

(1) *$X \in \mathcal{SN}$ and has the Hurewicz property,*
(2) *X has both the Rothberger and the Hurewicz property,*
(3) *X is $add(\mathcal{M})$-small.*

PROOF. See [**NSW**, Theorem 14]. □

THEOREM 2.11 (Scheepers, Weiss). *Assume that $X \in \mathcal{SN}$ and has the Hurewicz property. Then for every $Y \in \mathcal{SN}$ $X+Y$ is strongly measure zero.*

PROOF (SKETCH). By Theorem 2.10 X is $add(\mathcal{M})$-small. Using Theorem 2.4 it is quite easy to see that $add(\mathcal{M})$-smallness of X implies that $X \in \mathcal{M}^*$. Assume that $A \in \mathcal{M}$. Then $X+Y+A = X+A+Y$, and since $X+A$ is meager, $X+A+Y \neq 2^\omega$. □

We call a family $\mathcal{F}$ of subsets of $X \subseteq 2^\omega$ an *ω-cover* of X if every finite subset of X can be covered by a set from $\mathcal{F}$. Next we present the family of the smallest sets in the hierarchy of subsets of 2^ω defined in terms of topological selections (see [**GN**]).

DEFINITION 2.12 (Gerlits, Nagy). We shall say that $X \subseteq 2^\omega$ is a *γ-set* if for every open ω-cover $\mathcal{F}$ of X one can choose a sequence $\{D_n\}_{n\in\omega} \in \mathcal{F}$ such that each $x \in X$ is in all but finitely many of the sets D_n.

It is easy to prove that every γ-set has both the Rothberger and the Hurewicz property. Thus we obtain the following theorem.

THEOREM 2.13 (Galvin, Miller). *Every γ-set is in $\mathcal{M}^*$.*

PROOF. See [**GM**, Theorem 6]. □

It was tempting to conjecture that all γ-sets belong to $\mathcal{N}^*$. It turns out that this conjecture is consistently false (see [**BR**]).

THEOREM 2.14 (Bartoszyński, Recław). *Assume CH. Then there are a γ-set X and $G \in \mathcal{N}$ such that $X+G$ is not null.*

PROOF (SKETCH). Let $\{k_n\}_{n\in\omega}$ be an increasing sequence of natural numbers such that the set
$$G = \{x \in 2^\omega : \exists_n^\infty \; x \restriction [k_n, k_{n+1}) \equiv \mathbb{O}\}$$
has measure zero, and let $\{y_\alpha\}_{\alpha<\mathfrak{c}}$ be a bijective enumeration of $G^c = 2^\omega \setminus G$. We construct by induction on α a γ-set $X = Q \cup X'$ as follows (see Lemma 2.2 in [**BR**]). Suppose that for $\alpha < \lambda < \mathfrak{c}$ is already given, so that
$$\forall_{\alpha<\lambda} \, \exists_n^\infty \; x_\alpha \restriction [k_n, k_{n+1}) \equiv \mathbb{1},$$
$$\forall_{\alpha<\lambda} \, \forall_n^\infty \; x_{\alpha+1}(n) \le x_\alpha(n),$$
and
$$\forall_{\kappa<\lambda}, \; \kappa \in \mathrm{Lim}, \; \forall_{\alpha<\kappa} \, \forall_n^\infty \; x_\kappa(n) \le x_\alpha(n).$$
By applying CH we find x_λ such that
$$\exists_n^\infty \; x_\lambda \restriction [k_n, k_{n+1}) \equiv \mathbb{1},$$
and
$$\forall_{\alpha<\lambda} \, \forall_n^\infty \; x_\lambda(n) \le x_\alpha(n).$$

We can choose x_λ, so that

$$\exists_n^\infty\ x_\lambda \upharpoonright [k_n, k_{n+1}) = y_\lambda \upharpoonright [k_n, k_{n+1}).$$

It is clear that $G^c \subseteq X + G$. □

Recall that for $m \in \omega$, $T \subseteq 2^m$ $\mu(T) = \frac{|T|}{2^m}$, and $T_0, \ldots, T_k \subseteq 2^m$ are said to be stochastically independent if for every $j \leq k$

$$\mu\left(\bigcap_{i \leq j} T_i\right) = \mu(T_0) \cdot \ldots \cdot \mu(T_j).$$

LEMMA 2.15 (Pawlikowski). *Let $m \geq n + 2^n k$, $k, m, n \in \omega$. Then there exists $T \subseteq 2^m$ with measure $\mu(T) = 2^{-k}$ such that for all $\langle \sigma_i, \tau_i \rangle \in 2^n \times 2^{[n,m)}$, $i \in I$ with σ_i distinct the sets $T + \langle \sigma_i, \tau_i \rangle$ are stochastically independent.*

PROOF. See the proof of Lemma 0 in [**P**]. □

THEOREM 2.16 (Shelah). *If $X \in \mathcal{N}^*$, then for every sufficiently fast increasing function $f \in \omega^{\omega\uparrow}$, there is a sequence $\{A_n\}_{n\in\omega}$, $A_n \subseteq 2^{[f(n), f(n+1))}$, $|A_n| \leq 2^n$ for every $n \in \omega$ such that*

$$X \subseteq \{x \in 2^\omega :\ \forall_n^\infty\ x \upharpoonright [f(n), f(n+1)) \in A_n\}.$$

PROOF (SKETCH). Our technique is the same as in the proof of (♠) from [**P**]. Let $\{a_n\}_{n\in\omega}$, $\{b_n\}_{n\in\omega}$ be two increasing sequences of integers with $a_n < b_n < a_{n+1}$ for $n \in \omega$. Each number $b_i - a_i$ and a_{i+1} play the role of n and m, respectively, from Lemma 2.15. Each set T_i with $\mu(T_i) = 1/2^i$ and appearing below plays the role of a set T from Lemma 2.15. Put

$$A = \{x \in 2^\omega :\ \exists_n^\infty\ x \upharpoonright [a_n, a_{n+1}) \in T_n\}.$$

Suppose that $X \in \mathcal{N}^*$. Then $X + A$ is of measure zero, thus there exists an open set G, $\mu(G) < 1$ such that $X + A \subseteq G$. We can assume that for every $\tau \in 2^{<\omega}$, if $[\tau] \nsubseteq G$, then $\mu(\tau \cap (2^\omega \setminus G)) > 0$. By Baire's Category Theorem for every $x \in X$ there are $n_x \in \omega$ and τ_x, $[\tau_x] \nsubseteq G$, so that

$$\left(\bigcup_{n \geq n_x} (x + T_n)\right) \cap ([\tau_x] \setminus G) = \emptyset.$$

Define for every $n \in \omega$ and $[\tau] \nsubseteq G$

$$K_n^\tau = \{x \upharpoonright [a_n, b_n) :\ x \in X, \text{ and } (x + T_n) \cap [\tau] = \emptyset\}.$$

We have that for every $x \in X$, $x \upharpoonright [a_n, b_n) \in K_n^{\tau_x}$, where $n \geq n_x$. If we let $\{x_{k,n}^\tau :\ k < |K_n^\tau|\}$ be a list of all x's such that $x \upharpoonright [a_n, b_n)$ are distinct and give the whole set K_n^τ, then

$$[\tau] \setminus G \subseteq \bigcap_{n\in\omega} \left(2^\omega \setminus \bigcup_{k < |K_n^\tau|} (x_{k,n}^\tau + T_n)\right).$$

Then by the stochastic independence condition from Lemma 2.15 we obtain that

$$\prod_{n\in\omega} \left(1 - \frac{1}{2^n}\right)^{|K_n^\tau|} > 0,$$

and by Euler's formula this implies that

$$\sum_{n\in\omega}\frac{|K_n^\tau|}{2^n} < +\infty.$$

This gives us the required cardinality of the set

$$\{x \restriction [a_n, b_n) : x \in X\} \quad \text{for } n \in \omega,$$

and by applying the same argument we obtain similar bounds on the number of elements in every set of the form $\{x \restriction [b_n, a_{n+1}) : x \in X\}$. □

Theorem 2.16, which in fact gives the characterization of null additive sets (see [**BJ**, Theorem 2.7.18]), provides the proof of the following deep theorem (see [**BJ**, Theorem 2.7.1]).

THEOREM 2.17 (Shelah). *Every null additive set is meager additive.*

PROOF. By a straightforward application of Theorem 2.16 one can easily see that if $X \in \mathcal{N}^*$, then X satisfies the combinatorial formula from Theorem 2.4. □

The next observation shows that the algebraic union of two "small" subsets of 2^ω can be equal to the whole 2^ω (see [**Sier2**]).

THEOREM 2.18 (Sierpiński). *Assume CH. Then there are $X, Y \in \mathcal{SN}$ such that $X + Y = 2^\omega$.*

PROOF. Suppose that $\{F_\alpha\}_{\alpha<\mathfrak{c}}$ enumerates all F_σ meager sets and $\{z_\alpha\}_{\alpha<\mathfrak{c}}$ is a bijective enumeration of 2^ω. We construct X and Y by induction. Assume that for $\kappa < \mathfrak{c}$ we have already constructed $X_\kappa = \{x_\alpha\}_{\alpha<\kappa}$ and $Y_\kappa = \{y_\alpha\}_{\alpha<\kappa}$, so that $\{z_\alpha\}_{\alpha<\kappa} \subseteq X_\kappa + Y_\kappa$. Let

$$G_\kappa = 2^\omega \setminus \left(\bigcup_{\alpha<\kappa+1} F_\alpha \cup X_\kappa \cup Y_\kappa \right).$$

Clearly G_κ is a comeager set. Thus we can pick $x_\kappa \in G_\kappa$ and $y_\kappa \in G_\kappa + z_\kappa$ such that $x_\kappa + y_\kappa = z_\kappa$. Put $X_{\kappa+1} = X_\kappa \cup \{x_\kappa\}$ and $Y_\kappa = Y_\kappa \cup \{y_\kappa\}$. Then $X = \{x_\alpha\}_{\alpha<\mathfrak{c}}$ and $Y = \{y_\alpha\}_{\alpha<\mathfrak{c}}$ defined by the above procedure are two $\aleph_1$-Luzin sets, which satisfy $X + Y = 2^\omega$. □

By a similar argument we can obtain $X, Y \in \mathcal{SM}$ such that $X + Y = 2^\omega$. With smaller sets the situation is different, and it can be illustrated by the following very simple fact.

PROPOSITION 2.19. *If $X, Y \in \mathcal{M}^*$, then $X + Y \in \mathcal{M}^*$, and if $X, Y \in \mathcal{N}^*$, then $X + Y \in \mathcal{N}^*$.*

PROOF. Let $A \in \mathcal{M}$. Then $(X + Y) + A = X + (Y + A)$, and since the second set is meager, we have that $X + Y \in \mathcal{M}^*$. We prove the measure case in the same way. □

REMARK. By using an argument from above we can easily prove that for $X \in \mathcal{N}^*$ and $Y \in \mathcal{SM}$ $X + Y$ is an $\mathcal{SM}$ set. A category counterpart of this fact was already mentioned in Theorem 2.11. For analogous results in which the addition is replaced by the cartesian product we refer to [**TW**] and [**Z2**].

We now consider the algebraic sum of a subset of 2^ω with a set that belongs to the intersection ideal $\mathcal{M} \cap \mathcal{N}$. Recall that $\mathcal{M} \cap \mathcal{N}$ is a strictly larger σ-ideal than $\mathcal{E}$ (see [**BJ**, page 73]).

Suppose that I and J are σ-ideals of subsets of 2^ω with $I \subseteq J$. The following is a generalization of the notion of $\mathcal{M}^*$, $\mathcal{N}^*$ and $\mathcal{E}^*$ that appeared earlier.

DEFINITION 2.20. We shall say $X \subseteq 2^\omega$ is I additive, or $X \in I^*$, if $X + A \in I$ for any set $A \in I$, and $X \in (I, J)^*$ if for every set $A \in I$, $X + A \in J$.

The next results with detailed proofs can be found in [**W2**] and [**W3**].

Assume that $\to$ denotes the inclusion, and $\not\leftarrow$ means that the reverse inclusion cannot be proved in ZFC.

PROPOSITION 2.21.

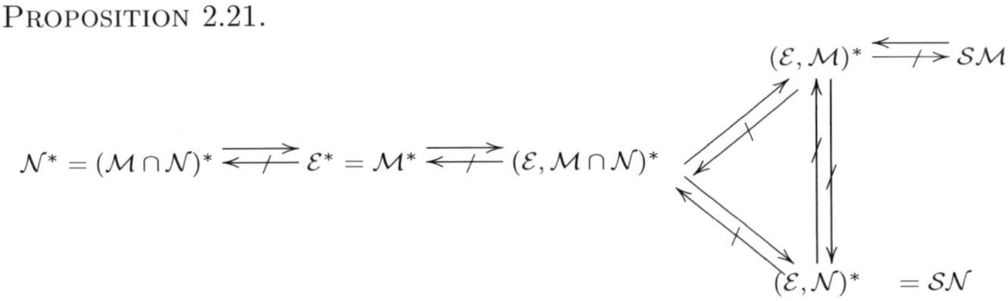

PROOF. Most inclusions and crossed arrows follow from the definitions, earlier considerations in this article and from Theorem 2.23, 2.24, 2.25, 2.26 below. The proof of $\mathcal{SM} \not\leftarrow (\mathcal{E}, \mathcal{M})^*$ can be found in [**P**], page 182. □

QUESTION 2.22. Is it consistent with ZFC that every element of the class $(\mathcal{E}, \mathcal{M})^*$ is at most countable?

THEOREM 2.23. *Assume CH. Then there are a γ-set X and $A \in \mathcal{M} \cap \mathcal{N}$ such that $X + A \notin \mathcal{M} \cap \mathcal{N}$.*

PROOF (SKETCH). Let

$$A' = \{x \in 2^\omega : \exists_n^\infty x \restriction [k_n, k_{n+1}) \equiv \mathbb{O}\},$$

where $\{k_n\}_{n \in \omega}$ is an increasing sequence of natural numbers such that $\mu(A') = 0$. Put

$$A = A' \cap \{x \in 2^\omega : \forall_n^\infty\ x \restriction [k_n, k_{n+1}) \not\equiv \mathbb{1}\},$$

and let $\{y_\alpha\}_{\alpha < \mathfrak{c}}$ be a bijective enumeration of the following set.

$$B = \{x \in 2^\omega : \forall_n^\infty\ (x \restriction [k_n, k_{n+1}) \not\equiv \mathbb{O}, \text{ and } x \restriction [k_n, k_{n+1}) \not\equiv \mathbb{1})\}.$$

Obviously, $\mu(B) = 1$.

We choose x_λ for $\lambda \in \mathrm{Lim}$ in a similar way to this in the proof of Theorem 2.14 of Bartoszyński and Recław, so that it satisfies

$$\exists_n^\infty\ x_\lambda \restriction [k_n, k_{n+1}) \equiv \mathbb{1}, \qquad \forall_{\alpha<\lambda} \forall_n^\infty\ x_\lambda(n) \leq x_\alpha(n)$$

and additionaly: there is an infinite $a \subseteq \omega$, with $\omega \setminus a$ also infinite such that

$$\forall_{n \in a}\ x_\lambda \restriction [k_n, k_{n+1}) = y_\lambda, \qquad \forall_{n \in \omega \setminus a}\ (x_\lambda \restriction [k_n, k_{n+1}) \equiv \mathbb{O} \text{ or } x_\lambda \restriction [k_n, k_{n+1}) \equiv \mathbb{1}).$$

As a consequence we obtain that

$$B \subseteq X + A.$$ □

THEOREM 2.24. $(\mathcal{M} \cap \mathcal{N}, \mathcal{M})^* \subseteq \mathcal{E}^* = \mathcal{M}^*$.

PROOF (SKETCH). By Theorem 2.24 from [**BJ**] we know that every meager set is contained in the set $B_{f,y}$ for some $f \in \omega^{\omega\uparrow}$ and $y \in 2^\omega$, where

$$B_{f,y} = \{x \in 2^\omega : \forall_n^\infty \; x \restriction [f(n), f(n+1)) \neq y \restriction [f(n), f(n+1))\}.$$

For $f \in \omega^\omega$, which satisfies $f(n+1) \geq f(n) + n$ for $n \in \omega$ we put

$$H_f = \{x \in 2^\omega : \exists_n^\infty \; x \restriction [f(n), f(n+1)) \equiv \mathbb{1}\},$$

and we define

$$\overline{B}_{f,\mathbb{0}} = B_{f,\mathbb{0}} \cap H_f.$$

If $X \in \mathcal{M}^*$, then $X + \overline{B}_{f,\mathbb{0}} \subseteq B_{g,y}$ for some $g \in \omega^{\omega\uparrow}$ and $y \in 2^\omega$.

Proceeding as in the proof of Lemma 2.7.5 from [**BJ**] we show that X fullfills the characterization from Theorem 2.4 above. □

THEOREM 2.25. $(\mathcal{M} \cap \mathcal{N})^* \subseteq \mathcal{N}^*$.

PROOF (SKETCH). We use the same terminology and notation as in the proof of Theorem 2.16. Assume that $X \in (\mathcal{M} \cap \mathcal{N})^*$. By Theorem 2.24 $X \in \mathcal{M}^*$. Let

$$A' = A \cap \{x \in 2^\omega : \forall_n^\infty \; x \restriction [a_n, a_{n+1}) \neq h \restriction [a_n, a_{n+1})\},$$

where $h \in 2^\omega$ is such that $A' \neq \emptyset$. Then $X + A' \subseteq G$, where G is open, and $\mu(G) < 1$. By De Morgan law

$$\bigcap_{x \in X} ((x + (2^\omega \setminus A)) \cup (x + B)) \supseteq [\tau] \setminus G,$$

where $[\tau]$ is any basic clopen set satisfying $\mu([\tau] \setminus G) > 0$, and

$$B = \{x \in 2^\omega : \exists_n^\infty \; x \restriction [a_n, a_{n+1}) = h \restriction [a_n, a_{n+1})\}.$$

Since

$$\bigcap_{x \in X} ((x + (2^\omega \setminus A)) \cup (x + B)) \subseteq \bigcap_{x \in X} (x + (2^\omega \setminus A)) \cup \bigcup_{x \in X} (x + B),$$

and the union in the latter formula is of measure zero by the characterization of sets in $\mathcal{M}^*$, we obtain that for every basic clopen set $[\tau]$, with $\mu([\tau] \setminus G) > 0$ there is $a_\tau \subseteq ([\tau] \setminus G)$, $\mu(a_\tau) > 0$

$$\left(\bigcup_{x \in X} (x + A) \right) \cap a_\tau = \emptyset.$$

We can now repeat the main argument from the proof of Theorem 2.16 to find the similar bounds on the cardinalities of sets of the form

$$\{x \restriction [a_n, b_n) : x \in X\} \quad \text{and} \quad \{x \restriction [b_n, a_{n+1}) : x \in X\}$$

for $n \in \omega$. □

Notice that Theorem 2.25 completes the list of combinatorial characterizations of I additive sets, where I is one of the following most common σ-ideals in 2^ω that is $\mathcal{M}, \mathcal{N}, \mathcal{E}$ and $\mathcal{M} \cap \mathcal{N}$.

THEOREM 2.26. *It is consistent with ZFC that* $\mathcal{E}^* = \mathcal{M}^* \not\supseteq (\mathcal{E}, \mathcal{M} \cap \mathcal{N})^*$.

PROOF (SKETCH). Let V be a model of $ZFC + CH$. By applying methods from the proof of Theorem 2.18 we construct two sets $X, Y \in \mathcal{SM}$ such that $X + Y = 2^\omega$. Assume that $V[c]$ is a model of ZFC obtained by adding Cohen real. Since $2^\omega \cap V$ is non meager in $V[c]$, we have that X and Y are not meager additive. Next we prove that for every $F \in \mathcal{E} \cap V[c]$ $X + F \in \mathcal{M}$. This observation follows easily from the fact that each $F \in \mathcal{E} \cap V[c]$ is contained in a G_δ null set coded in V. As $X \in \mathcal{SN}$, $X + F \in \mathcal{N}$ for every $F \in \mathcal{E}$. Thus $X \in (\mathcal{E}, \mathcal{M} \cap \mathcal{N})^*$. □

As a consequence of Theorem 2.23 we obtain that $(\mathcal{M} \cap \mathcal{N}, \mathcal{M})^* \not\to (\mathcal{M} \cap \mathcal{N}, \mathcal{N})^*$. It seems natural to ask whether the reverse implication holds true.

QUESTION 2.27. $(\mathcal{M} \cap \mathcal{N}, \mathcal{N})^* \to (\mathcal{M} \cap \mathcal{N}, \mathcal{M})^*$?

In the last part of Section 2, we turn attention to a new result related to our discussion of $\mathcal{SN}$ and $\mathcal{SM}$ sets, which recently appeared in [**HL**].

DEFINITION 2.28. For a family I of subsets of 2^ω, which does not necessarily form an ideal, we shall say that $X \in I_*$ if for every $A \in I$ $X + A \neq 2^\omega$.

We use the notation I_* to distinguish this family from the class of I additive sets. Obviously, $\mathcal{SN} = \mathcal{M}_*$, and $\mathcal{SM} = \mathcal{N}_*$. By an easy analysis of the definition of I_* family we obtain that $I \subseteq (I_*)_* \stackrel{\text{def}}{=} I_{**}$, and $I_* = (I_{**})_* \stackrel{\text{def}}{=} I_{***}$.

THEOREM 2.29 (Horbaczewska, Lindner). *The following two conditions are equivalent.*

1. *For every* $A \notin I$ $(I \cup \{A\})_* \neq I_*$.
2. $I = I_{**}$.

PROOF. See Theorem 3 in [**HL**]. □

The reasoning from the proof of the fact (due to the author of this article), which says that under CH $J_* \not\subseteq I_*$ for any two Borel supported and + invariant σ-ideals I, J satisfying $I \not\subseteq J$, yields the following observation.

THEOREM 2.30 (Horbaczewska, Lindner). *Suppose that CH holds. Then for every σ-ideal I, which is + invariant and with* $\operatorname{cof}(I) \leq \mathfrak{c}$ *we have that*

$$(I \cup \{A\})_* \not\subseteq I_*$$

for every $A \notin I$.

PROOF. See Theorem 9 in [**HL**]. □

It is now easy to draw consequences of our previous considerations (see Corollary 12 in [**HL**]).

COROLLARY 2.31 (Horbaczewska, Lindner). *Assume that CH holds. Then* $\mathcal{SN}_* = \mathcal{M}$, *and* $\mathcal{SM}_* = \mathcal{N}$.

PROOF. Apply Theorem 2.29 and 2.30. □

Since $\mathcal{M}$ and $\mathcal{N}$ are contained in $([2^\omega]^\omega)_*$, we have that in the Laver model for the Borel Conjecture (see [**L**]) $\mathcal{SN}_* \not\subseteq \mathcal{M}$, and in the Cohen model for the Dual Borel Conjecture (see [**C**]) $\mathcal{SM}_* \not\subseteq \mathcal{N}$.

3

In this section, we investigate Ramseyan properties of $\mathcal{SN}$ and $\mathcal{SM}$ sets and their relatives. We concentrate on the two most common properties by considering Marczewski s_0-sets and completely Ramsey null sets.

Following Marczewski (see [**Mar**]) we define an s_0-set in 2^ω.

DEFINITION 3.1. We say that $X \subseteq 2^\omega$ is *an s_0-set* (or an *s_0-Marczewski set*) if for every perfect set P in 2^ω there exists a perfect set $P' \subseteq P$ such that $X \cap P' = \emptyset$.

DEFINITION 3.2. A set $X \subseteq 2^\omega$ is called a *perfectly meager set* (or a $P\mathcal{M}$) set if for every perfect set P in 2^ω, $X \cap P$ is meager in the relative topology of P.

$P\mathcal{M}$ sets are also called *always first category sets*, and are sometimes denoted by AFC. A smaller class of sets, namely sets, which are *perfectly meager in the transitive sense* we define in the following way, and we denote them by $P\mathcal{M}'$.

DEFINITION 3.3. We say that $X \in P\mathcal{M}'$ if there is for every perfect set P a sequence $\{F_n\}_{n\in\omega}$ of closed sets in 2^ω such that $X \subseteq \bigcup_{n\in\omega} F_n$ and for every $t \in 2^\omega$ $(\bigcup_{n\in\omega} F_n + t) \cap P$ is meager in the relative topology of P.

THEOREM 3.4 (Nowik). *It is a theorem of ZFC that the class $P\mathcal{M}'$ has uncountable representatives.*

PROOF. See [**N**, Conclusion 6]. □

QUESTION 3.5 (Scheepers). Does ZFC prove that $P\mathcal{M}' \not\subseteq P\mathcal{M}$?

The following result provides a partial solution to M. Scheeper's question.

THEOREM 3.6 (Weiss). *Assume $\mathfrak{c} \leq \aleph_2$. Then there is $X \in P\mathcal{M}$, which does not belong to $P\mathcal{M}'$.*

PROOF. See Theorem 4 in [**W1**]. □

THEOREM 3.7 (Nowik, Scheepers, Weiss). *Suppose that $X \in \mathcal{SN}$, and Y is a $P\mathcal{M}'$ set. Then $X + Y$ is an s_0-set.*

PROOF (SKETCH). Let P be a perfect set, and assume that $Y \subseteq \bigcup_{n\in\omega} F_n$, where for every $n \in \omega$ and $t \in 2^\omega$ $(F_n + t) \cap P$ is closed nowhere dense in the relative topology of P. Suppose also that $F_n \subseteq F_{n+1}$ for $n \in \omega$. Using compactness of 2^ω find $\delta_0 > 0$ such that for every basic open set I with $\operatorname{diam}(I) \leq \delta_0$ there are disjoint perfect set $P_0, P_1 \subseteq P$, which satisfy

$$(I + F_0) \cap (P_0 \cup P_1) = \emptyset.$$

By repeating this procedure infinitely many times we obtain a sequence $\{\delta_n\}_{n\in\omega}$ with each $\delta_n > 0$ such that for any sequence $\{I_n\}_{n\in\omega}$ of basic open sets with every $\operatorname{diam}(I_n) \leq \delta_n$ there is a sequence $\{P_s\}_{s\in 2^{<\omega}}$ of perfect subsets of P satisfying the following conditions

(1) $P_{s^\frown 0}, P_{s^\frown 1} \subseteq P_s$ for every $s \in 2^{<\omega}$, and are disjoint.
(2) For every $n \in \omega$

$$(I_n + F_n) \cap \left\{ \bigcup_{s\in 2^{<\omega}} P_s : \text{ length } (s) = n+1 \right\} = \emptyset.$$

If for such a sequence $\{I_n\}_{n\in\omega}$

$$X \subseteq \bigcap_{m\in\omega} \bigcup_{n\geq m} I_n, \quad \text{then } X+Y \subseteq \bigcup_{n\in\omega}(I_n + F_n),$$

and the latter set is disjoint with a perfect set $P' \subseteq P$, where

$$P' = \bigcap_{n\in\omega} \left\{ \bigcup_{s\in 2^{<\omega}} P_s : \text{length}\,(s) = n+1 \right\}. \qquad \square$$

THEOREM 3.8. *The class $\mathcal{SM}$ is included in $P\mathcal{M}'$.*

PROOF (SKETCH). Suppose that $X \in \mathcal{SM}$, and P is a perfect set. Let $\{I_n\}_{n\in\omega}$ be a bijective enumeration of all basic open sets in 2^ω. If $I_n \cap P$ is perfect, we choose A_n to be a perfect measure zero set that $(I_n \cap P) + A_n = 2^\omega$ (use for instance Corollary 2 from [**NW3**]). Let G be a G_δ measure zero set such that $\bigcup_{n\in\omega} A_n \subseteq G$. Then for some $t \in 2^\omega$ $X \cap (G+t) = \emptyset$, thus $X \subseteq F$, where $F = 2^\omega \setminus (G+t)$. Next we argue that F is a required F_σ set. $\square$

As a corollary we obtain the following theorem (see [**NSW**]).

THEOREM 3.9 (Nowik, Scheepers, Weiss). *If $X \in \mathcal{SN}$, and $Y \in \mathcal{SM}$, then $X+Y$ is an s_0-set.*

PROOF. Apply Theorem 3.7 and 3.8. $\square$

THEOREM 3.10 (Nowik, Weiss). *Assume CH. Then there are $X \in \mathcal{SN}$ and $Y \in P\mathcal{M}$ such that $X+Y$ is not an s_0-set.*

PROOF (SKETCH). Let C, D be disjoint perfect subsets of 2^ω with the following property (see [**NW1**, Claim 1] and [**R**, Lemma 1])

$$(C+C) \cap (D+D) = \{0\}.$$

By CH there is a γ-set $X \subseteq C$ of cardinality $\mathfrak{c}$ (see [**GM**, Theorem 1]. Let $\{d_x\}_{x\in X}$ be a bijective enumeration of D, and define $Y = \{x + d_x : x \in X\}$. Then $Y \in P\mathcal{M}$ as a continous one to one inverse image of a perfectly meager set X (see [**R**, Lemma 1]), and $D \subseteq X+Y$. $\square$

The content of Theorem 3.9 was motivated by Problem 2 from [**S**] and by the following result of M. Scheepers.

THEOREM 3.11 (Scheepers). *If X is an $\aleph_1$-Luzin set, and $Y \in \mathcal{M}$ is an s_0-set, then $X+Y$ is an s_0-set. Analogously, if X is an $\aleph_1$-Sierpiński set, and $Y \in \mathcal{N}$ is an s_0-set, then $X+Y$ is an s_0-set.*

PROOF. We apply the following fact (see Theorem 11 in [**S**]). If $A \in \mathcal{M}$, $B \in \mathcal{N}$, and P is perfect, then there is a perfect $P' \subseteq P$ such that $A + P' \in \mathcal{M}$, and $B + P' \in \mathcal{N}$. We prove the category part of Theorem 3.11. The proof of the measure case is the same. Let X be an $\aleph_1$-Luzin set and suppose that P is perfect. In P find a perfect subset P' such that $Y + P'$ is meager, where Y is a given meager set with the s_0-property. Then $X \cap ((Y + P')) = X'$ is a countable set. We have that

$$X + Y = [(X \setminus X') + Y] \cup [X' + Y].$$

Clearly, the first set in the above union is disjoint with P', and the second is an s_0-set. Thus $X+Y$ is an s_0-set. $\square$

In [**KW**], the authors introduce a notion of perfectly null sets, which is a possible counterpart to perfectly meager sets.

Let P be a perfect set with $P = [T_p]$, where T_p is a perfect tree in $2^{<\omega}$. A sequence $s \in T_p$ is called a *branching point* of P if $s^\frown 0, s^\frown 1 \in T_p$. We say that a branching point is *on level* $i \in \omega$ if there are i branching point of P below it. We denote by $\mathrm{Split}_i(P)$ the set of all branching points of P on level i, and

$$\mathrm{Split}(P) = \bigcup_{i\in\omega} \mathrm{Split}_i(P).$$

Assume that for a perfect P $h_P : 2^\omega \to P$ is the homeomorphism induced by the order isomorphism of $2^{<\omega}$ and $\mathrm{Split}(P)$.

DEFINITION 3.12. Suppose that P is a perfect set, and $A \subseteq P$ is such that $h_P^{-1}(A)$ is measurable in 2^ω. We define the *canonical measure* $\mu_P(A) = \mu(h_P^{-1}(A))$.

DEFINITION 3.13. An $A \subseteq 2^\omega$ is said to be *perfectly null* if for any perfect set P $\mu_P(A \cap P) = 0$.

We denote the family of all perfectly null sets by $\mathcal{PN}$.

Let $A \subseteq 2^\omega$. We shall say that A is *universally null*, or $A \in \mathcal{UN}$, if for every finite diffused Borel measure μ on 2^ω there exists a Borel set B containing A and such that $\mu(B) = 0$. Assume now that for a perfect set P a measure λ is such that $\lambda(B) = \mu_P(B \cap P)$ for any Borel set $B \subseteq 2^\omega$. It is clear that for a universally null set A every measure λ of this form vanishes on A. Hence $\mathcal{UN} \subseteq \mathcal{PN}$.

QUESTION 3.14. Is there a model of ZFC, in which $\mathcal{UN} \not\subseteq \mathcal{PN}$?

The following notion was motivated by the definition of $\mathcal{PM}'$ sets.

DEFINITION 3.15. We say that a set X is *perfectly null in the transitive sense*, or $X \in \mathcal{PN}'$, if there exists for every perfect set P a G_δ set G containing X such that for every $t \in 2^\omega$ $(G + t) \cap P$ is a null set with respect to μ_P.

QUESTION 3.16. Does ZFC prove that there are uncountable sets in $\mathcal{PN}'$?

The next theorem can be concluded easily from the definition of $\mathcal{PN}'$ sets (see Theorem 3.3 in [**KW**]).

THEOREM 3.17. *$\mathcal{SN} \subseteq \mathcal{PN}'$.*

PROOF. Suppose that $X \in \mathcal{SN}$, and P is perfect. Let $\{a_n\}_{n\in\omega}$ be a sequence such that $a_n = \max\{\mathrm{length}(s) : s \in \mathrm{Split}_n(P)\}$ for $n \in \omega$. Pick a sequence $\{A_n\}_{n\in\omega}$ of basic open sets, so that

$$X \subseteq \bigcap_{m\in\omega} \bigcup_{n\geq m} A_n = \{x \in 2^\omega : \exists_n^\infty \ x \in A_n\},$$

and $\mathrm{diam}(A_n) \leq \frac{1}{2^{a_n+1}}$ for $n \in \omega$. For $t \in 2^\omega$ every $B_n = (A_n + t) \cap P$ satisfies $\mu_P(B_n) \leq \frac{1}{2^{n+1}}$. Thus

$$(X + t) \cap P \subseteq \bigcap_{m\in\omega} \bigcup_{n\geq m} B_n = \{x \in 2^\omega : \exists_n^\infty \ x \in B_n\},$$

and the latter set is μ_P null. □

QUESTION 3.18. Is there a set in $\mathcal{PN}'$ family, which is not strongly null?

If $\mu_P(A) = 0$ for $A \subseteq P$, where P is perfect, then $2^\omega \setminus h_P^{-1}(A)$ contains a perfect set, hence every member of $P\mathcal{N}$ family is an s_0-set.

QUESTION 3.19. Is it true that for every $X \in P\mathcal{N}'$ and every $Y \in \mathcal{SM}$ $X+Y$ is an s_0-set?

If $s \in [\omega]^{<\omega}$, $S \in [\omega]^\omega$, and $\max(s) < \min(S)$, we define an *Ellentuck set* (see Definition 3 in [**NW2**])

$$\langle s, S\rangle = \{A \in [\omega]^\omega : s \subseteq A \subseteq s \cup S\}.$$

Depending on the context $\langle s, S\rangle$ is often conflated with $\{\chi(x) :\ x \in \langle s, S\rangle\}$, where $\chi(x)$ stands for the characteristic function of x.

DEFINITION 3.20. We shall say that $X \subseteq 2^\omega$ is *completely Ramsey null* (or cr_0) if for every $\langle s, S\rangle$ there is $\langle s, T\rangle \subseteq \langle s, S\rangle$ such that

$$\{\chi(x) :\ x \in \langle s, T\rangle\} \cap X = \emptyset.$$

THEOREM 3.21 (Nowik, Weiss). *If $X \in \mathcal{SM}$, then X is a cr_0 set.*

PROOF. We apply the below lemma and an argument, which is very much like that of Theorem 3 in [**NW2**].

LEMMA 3.22 (Lorentz). *If H is a finite group with the group operation $\oplus$, then for every $H_1 \subseteq H$ there exists $H_2 \subseteq H$ with $|H_2| \leq |H| \cdot \frac{\log|H_1|+1}{|H_1|}$ such that*

$$H_1 \oplus H_2 = \{h_1 \oplus h_2 :\ h_1 \in H_1,\ h_2 \in H_2\} = H.$$

PROOF. See [**BJ**, Lemma 3.3.10]. □

Let $s \in [\omega]^{<\omega}$, $S \in [\omega]^\omega$, $\max(s) < \min(S)$. Suppose that $\{I_n\}_{n\in\omega}$ is a partition of ω into finite disjoint intervals with $\max(I_n) + 1 = \min(I_{n+1})$ and satisfying

$$\frac{\log|I_n \cap S| + 1}{|I_n \cap S|} \leq \frac{1}{2^n} \quad \text{for } n \in \omega.$$

For $n \in \omega$ let $H_1^n = \{\chi(a) \restriction I_n :\ a \in I_n \cap S\}$, and let H_2^n be such that $H_1^n \oplus H_2^n = 2^{I_n}$ with

$$|H_2^n| \leq \frac{\log|H_1^n| + 1}{|H_1^n|} \cdot 2^{|I_n|}.$$

Put

$$A = \{x \in 2^\omega : \exists_n^\infty\ x \restriction I_n \in H_2^n\}.$$

Clearly, $X + A \neq z$ for some $z \in 2^\omega$ as $X \in \mathcal{SM}$, and $A \in \mathcal{N}$. For $n \in \omega$ let $z \restriction I_n = h_1^n + h_2^n$, where $h_1^n \in H_1^n$, $h_2^n \in H_2^n$. It is easy to check that $X \cap \langle s, S'\rangle = \emptyset$, where $S' \subseteq S$ is such that $S' \restriction I_n = h_1^n$ for $n \in \omega$. □

The following sets appeared in [**NW3**] and indirectly in [**Rep**].

DEFINITION 3.23. A set $X \subseteq 2^\omega$ has the (T') *property* if there is $g \in (\omega \setminus \{0\})^\omega$ such that for every partition $\{I_n\}_{n\in\omega}$ of ω into finite, disjoint intervals with $\max(I_n) + 1 = \min(I_{n+1})$ one can find an increasing sequence $\{n_k\}_{k\in\omega}$ and $\{H_k\}_{k\in\omega}$, $H_k \subseteq 2^{I_{n_k}}$, $|H_k| \leq g(n_k)$ for $k \in \omega$, so that

$$X \subseteq \{x \in 2^\omega : \forall_k^\infty\ x \restriction I_{n_k} \in H_k\}.$$

Using similar argument as in the proof of Theorem 3.21 one can show the following more general theorem.

THEOREM 3.24 (Nowik, Weiss). *For every $X \in \mathcal{SM}$ and every Y with the (T') property $X+Y$ is a completely Ramsey null set.*

PROOF. See [**NW3**, Theorem 2.2]. □

The proof that the class of sets with the (T') property is contained in $\mathcal{M}^*$ can be handled in the following way (see Proposition 3.7 in [**NW3**]).

THEOREM 3.25. $(T') \subseteq \mathcal{M}^*$.

PROOF (SKETCH). First notice that in the definition of the (T') property we may suppose without loss of generality that $g(n) = n$ for $n \in \omega$ (see Proposition 3.1 and 3.2 in [**NW3**]). Let X be a (T') set and $\{I_n\}_{n\in\omega}$ a partition of ω with $\max(I_n)+1 = \min(I_n)$. Put $I'_0 = I_0$, $I'_1 = I_1 \cup I_2$, $I'_3 = I_3 \cup I_4 \cup I_5$, etc. There are sequences $\{n_k\}_{k\in\omega}$, $\{H_k\}_{k\in\omega}$, with $H_k \subseteq 2^{I'_{n_k}}$ and $|H_k| \leq n_k + 1$ for $k \in \omega$ satisfying

$$X \subseteq \{x \in 2^\omega : \forall_k^\infty\ x \restriction I'_{n_k} \in H_k\}.$$

Suppose that $H_k = \{h_0^k, \ldots, h_{n_k}^k\}$ for $k \in \omega$. Let f be such that

$$f \restriction I'_{n_k} = h_0^k \restriction I_{m_{n_k}} \cup \cdots \cup h_{n_k}^k \restriction I_{m_{n_k}+n_k} \qquad \text{for } k \in \omega,$$

where $m_{n_k} = \frac{n_k(n_k+1)}{2}$, and let $f \equiv \mathbb{O}$ elsewhere. Clearly, X satisfies the condition from Theorem 2.4 with y being equal to f. □

Below we prove that Theorem 3.21 does not necessarily hold for meager additive sets.

LEMMA 3.26. *For $M, M_1 \subseteq 2^\omega$,*

$$(M + M_1^c)^c = \{x \in 2^\omega : x + M \subseteq M_1\}.$$

PROOF. See [**NW3**, Lemma 4.1]. □

LEMMA 3.27. *Assume that $M \in \mathcal{M}$. Then there exists $M_1 \in \mathcal{M}$ such that $M + M_1^c$ is completely Ramsey null.*

PROOF. Let $\{I_n\}_{n\in\omega}$, a partition of ω into finite disjoint intervals with $\max(I_n)+1 = \min(I_{n+1})$ and $y \in 2^\omega$ be such that

$$M \subseteq \{x \in 2^\omega : \forall_n^\infty\ x \restriction I_n \neq y \restriction I_n\}.$$

Put $J_k = I_{2k} \cup I_{2k+1}$ for $k \in \omega$, and define

$$M_1 = \{x \in 2^\omega : \forall_k^\infty\ x \restriction J_k \neq y \restriction J_k\}.$$

Let $S' \in \langle s, S\rangle$ be such that

$$\forall_k^\infty\ |J_k \cap S'| \leq 1.$$

Suppose that $m \in M$, and $m' \in M_1^c$. Pick $k_0 \in \omega$, so that

$$m \restriction I_{2k_0+i} \neq y \restriction I_{2k_0+i}$$

for $i \in 2$, and $m' \restriction J_{k_0} = y$.
Then $(m + m') \restriction J_k \neq S' \restriction J_k$. □

THEOREM 3.28. *If CH holds, then there exists $X \in \mathcal{M}^*$, which is not cr_0.*

PROOF. We follow directly the proof of Theorem 4.3 in [**NW3**]. Assume that $\{\mathcal{U}_\alpha\}_{\alpha<\aleph_1}$ enumerates bijectively $[\omega]^\omega$, and let $\{M_\alpha\}_{\alpha<\aleph_1}$ be a bijective enumeration of all F_σ meager sets. For each M_α choose a meager set $M_{\alpha,1}$ as in Lemma 3.27 and pick

$$x_\xi \in [\mathcal{U}_\xi]^\omega \setminus \bigcup_{\alpha<\xi} (M_\alpha + M^c_{\alpha,1}) \quad \text{for } \xi < \aleph_1.$$

This can be done as the union in the above formula is completely Ramsey null. Clearly, $X = \{x_\xi\}_{\xi<\aleph_1}$ is not a completely Ramsey null set as it captures some member of $[\mathcal{U}_\alpha]^\omega$ for $\alpha < \aleph_1$. We have that for $\xi < \aleph_1$

$$X + M_\xi = [\{x_\alpha\}_{\alpha\leq\xi} + M_\xi] \cup [\{x_\alpha\}_{\alpha>\xi} + M_\xi] \subseteq [\{x_\alpha\}_{\alpha\leq\xi} + M_\xi] \cup M_{\xi,1}$$

as by Lemma 3.26, $\{x_\alpha\}_{\alpha>\xi} + M_\xi \subseteq M_{\xi,1}$. Thus $X + M_\xi \in \mathcal{M}$. □

Theorem 3.28 implies that $\mathcal{SN}$ do not have to be cr_0. This fact was proved earlier by J. Brown (see Example 6 in [**Bro**]).

References

[B] T. Bartoszynski, *Remarks on small sets of reals*, Proc. Amer. Math. Soc. **131** (2003), no. 2, 625–630, DOI 10.1090/S0002-9939-02-06567-X. MR1933355

[Bor] E. Borel, *Sur la classification des ensembles de mesure nulle* (French), Bull. Soc. Math. France **47** (1919), 97–125. MR1504785

[Bro] J. B. Brown, *The Ramsey sets and related sigma algebras and ideals*, Fund. Math. **136** (1990), no. 3, 179–185, DOI 10.4064/fm-136-3-179-185. MR1095690

[BJ] T. Bartoszyński and H. Judah, *Set theory*, A K Peters, Ltd., Wellesley, MA, 1995. On the structure of the real line. MR1350295

[BNW] T. Bartoszynski, A. Nowik, and T. Weiss, *Strongly meager sets can be quite big*, Israel J. Math. **139** (2004), 237–251, DOI 10.1007/BF02787551. MR2041793

[BR] T. Bartoszyński and I. Recław, *Not every γ-set is strongly meager*, Set theory (Boise, ID, 1992), Contemp. Math., vol. 192, Amer. Math. Soc., Providence, RI, 1996, pp. 25–29, DOI 10.1090/conm/192/02346. MR1367132

[BS] T. Bartoszynski and S. Shelah, *Strongly meager sets do not form an ideal*, J. Math. Log. **1** (2001), no. 1, 1–34, DOI 10.1142/S0219061301000028. MR1838340

[Bu] L. Bukovský, *The structure of the real line*, Instytut Matematyczny Polskiej Akademii Nauk. Monografie Matematyczne (New Series) [Mathematics Institute of the Polish Academy of Sciences. Mathematical Monographs (New Series)], vol. 71, Birkhäuser/Springer Basel AG, Basel, 2011. MR2778559

[C] T. J. Carlson, *Strong measure zero and strongly meager sets*, Proc. Amer. Math. Soc. **118** (1993), no. 2, 577–586, DOI 10.2307/2160341. MR1139474

[F] D. H. Fremlin, *Measure theory. Vol. 5. Set-theoretic measure theory. Part II*, Torres Fremlin, Colchester, 2015. Corrected reprint of the 2008 original. MR3723041

[GM] F. Galvin and A. W. Miller, *γ-sets and other singular sets of real numbers*, Topology Appl. **17** (1984), no. 2, 145–155, DOI 10.1016/0166-8641(84)90038-5. MR738943

[GMS] F. Galvin, J. Mycielski, and R. Solovay, *Abstract a-280.*, Notices of the American Mathematical Society, vol. **26** (1979).

[GN] J. Gerlits and Zs. Nagy, *Some properties of $C(X)$. I*, Topology Appl. **14** (1982), no. 2, 151–161, DOI 10.1016/0166-8641(82)90065-7. MR667661

[HL] G. Horbaczewska and S. Lindner, *On sets which can be moved away from sets of a certain family*, J. Math. Anal. Appl. **472** (2019), no. 1, 231–237, DOI 10.1016/j.jmaa.2018.11.020. MR3906370

[HZ] M. Hrušák and O. Zindulka, *Strong measure zero in Polish groups.*, this volume, 2019.

[JMSSz] W. Just, A. W. Miller, M. Scheepers, and P. J. Szeptycki, *The combinatorics of open covers. II*, Topology Appl. **73** (1996), no. 3, 241–266, DOI 10.1016/S0166-8641(96)00075-2. MR1419798

[K] M. Kysiak, *On Erdős–Sierpiński duality between Lebesgue measure and Baire category*, (in Polish), Master's thesis (2000).

[KW] M. Korch and T. Weiss, *On the class of perfectly null sets and its transitive version*, Bull. Pol. Acad. Sci. Math. **64** (2016), no. 1, 1–20, DOI 10.4064/ba8029-5-2016. MR3550609

[L] R. Laver, *On the consistency of Borel's conjecture*, Acta Math. **137** (1976), no. 3-4, 151–169, DOI 10.1007/BF02392416. MR422027

[M] A. W. Miller, *Special subsets of the real line*, Handbook of set-theoretic topology, North-Holland, Amsterdam, 1984, pp. 201–233. MR776624

[Mar] E. Szpilrajn-Marczewski, *Sur deux propriétés des classes d'ensembles* (French), Fund. Math. **33** (1945), 303–307, DOI 10.4064/fm-33-1-303-307. MR15448

[MF] A. W. Miller and D. H. Fremlin, *On some properties of Hurewicz, Menger, and Rothberger*, Fund. Math. **129** (1988), no. 1, 17–33, DOI 10.4064/fm-129-1-17-33. MR954892

[N] A. Nowik, *Remarks about a transitive version of perfectly meager sets*, Real Anal. Exchange **22** (1996/97), no. 1, 406–412. MR1433627

[NSW] A. Nowik, M. Scheepers, and T. Weiss, *The algebraic sum of sets of real numbers with strong measure zero sets*, J. Symbolic Logic **63** (1998), no. 1, 301–324, DOI 10.2307/2586602. MR1610427

[NW1] A. Nowik and T. Weiss, *Not every Q-set is perfectly meager in the transitive sense*, Proc. Amer. Math. Soc. **128** (2000), no. 10, 3017–3024, DOI 10.1090/S0002-9939-00-05355-7. MR1664434

[NW2] A. Nowik and T. Weiss, *Strongly meager sets and their uniformly continuous images*, Proc. Amer. Math. Soc. **129** (2001), no. 1, 265–270, DOI 10.1090/S0002-9939-00-05499-X. MR1694343

[NW3] A. Nowik and T. Weiss, *On the Ramseyan properties of some special subsets of 2^ω and their algebraic sums*, J. Symbolic Logic **67** (2002), no. 2, 547–556, DOI 10.2178/jsl/1190150097. MR1905154

J. Pawlikowski, *A characterization of strong measure zero sets*, Israel Journal of Mathematics **93** (1996), 171–183.

[P] J. Pawlikowski, *A characterization of strong measure zero sets*, Israel J. Math. **93** (1996), 171–183, DOI 10.1007/BF02761100. MR1380640

[R] I. Recław, *Some additive properties of special sets of reals*, Colloq. Math. **62** (1991), no. 2, 221–226, DOI 10.4064/cm-62-2-221-226. MR1142923

[Rep] M. Repický, *A family of permitted trigonometric thin sets*, Proc. Amer. Math. Soc. **125** (1997), no. 1, 137–144, DOI 10.1090/S0002-9939-97-03516-8. MR1343721

[S] M. Scheepers, *Additive properties of sets of real numbers and an infinite game*, Quaestiones Math. **16** (1993), no. 2, 177–191. MR1234464

[Sier1] W. Sierpiński, *Sur un ensemble non denombrable, donc toute image continue est de mesure nulle.*, Fund. Math., vol. **11** (1928), 301–304.

[Sier2] W. Sierpiński, *Sur le produit combinatoire de deux ensemble jouissant de la propriété C.*, Fund. Math., vol. **24**, 48–50.

[TW] B. Tsaban and T. Weiss, *Products of special sets of real numbers*, Real Anal. Exchange **30** (2004/05), no. 2, 819–835. MR2177439

[W1] T. Weiss, *On perfectly meager sets in the transitive sense*, Proc. Amer. Math. Soc. **130** (2002), no. 2, 591–594, DOI 10.1090/S0002-9939-01-06073-7. MR1862142

[W2] T. Weiss, *A note on the intersection ideal $\mathcal{M} \cap \mathcal{N}$*, Comment. Math. Univ. Carolin. **54** (2013), no. 3, 437–445. MR3090421

[W3] T. Weiss, *More remarks on the intersection ideal $\mathcal{M} \cap \mathcal{N}$*, Comment. Math. Univ. Carolin. **59** (2018), no. 3, 311–316, DOI 10.14712/1213-7243.2015.252. MR3861554

[Z1] O. Zindulka, *Strong measure zero and meager-additive sets through the prism of fractal measures*, Comment. Math. Univ. Carolin. **60** (2019), no. 1, 131–155, DOI 10.14712/1213-7243.2015.277. MR3946667

[Z2] O. Zindulka, *Meager-additive sets in topological groups*, preprint (2019).

Institute of Mathematics, College of Science, Cardinal Stefan Wyszyński University, Dewajtis 5, 01-815 Warsaw, Poland

Email address: `tomaszweiss@o2.pl`

Contemporary Mathematics
Volume **755**, 2020
https://doi.org/10.1090/conm/755/15216

Borel Conjecture, dual Borel Conjecture, and other variants of the Borel Conjecture

Wolfgang Wohofsky

ABSTRACT. This survey article is about the Borel Conjecture and several variants (which are inspired by the Galvin-Mycielski-Solovay characterization of strong measure zero) such as the dual Borel Conjecture and the Marczewski Borel Conjecture. We discuss their status in various models of ZFC, and provide the values of the cardinal characteristics in Cichoń's diagram.

In particular, we prove necessary conditions for the (dual) Borel Conjecture, and give a detailed proof of Laver's theorem that the Borel Conjecture holds in the Laver model and of Carlson's theorem that the dual Borel Conjecture holds in the Cohen model. Moreover, we give an informal overview of the construction for the model of ZFC in which both the Borel Conjecture and the dual Borel Conjecture hold. We also discuss the notion of very meager (a weakening of strongly meager), and we demonstrate that the (dual) Borel Conjecture is consistent with a projective well-order of the reals.

1. Introduction, basic definitions, and general facts

The aim of this survey article is to present several concepts and results related to the Borel Conjecture and its variants, thereby celebrating the 100th birthday of Borel's conjecture from his 1919 paper [**Bor19**].

To mention a few: we present detailed proofs of the consistency of the Borel Conjecture (see Section 2) and the dual Borel Conjecture (see Section 3), and give an overview of the proof from [**GKSW14**] which shows that it is consistent that the Borel Conjecture and the dual Borel Conjecture hold simultaneously (see Section 4); moreover, we discuss several other variants of the Borel Conjecture; for instance, we summarize work from [**BW**] showing that the Marczewski Borel Conjecture (based on the Marczewski ideal s_0) is consistent (see Section 7).

In this introductory section, we discuss the relevant concepts which are needed later, and provide several basic facts. Moreover, we give references to the literature (see Section 1.7) and annotated contents of this paper (see Section 1.8) as well as a table listing all models mentioned later on (see Table 1 on page 151). More specifically, this section is structured as follows.

In Section 1.1, we recall the notion of *strong measure zero* and its basic properties, and quote from the 1919 paper [**Bor19**], in which Borel conjectured that all

2010 *Mathematics Subject Classification.* Primary 03E35; secondary 03E17.

The author gratefully acknowledges the following partial support by the Austrian Science Fund (FWF): Y 1012 (START) and I 4039 (International Project).

strong measure zero sets are countable: this was the birth of the *Borel Conjecture*, which turned out to be independent from ZFC. In Section 1.2, we consider the *Cantor space* 2^ω, in which we work for the rest of the paper.

In Section 1.3, we present the *Galvin-Mycielski-Solovay theorem*, saying that a set is strong measure zero if and only if it can be translated away from every meager set (for much more, see the survey article [**HZ20**] by Michael Hrušák and Ondřej Zindulka). In Section 1.4, we give the definition of *strongly meager*, obtained by replacing "meager" by "measure zero" in the Galvin-Mycielski-Solovay characterization of strong measure zero, and discuss the *dual Borel Conjecture.* In Section 1.5, we generalize both the Galvin-Mycielski-Solovay characterization of strong measure zero and the definition of strongly meager by introducing the concept of *$\mathcal{I}$-shiftable*, and discuss the associated *Borel Conjecture with respect to $\mathcal{I}$.*

In Section 1.6, we recall well-known *cardinal characteristics* of the continuum, provide general constructions of $\mathcal{I}$-shiftable sets under certain cardinal characteristics assumptions, and recall *Cichoń's diagram.*

In Section 1.7, we talk about prerequisites for reading this paper, provide an annotated list of books and survey articles, and comment on our forcing notation. In Section 1.8, we give annotated contents, as well as a list of all models mentioned in this paper (see Table 1).

1.1. Strong measure zero sets and the Borel Conjecture (BC). Let $\mathbb{R}$ denote the set of real numbers (i.e., the classical real line). For an interval $I \subseteq \mathbb{R}$, let $\lambda(I)$ denote its length (or, equivalently, its Lebesgue measure).

The following definition is implicit in [**Bor19**]:

Definition 1.1 (Borel, 1919). A set $X \subseteq \mathbb{R}$ is *strong measure zero* if for every sequence $\langle \varepsilon_k \colon k \in \omega \rangle$ of positive reals there exists a sequence $\langle I_k \colon k \in \omega \rangle$ of intervals such that $\lambda(I_k) \leq \varepsilon_k$ for each $k \in \omega$ and $X \subseteq \bigcup_{k\in\omega} I_k$.

Recall that $X \subseteq \mathbb{R}$ is (Lebesgue) measure zero if and only if for every $\varepsilon > 0$ there exists a sequence $\langle I_k \colon k \in \omega \rangle$ of intervals such that $\sum_{k\in\omega} \lambda(I_k) \leq \varepsilon$ and $X \subseteq \bigcup_{k\in\omega} I_k$; therefore, each strong measure zero set is measure zero (just let $\varepsilon_k := \varepsilon/2^{k+1}$).

Moreover, it is clear that each countable set is strong measure zero. It also follows quite easily from the definition that the union of countably many strong measure zero sets is strong measure zero (i.e., the collection of strong measure zero sets forms a σ-ideal; see [**BJ95**, Lemma 8.1.2(3)]).

No perfect set – and hence no uncountable Borel set (since any such set contains a perfect set) – is strong measure zero (see [**BJ95**, Lemma 8.1.4 and Corollary 8.1.5]); in particular (since there is a perfect set of measure zero), there are measure zero sets which are not strong measure zero.

So the collection of strong measure zero sets is a typical example for a collection of *special (or pathological) sets of real numbers* (see [**Mil84**] for more): there is no "nice" set (such as a Borel set) which belongs to a collection of this kind (unless the set is countable).

The question arises whether there are any *uncountable* strong measure zero sets.

In his 1919 paper [**Bor19**], Émile Borel observed that *each countable set is strong measure zero*, and *conjectured the converse*, i.e., that the only strong measure zero sets are the countable sets of reals:

> *Un ensemble énumérable a une mesure asymptotique inférieure à toute série donnée à l'avance; la réciproque me paraît exacte, mais je n'en possède pas de démonstration entièrement satisfaisante.*
>
> A countable set has asymptotic measure lower than any series given in advance; the converse appears true to me, but I don't have an entirely satisfactory proof thereof.

His conjecture turned out to be unprovable (in ZFC) but nevertheless became a well-known and extensively studied mathematical statement, usually referred to[1] as *Borel's conjecture*, or the Borel Conjecture:

DEFINITION 1.2. The *Borel Conjecture (BC)* is the assertion that there is *no uncountable* strong measure zero set.

In fact, the failure of the Borel Conjecture is consistent with ZFC: under the *continuum hypothesis* (CH) – the statement that the set of real numbers has the least possible size (i.e., $2^{\aleph_0} = \aleph_1$) –, it is (by *modern standards*) relatively easy to construct an uncountable strong measure zero set (see[2] Theorem 2.1), and hence the Borel Conjecture fails; this was first proved by Sierpiński in his 1928 paper [**Sie28**] by showing that every *Luzin*[3] *set* (a set of reals whose intersection with any meager set is countable) has strong measure zero; only about a decade later, Gödel established the consistency of CH (by showing that CH holds in his constructible universe L), thereby completing the arguments which altogether yield that the Borel Conjecture cannot be proved in ZFC.

On the other hand, the Borel Conjecture is consistent with ZFC (i.e., it cannot be disproved in ZFC either): in his celebrated 1976 paper [**Lav76**], Laver was able to construct a model of ZFC satisfying the Borel Conjecture (see Theorem 2.6), thereby showing that BC is actually independent from ZFC; most of Section 2 is dedicated to a detailed proof of this result.

1.2. Strong measure zero in the Cantor space 2^ω. The notion of strong measure zero can be generalized to any metric space (see [**BJ95**, Definition 8.1.6]): it is defined as in Definition 1.1 above, but instead of intervals I_n of length at most ε_n, arbitrary sets (or balls) B_n of *diameter* at most ε_n are used.

In particular, we can consider the notions of strong measure zero and Borel Conjecture in any *Polish space*, such as the classical real line $\mathbb{R}$, the unit interval $[0, 1]$, the circle S^1, the Cantor space 2^ω, the Baire space ω^ω, etc.; in fact, the respective Borel Conjecture is equivalent for all metric spaces (see [**BJ95**, Theorem

[1]Since the Borel Conjecture is not really a "conjecture" (any more), it is disputable whether the word "conjecture" should be used for a statement which turned out to be independent from ZFC; this applies even more to expressions such as "dual Borel Conjecture" (see Definition 1.7), "Marczewski Borel Conjecture" (see Definition 7.6), etc., especially because neither Borel nor Marczewski ever considered (let alone "conjectured") any of the respective statements.

Nevertheless, the expression "Borel Conjecture" became generally accepted and widely used, and the same holds true for "dual Borel Conjecture"; therefore we feel free to write "Borel Conjecture" as part of several expressions to denote the respective assertions (without "conjecture" being understood in its original meaning).

[2]. . . or Corollary 1.11 for a more general approach.

[3]According to [**CP04**, page xii], Luzin sets have been constructed by Luzin in 1914 (see [**Luz14**]), and also a year earlier by Mahlo (see [**Mah13**]).

8.1.8]), so we can freely choose our favorite version of "the reals" (as far as the Borel Conjecture is concerned).

For technical reasons, we are going to use (in the following, and throughout the rest of the paper) the Cantor space 2^ω as our "official" version of the reals. It is a compact, zero-dimensional (hence totally disconnected) Polish space with nice combinatorial properties, and the following[4] (complete) metric d on 2^ω: given $x, y \in 2^\omega$ with $x \neq y$, let $d(x,y) := 2^{-n}$, where

$$n := \min\{i \in \omega \colon x(i) \neq y(i)\}.$$

Given $s \in 2^{<\omega}$, let

$$[s] := \{x \in 2^\omega : x \supseteq s\}$$

denote the corresponding *basic clopen* set in 2^ω. Note that $\{[s] : s \in 2^{<\omega}\}$ forms a basis for the topology on 2^ω. Further note that the diameter of $[s]$ (with respect to the above metric d) is 2^{-n}, where $n := |s|$ is the length of s; using this, it is easy to see that being strong measure zero in 2^ω can be rephrased as follows:

Definition 1.3. A set $X \subseteq 2^\omega$ is *strong measure zero* ($X \in \mathcal{SN}$) if for every sequence $\langle n_k \colon k \in \omega \rangle$ of natural numbers there exists a sequence $\langle s_k \colon k \in \omega \rangle \subseteq 2^{<\omega}$ such that $|s_k| \geq n_k$ for each $k \in \omega$ and $X \subseteq \bigcup_{k \in \omega} [s_k]$.

Note that $\mathcal{SN}$ is supposed to abbreviate "strongly null" (which is an alternative expression for "strong measure zero").

Let μ be the *Lebesgue measure*, i.e., the usual product measure on 2^ω; in particular, we have

$$\mu([s]) = 2^{-|s|} \tag{1.1}$$

for $s \in 2^{<\omega}$; moreover, let

$$\mathcal{N} := \{H \subseteq 2^\omega : \mu(H) = 0\}$$

denote the σ-ideal of *(Lebesgue) measure zero* sets (or *null* sets) on the Cantor space 2^ω.

Observe that measure coincides with diameter in case of basic clopen sets $[s]$ (see (1.1)); consequently, it is easy to show that each strong measure zero set in 2^ω is indeed (Lebesgue) measure zero:

$$\mathcal{SN} \subseteq \mathcal{N}.$$

1.3. The Galvin-Mycielski-Solovay theorem (and 2^ω as a group). In the following, we discuss a key fact by Galvin-Mycielski-Solovay (see Theorem 1.4) which gives an equivalent characterization of being strong measure zero, involving meager sets and translations (for the original announcement of the result – without proof – consult [**GMS79**]; quite recently, another paper [**GMS17**] by the same authors appeared, this time containing a proof of their result); it opens up the possibility to replace the ideal of meager sets by any (ideal) $\mathcal{I} \subseteq \mathcal{P}(2^\omega)$, which gave rise to several new notions such as strongly meager (see Definition 1.6) and the dual Borel Conjecture (see Definition 1.7), or – more generally – $\mathcal{I}$-shiftable (see Definition 1.8) and the Borel Conjecture with respect to $\mathcal{I}$ (see Definition 1.9), e.g., the Borel Conjecture with respect to the Marczewski ideal s_0 (see Definition 7.6).

[4]In fact, for a compact metric space, being strong measure zero does not depend on the metric.

Let us mention that there is another survey paper in this volume (by Michael Hrušák and Ondřej Zindulka) which gives a detailed account on various aspects of the GMS theorem (see [**HZ20**, Section 3]).

The Galvin-Mycielski-Solovay theorem uses the *topological* as well as the *algebraic* structure of 2^ω: for $x, y \in 2^\omega$, let $x + y$ denote their bitwise sum modulo 2, i.e.,

$$(x + y)(i) := x(i) + y(i) \bmod 2$$

for each $i \in \omega$. Note that $(2^\omega, +)$ is an abelian *Polish group*; moreover, $-x = x$ for each $x \in 2^\omega$, so there is no difference between addition and subtraction.

For $x \in 2^\omega$ and $Y \subseteq 2^\omega$, let $x + Y$ denote the set Y "translated by x", i.e.,

$$x + Y := \{x + y : y \in Y\};$$

for $X, Y \subseteq 2^\omega$, let

$$X + Y := \bigcup_{x \in X} (x + Y)$$

denote the *Minkowski sum* of X and Y.

Recall that a set in a topological space is *nowhere dense* if the interior of its closure[5] is empty (in particular, $Y \subseteq 2^\omega$ is nowhere dense if for each $s \in 2^{<\omega}$ there is a $t \in 2^{<\omega}$ such that $t \supseteq s$ and $[t] \cap Y = \emptyset$); a set M is *meager* if there are countably many (closed) nowhere dense sets $\langle Y_n \rangle_{n \in \omega}$ such that $M \subseteq \bigcup_{n \in \omega} Y_n$ (i.e., the collection of meager sets is the σ-ideal generated from the ideal of nowhere dense sets). Let

$$\mathcal{M} := \{M \subseteq 2^\omega : M \text{ is meager}\}$$

denote the σ-ideal of meager sets on the Cantor space 2^ω.

THEOREM 1.4 (Galvin-Mycielski-Solovay, 1979). *A set $X \subseteq 2^\omega$ is strong measure zero if and only if $X + M \neq 2^\omega$ for each meager set M.*

It is easy to check that $X + M \neq 2^\omega$ if and only if X can be "translated away" from M, i.e., there is a "translation real" $z \in 2^\omega$ such that $(z + X) \cap M = \emptyset$, or, alternatively, $X \cap (z + M) = \emptyset$.

So the above theorem actually says that a set X is strong measure zero if and only if it can be translated away from each meager set, or – by taking complements – if and only if every co-meager (dense G_δ) set contains a translate of X.

PROOF OF THEOREM 1.4. We only show the easy direction of the theorem; for the other direction, see[6] [**BJ95**, Theorem 8.1.16], [**Mil84**], [**HWZ16**], or [**GMS17**]; for additional commentary, see Remark 1.5 below.

Indeed, the direction from right to left is quite trivial. Let $X \subseteq 2^\omega$ such that $X + M \neq 2^\omega$ for each meager set M; we have to show that X is strong measure zero (see Definition 1.3). So fix a sequence $\langle n_k : k \in \omega \rangle$ of natural numbers; we will find a sequence $\langle s_k : k \in \omega \rangle \subseteq 2^{<\omega}$ with $|s_k| = n_k$ for all $k \in \omega$ such that $X \subseteq \bigcup_{k \in \omega} [s_k]$.

Let

$$\mathbb{Q} := \{x \in 2^\omega : x(n) = 0 \text{ for all but finitely many } n \in \omega\}$$

[5] So the closure of a nowhere dense set is nowhere dense as well, hence each nowhere dense set is contained in a closed nowhere dense set.

[6] Note that the "original paper" [**GMS79**] by Galvin, Mycielski, and Solovay just "announces" the result (without proof); their recent paper [**GMS17**] also contains a proof.

be the (countable dense) set of "rationals" in the Cantor space, and let $\{q_k : k \in \omega\}$ be an enumeration of $\mathbb{Q}$. Moreover, let $s'_k := q_k \restriction n_k$ for each $k \in \omega$, and let

$$U := \bigcup_{k\in\omega} [s'_k].$$

Note that U is open dense: it is *open* as the union of the clopen (hence in particular open) sets $[s'_k]$, and it is *dense* since it contains the dense set $\mathbb{Q}$ as a subset.

Therefore, its complement $M := 2^\omega \setminus U$ is (closed) nowhere dense, hence in particular meager, so by assumption, $X + M \neq 2^\omega$; in other words, there exists a $z \in 2^\omega$ such that $X \cap (z + M) = \emptyset$, yielding

$$X \subseteq (z + U),$$

which gives a legitimate cover of X (just let $s_k := (z + q_k) \restriction n_k$). □

Note that Theorem 1.4 remains true if "meager" is replaced by "nowhere dense"; in fact, the following three properties are equivalent:

(1) X is strong measure zero,
(2) X is meager-shiftable (see also Definition 1.8),
(3) X is (nowhere dense)-shiftable;

the above proof of the easy direction of the theorem (i.e., (2) → (1)) indeed shows (3) → (1), whereas (1) → (2) is the hard direction of the theorem, and (2) → (3) is obvious.

REMARK 1.5. Actually, there would be a lot to say about (the proof of) the "difficult" direction of the Galvin-Mycielski-Solovay theorem (abbreviated by GMS), i.e., about the (various) strategies to prove that any strong measure zero set can be translated away from each meager set.

However, it is not really relevant for the theme of this paper, and – more importantly – there is the survey paper [**HZ20**] by Michael Hrušák and Ondřej Zindulka which discusses the GMS theorem in detail; so let us just mention a few aspects (which are not mentioned there).

The usual proof of (the difficult direction of) GMS (as it can be found in [**BJ95**, Theorem 8.1.16] or [**Mil84**]) is – in the following sense – misleading: when analyzing the proof, it *seems* that *compactness* of the group $(2^\omega, +)$ is used in an essential way; we are going to argue that this connection between compactness and GMS is only apparent, even though it is also supported by the fact that the GMS theorem generalizes[7] to all Polish groups which are (locally) compact, but typically[8] fails[9] for Polish groups which are not locally compact, such as the Baer-Specker group $(\mathbb{Z}^\omega, +)$, or the group $(S^\infty, \circ)$ of permutations on ω (for the proofs and many more details, see [**Woh13**, Chapter 5], [**HWZ16**], [**HZ16**], and [**HZ20**]).

Nevertheless, there is strong evidence for the viewpoint that the "essential use" of compactness is only apparent, and actually less is sufficient to prove GMS:

[7]Given any Polish group $(G, +)$, it makes sense to talk about the question whether the GMS theorem is valid for $(G, +)$: first of all, the notion of strong measure zero is defined in any metric space (see the first paragraph of Section 1.2), and second, the notion of meager as well as translations are provided by the structure of the Polish group.

[8]In fact, it has only been proved for a certain subclass of the class of Polish groups which are not locally compact.

[9]...consistently (more precisely: under the assumption $\mathrm{cov}(\mathcal{M}) = 2^{\aleph_0}$, hence, in particular, under CH)

the notions of strong measure zero (and[10] hence the Borel Conjecture) can be generalized to the *higher Cantor space* 2^κ (for κ being an uncountable regular cardinal), basically by replacing ω (in the combinatorial definition of strong measure zero on 2^ω; see Definition 1.3) by[11] some regular $\kappa > \omega$; this was first done by Aapo Halko in his PhD thesis (see [**Hal96**]; for more on strong measure zero sets on 2^κ, including the associated Borel Conjecture, see also [**HS01**]); in the PhD thesis of the author of this paper (see [**Woh13**, Chapter 5]), and independently by Michał Korch in his PhD thesis (see [**Kor18**]), it has been shown that the GMS theorem is valid[12] for 2^κ provided that 2^κ is "compact"[13] (which is equivalent to κ being *weakly compact*); however, upcoming work of the author of this paper (see [**Woh**]) is going to show that compactness is not necessary: κ being *(strongly) inaccessible* is sufficient[14] for GMS on 2^κ; since the same arguments apply to $\kappa = \omega$ as well, we can describe the situation as follows: actually, not the "weak compactness of ω" (i.e., König's lemma on 2^ω), but only the "inaccessibility of ω" is used to prove the difficult direction of GMS on 2^ω.

Let us give one (last) comment about the[15] *Borel Conjecture on* κ. In [**HS01**], Halko and Shelah show that BC on κ is inconsistent for successor cardinals κ, but the question whether BC is consistent on inaccessible cardinals remains open; indeed, it is not possible to generalize the usual proof of the consistency of BC (see Theorem 2.6, showing that BC holds after a countable support iteration of Laver forcing) in a straightforward way: this is due to the fact that any kind of (generalized) *Laver forcing on* κ necessarily adds a κ-Cohen real (see [**KKLW**]), which inevitably destroys the Borel Conjecture on κ (analogous to the situation on ω; see Corollary 2.4 and the discussion afterwards); in fact, it is unclear how to add κ-dominating reals without adding κ-Cohen reals (if at all possible), which would be analogous to the usual procedure for getting BC on ω; however, on κ, it might be unnecessary to add κ-dominating reals; in [**HS01**], it has only been proved that the dominating number on κ being large is necessary for BC on κ (not the bounding number being large, as in the ω-case; compare with Theorem 2.3); so it is only mandatory (and might be sufficient) to perform an iteration adding κ-unbounded reals (without adding κ-Cohen reals) in order to obtain a model satisfying the Borel Conjecture on κ.

1.4. Strongly meager sets and the dual Borel Conjecture (dBC). Replacing "meager" by "(Lebesgue) measure zero" in the Galvin-Mycielski-Solovay

---

[10] For another, yet completely different, generalization of the Borel Conjecture to uncountable cardinals κ, see [**GS13**] by Galvin and Scheepers, as well as [**Sch19**] by Scheepers.

[11] A set $X \subseteq 2^\kappa$ is *strong measure zero on* 2^κ if for every sequence $\langle \alpha_i : i \in \kappa \rangle$ of ordinals less than κ there exists a sequence $\langle s_i : i \in \kappa \rangle \subseteq 2^{<\kappa}$ such that $|s_i| \geq \alpha_i$ for each $i \in \kappa$ and $X \subseteq \bigcup_{i \in \kappa} [s_i]$ (where $[s] := \{x \in 2^\kappa : x \supseteq s\}$).

[12] This means that a set $X \subseteq 2^\kappa$ is strong measure zero on 2^κ (see footnote 11) if and only if X can be translated away – with respect to bitwise addition modulo 2 – from each κ-meager set in 2^κ (where a set is *κ-meager* if it is the union of κ many nowhere dense sets in 2^κ; here, the notion of nowhere dense is based on the so-called *bounded topology on* 2^κ, which is generated by basic clopen sets of the form $[s]$, for $s \in 2^{<\kappa}$).

[13] ...meaning that every cover of 2^κ consisting of open sets (with respect to the bounded topology on 2^κ; see footnote 12) has a subcover of size strictly less than κ.

[14] ...and (consistently) there are counterexamples to GMS for successor cardinals, so κ being inaccessible is essentially equivalent to GMS on 2^κ.

[15] The *Borel Conjecture on* κ is the assertion that there is *no* strong measure zero set on 2^κ *of size* κ^+ (note that every subset of 2^κ of size at most κ is strong measure zero on 2^κ).

characterization of strong measure zero (see Theorem 1.4) yields the following notion "dual" to strong measure zero:

DEFINITION 1.6 (Příkrý). A set $X \subseteq 2^\omega$ is *strongly meager* ($X \in \mathcal{SM}$) if $X + H \neq 2^\omega$ for each (Lebesgue) measure zero set H.

In the literature, "strongly first category" (sfc) is sometimes used instead of "strongly meager" (as "first category" is an alternative expression for "meager").

As in the case of Theorem 1.4, $X + H \neq 2^\omega$ if and only if X can be "translated away" from H, i.e., there is a "translation real" $z \in 2^\omega$ such that $(z + X) \cap H = \emptyset$. So a set X is strongly meager if it can be translated away from each measure zero set, or – by taking complements – if every set of Lebesgue measure 1 contains a translate of X.

Actually, each countable set X is strongly meager: let H be any measure zero set, and note that $X + H = \bigcup_{x \in X}(x + H)$ is the union of countably many measure zero sets (due to the fact that being measure zero is *translation-invariant*), and hence in particular $X + H \neq 2^\omega$ (due to the fact that the measure zero sets form a *proper*[16] *σ-ideal*).

Moreover, each strongly meager set is indeed meager (as the name suggests):

$$\mathcal{SM} \subseteq \mathcal{M};$$

this is not entirely obvious (at least not so obvious as the analogous fact for strong measure zero), but can be shown as follows. Let X be strongly meager; we have to show that X is meager. Recall that the reals can be partitioned into a meager and a measure zero set (see [**BJ95**, Lemma 2.1.6]), i.e., let $2^\omega = M \,\dot\cup\, H$ be the disjoint union of the meager set M and the measure zero set H (sometimes called *Marczewski partition* of the reals). Since X is strongly meager, we know that $X + H \neq 2^\omega$, in other words, there is $z \in 2^\omega$ such that $(z + X) \cap H = \emptyset$, and hence $(z + X) \subseteq M$; since being meager is translation-invariant, it follows that X is meager.

In contrast to the strong measure zero sets, the strongly meager sets do not (always) form an ideal (let alone a σ-ideal): in fact, CH implies that there are two strongly meager sets whose union is not strongly meager (for the involved proof, see [**BS01**]); on the other hand, they consistently form a σ-ideal: it is consistent that a set is strongly meager if and only if it is countable (i.e., the dual Borel Conjecture holds; see Definition 1.7 and Theorem 3.10). More information can be found in Section 5, which is concerned with the notion of very meager: this is a natural weakening of strongly meager, and closed under countable unions.

No perfect set (and hence no uncountable Borel set) is strongly meager (see [**BJ95**, Corollary 3.3.11]); in particular (since there is a perfect set which is meager), there are meager sets which are not strongly meager.

The following is the dual version of the Borel Conjecture:

DEFINITION 1.7. The *dual Borel Conjecture (dBC)* is the assertion that there is *no uncountable* strongly meager set.

Recall that the Borel Conjecture is independent from ZFC (see the discussion after Definition 1.2); in fact, the dual Borel Conjecture turns out to be independent

---

[16] Here, *proper* means that the whole space 2^ω is not in the ideal.

from ZFC as well: under CH, there is[17] an uncountable strongly meager set (see Theorem 3.1), and hence the dual Borel Conjecture fails; on the other hand, the dual Borel Conjecture is consistent with ZFC (which was conjectured by Příkrý); actually, Carlson showed[18] in his 1993 paper [**Car93**] that the dual Borel Conjecture holds true in the Cohen model (see Theorem 3.10); most of Section 3 is dedicated to a detailed proof of this result.

1.5. $\mathcal{I}$-shiftable sets and other variants of the Borel Conjecture. The following definition generalizes both the Galvin-Mycielski-Solovay characterization of strong measure zero sets (see Theorem 1.4) and Příkrý's definition of strongly meager sets (see Definition 1.6).

Let $\mathcal{I} \subseteq \mathcal{P}(2^\omega)$ be arbitrary (typically $\mathcal{I}$ is going to be a translation-invariant σ-ideal).

DEFINITION 1.8. A set $X \subseteq 2^\omega$ is *$\mathcal{I}$-shiftable* if $X + Z \neq 2^\omega$ for each $Z \in \mathcal{I}$.

Note that a set is $\mathcal{I}$-shiftable if and only if it can be translated away from each set in $\mathcal{I}$ (i.e., given $Z \in \mathcal{I}$, there is a "translation real" $z \in 2^\omega$ such that $(z+X) \cap Z = \emptyset$). Recall that $\mathcal{M}$ is the σ-ideal of meager sets, and $\mathcal{N}$ is the σ-ideal of measure zero sets; then a set is strong measure zero if and only if it is $\mathcal{M}$-shiftable (see Theorem 1.4), and a set is strongly meager if and only if it is $\mathcal{N}$-shiftable (see Definition 1.6).

Further note (see the respective discussion for strongly meager sets above) that each countable set is $\mathcal{I}$-shiftable provided that $\mathcal{I}$ is a translation-invariant proper σ-ideal (in fact, it is enough to assume that $\mathcal{I}$ has the following property: the whole space 2^ω cannot be covered by countably many translates of sets from $\mathcal{I}$, or – equivalently – $\mathcal{I}$ is *contained* in a translation-invariant proper σ-ideal).

We define the following generalization of the (dual) Borel Conjecture:

DEFINITION 1.9. The *Borel Conjecture with respect to $\mathcal{I}$* (or *$\mathcal{I}$-BC*) is the assertion that there is *no uncountable* $\mathcal{I}$-shiftable set.

With this definition, the Borel Conjecture is the $\mathcal{M}$-BC, and the dual Borel Conjecture is the $\mathcal{N}$-BC.

This paper is mainly concerned with (models of) the Borel Conjecture and/or the dual Borel Conjecture, but we are also going to mention other (stronger) variants (see Section 4.13): the Borel Conjecture with respect to $\mathcal{M} \cap \mathcal{N}$, and the Borel Conjecture with respect to $\mathcal{E}$ (where $\mathcal{E}$ is the σ-ideal generated by the closed measure zero sets); we do not know whether they are consistent.

Moreover, we are going to discuss the Borel Conjecture with respect to s_0 (the σ-ideal of Marczewski null sets, where a set is Marczewski null if each perfect set has a perfect subset disjoint from it); see the final Section 7, where we present results from the joint paper [**BW**] with Jörg Brendle: the s_0-BC follows from CH (due to the key ZFC result that there is no s_0-shiftable set of size continuum; see Theorem 7.7), and it holds in the Cohen model.

---

[17] Analogous to the strong measure zero case, it is also – under CH – straightforward to construct a *Sierpiński set* (a set of reals whose intersection with any measure zero set is countable); moreover, any Sierpiński set is strongly meager (see [**BJ95**, Theorem 8.5.13], or [**Paw96b**]), but showing this is far more difficult than showing that any Luzin set is strong measure zero; so Theorem 3.1 provides a much easier argument that the dual Borel Conjecture fails under CH.

[18] Carlson proved the theorem several years earlier, but the result was published only in 1993; see also footnote 71.

1.6. Facts involving cardinal characteristics, and Cichoń's diagram. First of all (see Section 1.6.1), we are going to recall the four well-known cardinal characteristics of an ideal $\mathcal{I}$ (see Figure 1); subsequently (see Section 1.6.2), we provide some general constructions of (uncountable) $\mathcal{I}$-shiftable sets under certain cardinal characteristics assumptions (see Proposition 1.10 and Corollary 1.11, as well as Proposition 1.12); we then recall the bounding number $\mathfrak{b}$ and the dominating number $\mathfrak{d}$ (see Section 1.6.3); finally (see Section 1.6.4), we will discuss the well-known Cichoń's diagram (see Figure 2).

For more on cardinal characteristics of the continuum and Cichoń's diagram, see [**Bla10**] and [**BJ95**].

1.6.1. *Cardinal characteristics of (an ideal) $\mathcal{I}$.* From now on, assume that $\mathcal{I} \subseteq \mathcal{P}(2^\omega)$ is a non-empty collection of sets of reals which is closed under subsets (i.e., for each $Y \subseteq Z \in \mathcal{I}$, also $Y \in \mathcal{I}$), satisfies $2^\omega \notin \mathcal{I}$, and contains all singletons (i.e., $\{z\} \in \mathcal{I}$ for each $z \in 2^\omega$; note that – for $\mathcal{I}$ being closed under subsets – this is equivalent to $\bigcup \mathcal{I} = 2^\omega$).

Such an $\mathcal{I}$ can be thought of as a "smallness notion": a set of reals is considered to be small (in the sense of $\mathcal{I}$) if and only if it belongs to $\mathcal{I}$. In general, we do not require $\mathcal{I}$ to be an ideal (i.e., to be closed under unions), but proper σ-ideals such as $\mathcal{M}$ (the meager sets) and $\mathcal{N}$ (the null sets) are typical examples.

Recall the four well-known *cardinal characteristics* associated to $\mathcal{I}$: let

$$\operatorname{add}(\mathcal{I}) := \min\{|\mathcal{Y}| : \mathcal{Y} \subseteq \mathcal{I} \wedge \bigcup \mathcal{Y} \notin \mathcal{I}\}$$

be the *additivity* of $\mathcal{I}$ (i.e., the smallest number of sets from $\mathcal{I}$ whose union is not in $\mathcal{I}$),

$$\operatorname{cov}(\mathcal{I}) := \min\{|\mathcal{Y}| : \mathcal{Y} \subseteq \mathcal{I} \wedge \bigcup \mathcal{Y} = 2^\omega\}$$

be the *covering number* of $\mathcal{I}$ (i.e., the smallest number of sets from $\mathcal{I}$ needed to cover the whole space 2^ω),

$$\operatorname{non}(\mathcal{I}) := \min\{|Y| : Y \subseteq 2^\omega \wedge Y \notin \mathcal{I}\}$$

be the *uniformity* of $\mathcal{I}$ (i.e., the least size of an *$\mathcal{I}$-positive* set), and

$$\operatorname{cof}(\mathcal{I}) := \min\{|\mathcal{B}| : \mathcal{B} \subseteq \mathcal{I} \wedge \forall Y \in \mathcal{I}\ \exists B \in \mathcal{B}\ (Y \subseteq B)\}$$

be the *cofinality* of $\mathcal{I}$ (i.e., the least size of a *basis* $\mathcal{B}$ of $\mathcal{I}$).

There are several easy inequalities involving these cardinal characteristics (see Figure 1, where arrows denote "$\leq$"): by definition, $\operatorname{add}(\mathcal{I}) \geq \aleph_0$ for any ideal $\mathcal{I}$, and $\operatorname{add}(\mathcal{I}) \geq \aleph_1$ provided that $\mathcal{I}$ is a σ-ideal; it is easy to check (using $2^\omega \notin \mathcal{I}$) that $\operatorname{add}(\mathcal{I}) \leq \operatorname{cov}(\mathcal{I})$ as well as $\operatorname{non}(\mathcal{I}) \leq \operatorname{cof}(\mathcal{I})$; moreover, it is easy to check (using that $\mathcal{I}$ contains all singletons) that $\operatorname{add}(\mathcal{I}) \leq \operatorname{non}(\mathcal{I})$ as well as $\operatorname{cov}(\mathcal{I}) \leq \operatorname{cof}(\mathcal{I})$; finally, note that $\operatorname{cof}(\mathcal{I}) \leq 2^{\aleph_0}$ provided[19] that $\mathcal{I}$ has a basis consisting of Borel sets (since the are only continuum many Borel sets); however, it can happen that the cofinality of an ideal is strictly above the continuum (a natural example is the Marczewski ideal: in fact, $\operatorname{cof}(s_0) > 2^{\aleph_0}$ holds true in ZFC; see Theorem 7.4).

[19] [**BJ95**, Lemma 1.3.2] lists the facts mentioned in this paragraph, but forgets to mention any assumption justifying $\operatorname{cof}(\mathcal{I}) \leq 2^{\aleph_0}$.

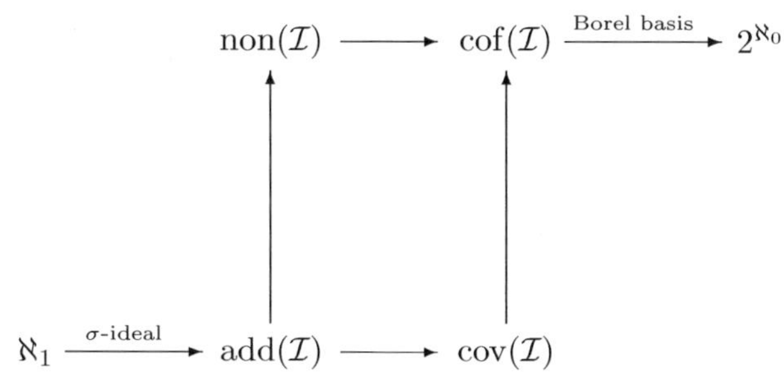

FIGURE 1. Cardinal characteristics of (an ideal) $\mathcal{I}$

1.6.2. *General constructions of $\mathcal{I}$-shiftable sets.* We start by proving the following rather easy (but very general) fact on constructing (uncountable) $\mathcal{I}$-shiftable sets (which in particular yields the failure of both BC and dBC under CH; see Corollary 1.11 and the discussion afterwards):

PROPOSITION 1.10. *Assume that $\mathcal{I} \subseteq \mathcal{P}(2^\omega)$ is* translation-invariant *(i.e., for each $z \in 2^\omega$ and $Z \in \mathcal{I}$, also $z + Z \in \mathcal{I}$), and*

$$\operatorname{cov}(\mathcal{I}) = \operatorname{cof}(\mathcal{I}) =: \lambda.$$

Then there exists an $\mathcal{I}$-shiftable set $X \subseteq 2^\omega$ with $|X| = \lambda$.

PROOF. Since $\operatorname{cof}(\mathcal{I}) = \lambda$, we can fix a basis $\langle B_\alpha : \alpha < \lambda \rangle \subseteq \mathcal{I}$ of $\mathcal{I}$, i.e., for each $Z \in \mathcal{I}$ there exists $\alpha < \lambda$ such that $Z \subseteq B_\alpha$. The idea is to construct the elements of $X = \{x_\alpha : \alpha < \lambda\}$ by induction, in such a way that $X + B_\alpha \neq 2^\omega$ for each $\alpha < \lambda$; to ensure this, we simultaneously construct an auxiliary sequence $\langle y_\alpha : \alpha < \lambda \rangle$ which is going to satisfy $y_\alpha \notin X + B_\alpha$ for each $\alpha < \lambda$.

More precisely, we proceed as follows. By induction on $\alpha < \lambda$, we construct two sequences $\langle x_\alpha : \alpha < \lambda \rangle$ and $\langle y_\alpha : \alpha < \lambda \rangle$ such that for each $\alpha < \lambda$, we have

(1) $x_\alpha \notin \bigcup_{\beta \leq \alpha}(y_\beta + B_\beta) \cup \{x_\beta : \beta < \alpha\}$, and
(2) $y_\alpha \notin \bigcup_{\beta < \alpha}(x_\beta + B_\alpha)$.

Assume that $\langle x_\beta \rangle_{\beta<\alpha}$ and $\langle y_\beta \rangle_{\beta<\alpha}$ have already been constructed; we want to find x_α and y_α satisfying (1) and (2) above.

Since $B_\alpha \in \mathcal{I}$, and $\mathcal{I}$ is translation-invariant, $x_\beta + B_\alpha \in \mathcal{I}$ for each $\beta < \alpha$, hence $\operatorname{cov}(\mathcal{I}) = \lambda$ together with $|\alpha| < \lambda$ yields

$$\bigcup_{\beta<\alpha} (x_\beta + B_\alpha) \neq 2^\omega,$$

and so we can pick $y_\alpha \in 2^\omega$ satisfying (2). Similarly as above, $y_\beta + B_\beta \in \mathcal{I}$ for each $\beta \leq \alpha$; since $\mathcal{I}$ contains each singleton, $\operatorname{cov}(\mathcal{I}) = \lambda$ yields

$$\bigcup_{\beta\leq\alpha} (y_\beta + B_\beta) \cup \{x_\beta : \beta < \alpha\} \neq 2^\omega,$$

so we can pick $x_\alpha \in 2^\omega$ satisfying (1). This finishes the construction of the two sequences.

Now let

$$X := \{x_\alpha : \alpha < \lambda\}.$$

Note that all the elements x_α are distinct by construction (see (1)), so $|X| = \lambda$. It remains to show that X is $\mathcal{I}$-shiftable; so fix $Z \in \mathcal{I}$; we have to show that $X + Z \neq 2^\omega$. Fix $\alpha < \lambda$ such that $Z \subseteq B_\alpha$. It suffices to show that $y_\alpha \notin x_\beta + B_\alpha$ for any $\beta \in \lambda$:

- if $\beta < \alpha$, then $y_\alpha \notin x_\beta + B_\alpha$ by (2), and
- if $\beta \geq \alpha$, then $x_\beta \notin y_\alpha + B_\alpha$ by (1), hence $y_\alpha \notin x_\beta + B_\alpha$,

and the proof is finished. □

It is easy to derive the following:

COROLLARY 1.11. *Let $\mathcal{I}$ be a translation-invariant σ-ideal with a Borel basis. Then CH implies that there exists an uncountable $\mathcal{I}$-shiftable set (i.e., $\mathcal{I}$-BC fails).*

PROOF. Since $\mathcal{I}$ is a σ-ideal and has a Borel basis,

$$\aleph_1 \leq \operatorname{add}(\mathcal{I}) \leq \operatorname{cov}(\mathcal{I}) \leq \operatorname{cof}(\mathcal{I}) \leq 2^{\aleph_0}$$

(see the discussion before Proposition 1.10 and Figure 1). By CH, we have $\operatorname{cov}(\mathcal{I}) = \operatorname{cof}(\mathcal{I}) = \aleph_1$, so Proposition 1.10 yields an $\mathcal{I}$-shiftable set of size $\aleph_1$. □

Since $\mathcal{M}$ and $\mathcal{N}$ are translation-invariant σ-ideals with a Borel[20] basis, the above corollary in particular yields the failure of both $\mathcal{M}$-BC (i.e., the Borel Conjecture) and $\mathcal{N}$-BC (i.e., the dual Borel Conjecture) under CH (see also Theorem 2.1 and Theorem 3.1).

Moreover, it is easy to show that there are uncountable $\mathcal{I}$-shiftable sets if the covering number of $\mathcal{I}$ is larger than $\aleph_1$; indeed, the following holds:

PROPOSITION 1.12. *Assume that $\mathcal{I} \subseteq \mathcal{P}(2^\omega)$ is translation-invariant. Then every $X \subseteq 2^\omega$ with $|X| < \operatorname{cov}(\mathcal{I})$ is $\mathcal{I}$-shiftable.*

PROOF. Let $X \subseteq 2^\omega$ with $|X| < \operatorname{cov}(\mathcal{I})$. Fix $Z \in \mathcal{I}$; we have to show that $X + Z \neq 2^\omega$.

Since $\mathcal{I}$ is translation-invariant, $x + Z \in \mathcal{I}$ for any $x \in 2^\omega$; by definition of $\operatorname{cov}(\mathcal{I})$, the whole space 2^ω cannot be covered by fewer than $\operatorname{cov}(\mathcal{I})$ many meager sets, so $|X| < \operatorname{cov}(\mathcal{I})$ yields

$$X + Z = \bigcup_{x \in X} (x + Z) \neq 2^\omega,$$

which finishes the proof. □

We provide the following corollary which is going to yield two of the (easy) inequalities of Cichoń's diagram below (see Figure 2, and Corollary 1.14):

COROLLARY 1.13. *Assume that $\mathcal{I}, \mathcal{J} \subseteq \mathcal{P}(2^\omega)$ are translation-invariant and* orthogonal *to each other (i.e., there are $A \in \mathcal{I}$ and $B \in \mathcal{J}$ with $A \cup B = 2^\omega$).*

Then $\operatorname{cov}(\mathcal{I}) \leq \operatorname{non}(\mathcal{J})$ holds true.

PROOF. Fix $A \in \mathcal{I}$ and $B \in \mathcal{J}$ with $A \cup B = 2^\omega$. Let $X \subseteq 2^\omega$ with $|X| < \operatorname{cov}(\mathcal{I})$; we have to show that $X \in \mathcal{J}$.

First note that X is $\mathcal{I}$-shiftable by Proposition 1.12; since $A \in \mathcal{I}$, we can fix (see the discussion after Definition 1.8) a real $z \in 2^\omega$ such that $(z + X) \cap A = \emptyset$.

[20] In fact, $\mathcal{M}$ has even an F_σ basis (since every meager set is covered by a countable union of *closed* nowhere dense sets), and $\mathcal{N}$ has even a G_δ basis (since any null set is – for every $\varepsilon > 0$ – covered by an *open* set of Lebesgue measure at most ε, hence any null set is contained in a null set which is the intersection of countably many open sets).

Consequently, $z + X \subseteq B$ (or, equivalently, $X \subseteq z + B$); since $B \in \mathcal{J}$ and $\mathcal{J}$ is translation-invariant, $X \in \mathcal{J}$ as well, as desired. $\square$

1.6.3. *Bounding number* $\mathfrak{b}$ *and dominating number* $\mathfrak{d}$. Recall that $\leq^*$ denotes *eventual domination* on ω^ω, i.e., given $f, g \in \omega^\omega$,

$$f \leq^* g \;:\Longleftrightarrow\; f(k) \leq g(k) \text{ for all but finitely many } k \in \omega.$$

We say that $F \subseteq \omega^\omega$ is *unbounded* if there is *no* single $g \in \omega^\omega$ such that $f \leq^* g$ for each $f \in F$, and we say that F is *dominating* if for every $h \in \omega^\omega$ there is an $f \in F$ with $h \leq^* f$. Let

$$\mathfrak{b} := \min\{|F| \colon F \subseteq \omega^\omega \;\wedge\; F \text{ is unbounded}\}$$

be the *(un)bounding number*, and

$$\mathfrak{d} := \min\{|F| \colon F \subseteq \omega^\omega \;\wedge\; F \text{ is dominating}\}$$

be the *dominating number*; it is easy to check that $\aleph_1 \leq \mathfrak{b} \leq \mathfrak{d} \leq 2^{\aleph_0}$.

1.6.4. *Cichoń's diagram.* Finally, we are going to recall and discuss "Cichoń's diagram", a well-known collection of (ZFC provable) relations between several cardinal characteristics of the continuum.

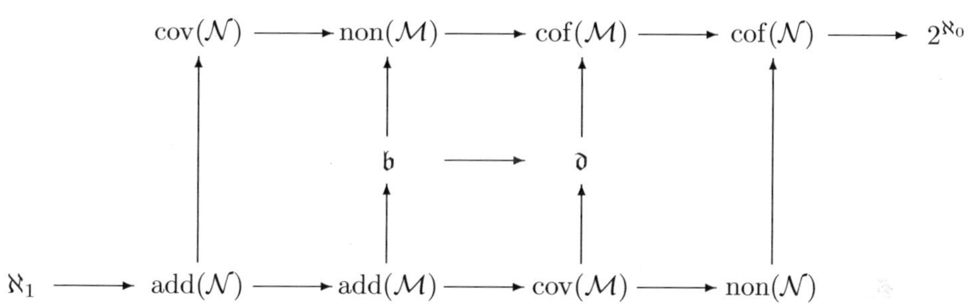

FIGURE 2. Cichoń's diagram

Cichoń's diagram (see Figure 2, where arrows denote "$\leq$") summarizes the ZFC provable inequalities between the following 10 cardinal characteristics: the additivity, the covering number, the uniformity, and the cofinality of both the ideal $\mathcal{M}$ of meager sets and the ideal $\mathcal{N}$ of measure zero sets, as well as $\mathfrak{b}$ and $\mathfrak{d}$, the two cardinal characteristics connected with $\leq^*$. Moreover, ZFC proves that

$$\text{add}(\mathcal{M}) = \min(\mathfrak{b}, \text{cov}(\mathcal{M})) \;\wedge\; \text{cof}(\mathcal{M}) = \max(\text{non}(\mathcal{M}), \mathfrak{d}). \tag{1.2}$$

For the proofs, we refer the reader to [**BJ95**, Chapter 2] which is devoted to establishing the ZFC relations expressed in Cichoń's diagram (some of them are very deep, e.g., $\text{add}(\mathcal{N}) \leq \text{add}(\mathcal{M})$).

Here, we only prove two inequalities (since they are easy, and the first of them appears in the proof of Theorem 2.5 and is relevant in the context of Theorem 3.30):

COROLLARY 1.14. *The following holds true in ZFC:*

- $\text{cov}(\mathcal{M}) \leq \text{non}(\mathcal{N})$
- $\text{cov}(\mathcal{N}) \leq \text{non}(\mathcal{M})$

PROOF. Since the ideal $\mathcal{M}$ of meager sets and the ideal $\mathcal{N}$ of measure zero sets are translation-invariant and orthogonal to each other (i.e., the reals can be partitioned into a meager and a measure zero set; see [**BJ95**, Lemma 2.1.6]), the two inequalities immediately follow from Corollary 1.13. □

Note that Cichoń's diagram trivializes[21] under CH: in this case, all involved cardinals are $\aleph_1$.

On the other hand, Cichoń's diagram (together with (1.2)) is everything one can prove in ZFC about the involved cardinal characteristics, at least when models of $2^{\aleph_0} = \aleph_2$ are concerned: it has been shown that for all assignments (there are actually 23 of them) of $\aleph_1$ and $\aleph_2$ to the cardinal characteristics which do not contradict Cichoń's diagram (and (1.2)), there is a model of ZFC realizing the assignment (see [**BJ95**, Chapter 7] for the construction of these models).

However, for continuum larger than $\aleph_2$, it is not clear whether every configuration not contradicting Cichoń's diagram[22] is consistent; in other words: are there any ZFC provable "relations" between three or more of the entries of Cichoń's diagram which have not been discovered yet (in particular, do not follow from the relations in Cichoń's diagram)?

Throughout the rest of the paper, we proceed as follows: whenever we discuss a certain model of ZFC (with $2^{\aleph_0} = \aleph_2$), we are going to present Cichoń's diagram displaying the values of all ten cardinal characteristics in the model; we follow the convention of [**BJ95**, Chapter 7] and put □ when the cardinal characteristic at the respective position is $\aleph_1$, and ■ when it is $\aleph_2$. As an example, have a look at Figure 5, displaying the values of the cardinal characteristics in Laver's model of the Borel Conjecture (see Theorem 2.6), or Figure 15, displaying the respective values in the Cohen model (which satisfies the dual Borel Conjecture; see Theorem 3.10).

1.7. Prerequisites, classical literature, and forcing notation. In the following, we discuss some prerequisites for reading this paper (see Section 1.7.1), provide an annotated list of references to books and survey articles related to the topic of this paper (see Section 1.7.2), and comment on the alphabet convention which we adopt for our notation of forcing (see Section 1.7.3).

1.7.1. *Prerequisites and forcing.* To understand this paper, a certain amount of knowledge about set theory is required: in particular, we assume that the reader is familiar with the basics of *(iterated) forcing*, including some experience with finite and countable support iterations of (tree) forcings.

Forcing is a technique to generate new models of ZFC with prescribed properties. It was developed by Paul Cohen, who used it in his 1963 proof of the independence of the continuum hypothesis (and the axiom of choice).

For details on forcing, we have to refer the reader to the literature (see Section 1.7.2 below).

Furthermore, it is certainly an advantage if the reader has already seen some tree forcings, and has heard a little bit about set theory of the reals and cardinal

[21] However, it is possible to define classes of small sets corresponding to the entries of Cichoń's diagram (see, e.g., [**PR95**]); there are ZFC provable inclusions between these classes corresponding to the relations in Cichoń's diagram; so, somehow, it *does* make sense to consider Cichoń's diagram under CH.

[22] By now, there are numerous papers dealing with the consistency of several of these configurations: in some of them, as many cardinals (in Cichoń's diagram) as possible are separated; see, e.g., [**GKS19**] or [**GKMS**], where also open questions are listed (see [**GKMS**, Section 5]).

characteristics of the continuum. However, we tried to introduce and discuss most of the relevant concepts in this introductory Section 1, as well as later on when needed, sometimes explicitly pointing to the literature.

1.7.2. *Annotated list of books and survey articles.* We provide an annotated list of references to several classical books and survey articles related to the topic of this paper (given in chronological order):

- [**Kun83**] Kunen's book "Set theory, an introduction to independence proofs" is *the* classical source for an *introduction to* the theory of *forcing*; for the new version, see [**Kun11**].
- [**Bau83**] Baumgartner's survey article gives an introduction to the technique of *iterated forcing* including *countable support.*
- [**Mil84**] Arnold Miller's survey article deals with *special subsets of the real line* (such as strong measure zero sets).
- [**Gol93**] Goldstern's paper "Tools for your forcing construction" is a good reference for *iterated forcing*; in particular, it excellently presents certain *preservation theorems* for *countable support* iteration of *proper forcing.*
- [**BJ95**] "Set theory: On the structure of the real line" by Bartoszyński and Judah is one of the standard references for the area of *set theory of the reals*; in particular, the book deals with the σ-ideals $\mathcal{M}$ and $\mathcal{N}$ of *meager* and *null* sets, respectively, and *Cichoń's diagram*; in its Chapter 8, numerous results on *strong measure zero* sets and related concepts are presented.
- [**Gol98**] Goldstern's article "A taste of proper forcing" gives a brief introduction to the concept of *proper forcing.*
- [**She98**] Shelah's book "Proper and improper forcing" contains numerous deep results on the theory of forcing, especially techniques for the *iteration of proper forcing,* including many very general *preservation theorems.*
- [**Jec03**] Jech's book "Set theory" gives an *encyclopedic view* of the state of the art in *set theory,* including the theory of *(iterated) forcing, (ultra)filters and ideals, Borel sets/codes,* etc.
- [**Bla10**] Blass' survey article gives a thorough overview of (combinatorial) *cardinal characteristics* of the continuum.
- [**Hal17**] Halbeisen's book "Combinatorial set theory" gives an introduction to *forcing,* and its connection with *cardinal characteristics* of the continuum.

1.7.3. *Alphabet convention for the notation of forcing.* Traditionally, there are two (contradictory) notations for interpreting a partial order as a forcing notion. We use the "Boolean" or "downwards" notation: if $(\mathbb{P}, \leq)$ is a forcing partial order, $q \leq p$ means "q extends p", "q is stronger than p", or "q has more information than p".

To avoid confusion, we employ the *alphabet convention* (see [**Gol98**]):

> Whenever two conditions are comparable, the notation is chosen so that the variable used for the stronger condition comes "lexicographically" later.

Consequently, we write, e.g., $q \leq p$ (for q stronger than p), but try to avoid expressions such as $p \leq r$ (for p stronger than r).

1.8. Annotated contents of this paper. For every section of this survey paper, we give a short description of its content.

We also provide a list of all models mentioned in this paper (see Table 1), for each model indicating the status of the Borel Conjecture and the dual Borel Conjecture, and pointing to the respective theorems, as well as to the figures displaying their Cichoń's diagrams.

Section 1 is this introductory section: it gives an overview of the concepts relevant for this paper, such as strong measure zero, Borel Conjecture, Galvin-Mycielski-Solovay theorem, strongly meager, dual Borel Conjecture, $\mathcal{I}$-shiftable, $\mathcal{I}$-BC, cardinal characteristics, Cichoń's diagram, etc.

Section 2 is concerned with strong measure zero sets and models of the Borel Conjecture: most of the section is dedicated to a proof that Laver's model [**Lav76**] satisfies the Borel Conjecture; we also discuss other models as well as necessary conditions for the Borel Conjecture.

Section 3 is concerned with strongly meager sets and models of the dual Borel Conjecture: most of the section is dedicated to a proof of Carlson's theorem [**Car93**] that the dual Borel Conjecture holds true in the Cohen model; we also discuss other models as well as necessary conditions for the dual Borel Conjecture.

Section 4 gives an overview of the proof of the following theorem: it is consistent that the Borel Conjecture and the dual Borel Conjecture hold simultaneously (this is joint work with Martin Goldstern, Jakob Kellner, and Saharon Shelah; see [**GKSW14**]); moreover, we discuss models for all combinations of BC and dBC being true or false, and stronger (perhaps inconsistent) versions of BC+dBC.

Section 5 discusses the notion of very meager sets and the respective stronger version of the dual Borel Conjecture, which holds both in the Cohen model and the model for BC+dBC presented in Section 4 (see [**SW16**]).

Section 6 contains unpublished joint work with Sy D. Friedman, showing that the Borel Conjecture (as well as the dual Borel Conjecture) is consistent with the existence of a projective well-order of the real numbers (see [**Woh13**, Chapter 4]).

Section 7 is concerned with s_0-shiftable sets and the Marczewski Borel Conjecture, summarizing joint work [**BW**] with Jörg Brendle: in ZFC, there are no s_0-shiftable sets of size continuum, and hence MBC holds under CH; moreover, MBC holds in the Cohen model.

2. Borel Conjecture (BC)

This section is concerned with strong measure zero sets and models of the Borel Conjecture. First, we recall well-known necessary conditions for the Borel Conjecture and argue why they are not sufficient (see Section 2.1). Subsequently, most of the section is dedicated to a detailed proof of Laver's theorem [**Lav76**] that the Borel Conjecture is consistent (see Theorem 2.6); at the end (see Section 2.7), we discuss alternative approaches and several models of the Borel Conjecture other than the original one constructed by Laver.

2.1. Necessary conditions for the Borel Conjecture. Before discussing the two well-known necessary conditions (involving cardinal characteristics from

TABLE 1. Status of BC and dBC in the models mentioned in this paper

theorem	Cichoń	name of model	BC	dBC	weak BC	weak dBC
Corollary 2.4	Figure 3	(necessary for BC)				
Theorem 2.5	Figure 4	Hechler $* \mathbb{B}_{\omega_1}$	no	?	?	?
Theorem 2.6	Figure 5	Laver	yes	?		yes
Theorem 2.23	Figure 6	Mathias	yes	?		?
Theorem 2.31	Figure 7	Laver-random	yes	no		no
Conjecture 2.32	Figure 8	Mathias-random	yes?	no		no
Theorem 2.34	Figure 9	Laver $* \mathbb{B}_{\omega_2}$	yes	no		no
Theorem 2.34	Figure 9	Mathias $* \mathbb{B}_{\omega_2}$	yes	no		no
Corollary 2.37	Figure 10	random	no	no	yes	no
Theorem 2.39	Figure 11	Corazza	no	no	yes	?
Theorem 2.40	Figure 12	$\mathfrak{h} < \mathrm{add}(s_0)$	?	?	?	?
Theorem 2.43	Figure 12	$\mathfrak{h} < \mathrm{add}(s_0) + \mathrm{BC}$	yes	?		?
Corollary 3.8	Figure 13	(necessary for dBC)				
Theorem 3.9	Figure 14	Cohen $* \mathbb{B}_{\omega_1}$	no	no	?	no
Theorem 3.10	Figure 15	Cohen	no	yes	no	
Theorem 3.28	Figure 16	Hechler	no	yes	no	
Theorem 3.29	Figure 16	MA(σ-centered)	no	yes	no	
Theorem 3.30	Figure 17	dBC w/o Cohens	?	yes	?	
Theorem 3.31	Figure 18	Miller	no	?	?	yes
Theorem 3.31	Figure 5	Laver (see above)				
Theorem 3.32	Figure 21	Sacks (see below)				
Question 3.33	Figure 5	Laver (see above)				
Theorem 4.1	Figure 19	BC+dBC	yes	yes		
Question 4.2	Figure 20	(nec. for BC+dBC)				
Corollary 4.3		CH	no	no	no	no
Theorem 4.4	Figure 21	Sacks	no	no	yes	yes
Theorem 4.4	Figure 21	Silver	no	no	yes	yes?
Theorem 4.5	Figure 10	random (see above)				
Theorem 4.6	Figure 22	Cohen-random	no	no	no	no
Theorem 4.7	Figure 23	amoeba	no	no	no	no
Theorem 4.8	Figure 23	MA	no	no	no	no
Theorem 5.11	Figure 15	Cohen (see above)				
Theorem 5.12	Figure 19	BC+dBC (s. ab.)				
Theorem 6.1	Figure 5	BC + proj. w.o.	yes	?		?
Theorem 6.3	Figure 16	dBC + proj. w.o.	no	yes	no	
Corollary 7.18		CH (see above)				
Theorem 7.19	Figure 15	Cohen (see above)				
Theorem 7.20	Figure 21	Sacks (see above)				

Cichoń's diagram) for the Borel Conjecture to hold (see Corollary 2.4), we first recall why $2^{\aleph_0} > \aleph_1$ is a necessary requirement; in other words, the Borel Conjecture necessarily fails under the continuum hypothesis.

In fact, this immediately follows from Corollary 1.11 and (the easy direction of) Theorem 1.4: since $\mathcal{M}$ is a translation-invariant σ-ideal and has a basis consisting of Borel sets, Corollary 1.11 yields an uncountable $\mathcal{M}$-shiftable set; this set, in turn, is strong measure zero by the easy direction of the Galvin-Mycielski-Solovay theorem (see Theorem 1.4).

Nevertheless, we also want to present a completely elementary direct proof (without using the notion of $\mathcal{M}$-shiftable or the Galvin-Mycielski-Solovay theorem, but only the "original" definition of strong measure zero):

THEOREM 2.1. *Assume CH. Then there exists an uncountable strong measure zero set.*

PROOF. In the definition of strong measure zero (see Definition 1.3), there are continuum many "challenges" $\langle n_k \colon k \in \omega \rangle$, so, by CH, we can enumerate them as $\langle \langle n_k^\alpha \colon k \in \omega \rangle \colon \alpha < \omega_1 \rangle$. As in the proof of Theorem 1.4, let $\{q_k : k \in \omega\}$ be an enumeration of $\mathbb{Q}$ (the set of rationals in the Cantor space).

For each $\alpha < \omega_1$, let[23] $s_{2k}^\alpha := q_k \restriction n_{2k}^\alpha$ for each $k \in \omega$, and let

$$U^\alpha := \bigcup_{k \in \omega} [s_{2k}^\alpha];$$

note that U^α is open, and, since $\mathbb{Q} \subseteq U^\alpha$, dense.

We now construct an uncountable strong measure zero set $X := \{x_\alpha \colon \alpha < \omega_1\}$ by induction, as follows: given $\alpha < \omega_1$, pick

$$x_\alpha \in \bigcap_{\beta \leq \alpha} U^\beta \setminus \{x_\beta \colon \beta < \alpha\};$$

this is possible, since $\bigcap_{\beta \leq \alpha} U^\beta \setminus \{x_\beta : \beta < \alpha\} = \bigcap_{\beta \leq \alpha} U^\beta \cap \bigcap_{\beta < \alpha} 2^\omega \setminus \{x_\beta\}$ is a countable intersection of open dense sets which is (dense, hence in particular) non-empty by the Baire Category Theorem.[24]

It remains to show that X is strong measure zero (note that X is of size $\aleph_1$). So fix a sequence $\langle n_k \colon k \in \omega \rangle$ of natural numbers; we are going to find a sequence $\langle s_k \colon k \in \omega \rangle \subseteq 2^{<\omega}$ with $|s_k| = n_k$ for all $k \in \omega$ such that $X \subseteq \bigcup_{k \in \omega} [s_k]$.

Fix $\alpha < \omega_1$ such that $\langle n_k^\alpha \colon k \in \omega \rangle$ equals the prescribed sequence. Then, by construction, $x_\gamma \in U^\alpha$ for each $\gamma \geq \alpha$, hence $X \setminus U^\alpha \subseteq \{x_\beta \colon \beta < \alpha\}$ is a countable[25] set; let $\{z_k \colon k \in \omega\}$ be such that $X \setminus U^\alpha \subseteq \{z_k \colon k \in \omega\}$, and let $s_{2k+1} := z_k \restriction n_{2k+1}$ for each $k \in \omega$, yielding $X \subseteq U^\alpha \cup \bigcup_{k \in \omega} [s_{2k+1}]$, as desired. □

We now show that the Borel Conjecture necessarily fails under certain assumptions about cardinal characteristics from Cichoń's diagram (see Section 1.6, and, in particular, Figure 2).

First of all, this is the case if $\mathrm{cov}(\mathcal{M})$ is large, i.e., if the reals cannot be covered by $\aleph_1$ many meager (or, equivalently, nowhere dense) sets:

THEOREM 2.2. *Assume $\mathrm{cov}(\mathcal{M}) > \omega_1$; then there exists an uncountable strong measure zero set. (In fact, each set of size strictly less than $\mathrm{cov}(\mathcal{M})$ is strong measure zero.)*

PROOF. This immediately follows from Proposition 1.12 and (the easy direction of) Theorem 1.4: let $X \subseteq 2^\omega$ with $|X| < \mathrm{cov}(\mathcal{M})$; since $\mathcal{M}$ is translation-invariant, we can apply Proposition 1.12 to derive that X is $\mathcal{M}$-shiftable (see Definition 1.8); by the easy direction of the Galvin-Mycielski-Solovay theorem (see Theorem 1.4), X is strong measure zero. □

[23] We "save" the other half of $\langle n_k^\alpha : k \in \omega \rangle$ for later, to cover countably many remaining points.

[24] This is basically the fact that $\mathrm{cov}(\mathcal{M}) > \aleph_0$ holds true in ZFC.

[25] Note that X is *concentrated on the rationals*, i.e., all but countably many points from X belong to any given open cover of the rationals. A set of this kind is always strong measure zero, as our proof shows.

Note that the above result can be rephrased as

$$\operatorname{cov}(\mathcal{M}) \leq \operatorname{non}(\mathcal{SN});$$

together with $\mathcal{SN} \subseteq \mathcal{N}$ (which trivially yields $\operatorname{non}(\mathcal{SN}) \leq \operatorname{non}(\mathcal{N})$), this reproves the first inequality of Corollary 1.14.

The following theorem shows that the Borel Conjecture necessarily fails if the bounding number $\mathfrak{b}$ is small, i.e., if there is an unbounded family (of functions in ω^ω) of size $\aleph_1$:

THEOREM 2.3. *Assume*[26] $\mathfrak{b} = \omega_1$; *then there exists an uncountable strong measure zero set.*

Note that this gives yet another proof of Theorem 2.1, i.e., the fact that there exists an uncountable strong measure zero set under CH (since CH implies $\mathfrak{b} = \omega_1$).

PROOF. By the assumption that $\mathfrak{b} = \omega_1$, there exists an unbounded family $F := \{f_\alpha : \alpha < \omega_1\} \subseteq \omega^\omega$, i.e., there is no $g \in \omega^\omega$ such that[27] $f_\alpha \leq^* g$ for each $\alpha < \omega_1$. It is easy to see (using the fact that each countable family of functions can be bounded by a single function) that without loss of generality we can assume that $f_\alpha \leq^* f_\beta$ for each $\alpha < \beta < \omega_1$; then, in particular, for each $g \in \omega^\omega$, we have

$$|\{\alpha < \omega_1 \colon f_\alpha \leq^* g\}| = \aleph_0. \tag{2.1}$$

As in the proof of Theorem 1.4, let $\mathbb{Q}$ denote the set of rationals in the Cantor space. Moreover, let $\iota : \omega^\omega \to 2^\omega \setminus \mathbb{Q}$ be the canonical bijection[28] defined by

$$f \mapsto 0^{(f(0))\frown}1^\frown 0^{(f(1))\frown}1^\frown \cdots \in 2^\omega \setminus \mathbb{Q},$$

where $0^{(\ell)}$ denotes "0 repeated ℓ times".

We claim that $\iota[F] \subseteq 2^\omega$ is strong measure zero (note that $\iota[F]$ is of size $\aleph_1$). So fix a sequence $\langle n_k \colon k \in \omega\rangle$ of natural numbers; we are going to find a sequence $\langle s_k \colon k \in \omega\rangle \subseteq 2^{<\omega}$ with $|s_k| = n_k$ for all $k \in \omega$ such that $\iota[F] \subseteq \bigcup_{k\in\omega}[s_k]$.

Let $\{q_k : k \in \omega\}$ be an enumeration of $\mathbb{Q}$, let[29] $s_{2k} := q_k \restriction n_{2k}$ for each $k \in \omega$, and let $U := \bigcup_{k\in\omega}[s_{2k}]$. To finish the proof it is enough to show that $\iota[F] \setminus U$ is countable:[30] then, let $\{z_k : k \in \omega\}$ be such that $\iota[F] \setminus U \subseteq \{z_k : k \in \omega\}$, and let $s_{2k+1} := z_k \restriction n_{2k+1}$ for each $k \in \omega$, yielding $\iota[F] \subseteq U \cup \bigcup_{k\in\omega}[s_{2k+1}]$, as desired.

In other words: it remains to show that $Y := \iota^{-1}(2^\omega \setminus U) \subseteq \omega^\omega$ intersects F in only countable many points. First note that Y is covered by the branches

---

[26] Actually, the proof of this theorem shows the following more general result: let $\mathfrak{b}$ be arbitrary, and assume that each set of size strictly less than $\mathfrak{b}$ is strong measure zero; then there exists a strong measure zero set of size $\mathfrak{b}$. (Moreover, the statement remains true if $\mathfrak{b}$ is replaced by $\mathfrak{d}$.) The proof is as the proof of Theorem 2.3, with the following two modifications: first, it is quite easy to check that there is an unbounded family $F := \{f_\alpha : \alpha < \mathfrak{b}\}$ with the property analogous to (2.1), i.e., $|\{\alpha < \mathfrak{b} : f_\alpha \leq^* g\}| < \mathfrak{b}$ for each $g \in \omega^\omega$ (and the same holds for $\mathfrak{d}$ replacing $\mathfrak{b}$), and second, $\iota[F]\setminus U$ then happens to be of size strictly less than $\mathfrak{b}$ (or, $\mathfrak{d}$, respectively), and hence is strong measure zero by assumption, and can therefore be covered by $\bigcup_{k\in\omega}[s_{2k+1}]$ (the "remaining half").

[27] Recall that $\leq^*$ denotes *eventual domination*, i.e., $f \leq^* g$ if $f(k) \leq g(k)$ for all but finitely many $k \in \omega$.

[28] In fact, ι is a homeomorphism of the spaces ω^ω and $2^\omega \setminus \mathbb{Q}$.

[29] As in Theorem 2.1, we "save" the other half of $\langle n_k \colon k \in \omega\rangle$ for later, to cover the remaining points.

[30] Note that (like X in the proof of Theorem 2.1) $\iota[F]$ is *concentrated on the rationals*, i.e., all but countably many points from $\iota[F]$ belong to any given open cover of the rationals.

of a *finitely splitting* subtree[31] of $\omega^{<\omega}$; more precisely, let $\tilde{\iota} : \omega^{<\omega} \to 2^{<\omega}$ be the mapping defined by $\tau \mapsto 0^{(\tau(0))\frown}1^\frown \cdots {}^\frown 0^{(\tau(|\tau|-1))\frown}1$, and observe that for each $\tau \in \omega^{<\omega}$, we have $[\tilde{\iota}(\tau^\frown \ell)] \subseteq U$ for all but finitely[32] many $\ell \in \omega$.

Hence we can inductively find a single $g \in \omega^\omega$ such that $f \leq g$ for all $f \in Y$; in particular, by (2.1), there are only countably many $\alpha < \omega_1$ with $f_\alpha \in Y$, i.e., $F \cap Y$ is indeed countable, as desired. □

Let us summarize the two necessary conditions involving the cardinal characteristics discussed above:

COROLLARY 2.4. *Assume BC. Then both* $\mathrm{cov}(\mathcal{M}) = \omega_1$ *and* $\mathfrak{b} > \omega_1$ *hold true.*

PROOF. Immediately follows from Theorem 2.2 and Theorem 2.3 above. □

Therefore, a forcing iteration (starting with a model of CH) yielding a model of the Borel Conjecture has to fulfill (at least) the following two requirements:

(1) it is *forbidden to add* (cofinally many) *Cohen reals* (since Cohen reals increase $\mathrm{cov}(\mathcal{M})$); so, in particular, it is not possible to use a finite support iteration (since finite support iterations always add Cohen reals at limit stages);
(2) it is *obligatory to add dominating reals* (in order to increase $\mathfrak{b}$).

Observe that Corollary 2.4 actually yields the following: in a model of BC satisfying $2^{\aleph_0} = \aleph_2$, we have $\mathrm{cov}(\mathcal{M}) = \omega_1$ and $\mathfrak{b} = \omega_2$, and hence the values of all but two cardinal characteristics of Cichoń's diagram (see Figure 2 on page 147) are determined; this is illustrated by Figure 3, where (following the convention of [**BJ95**, Chapter 7]) we put □ when the cardinal characteristic at the respective position is $\aleph_1$, and ■ when it is $\aleph_2$.

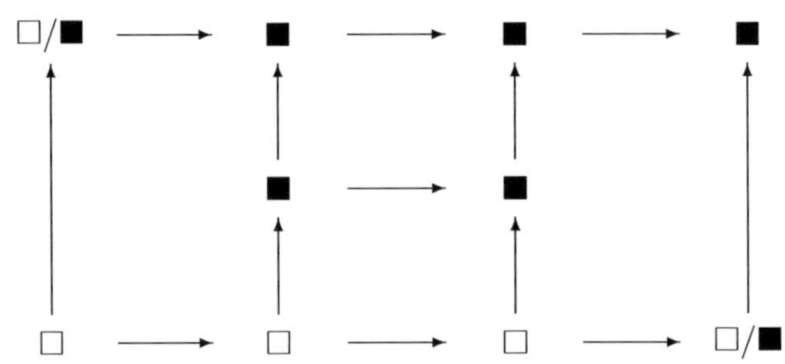

FIGURE 3. (□ $= \aleph_1$, ■ $= \aleph_2$) Cichoń's diagram in models of BC

It turns out that the two remaining cardinal characteristics of Cichoń's diagram (i.e., $\mathrm{cov}(\mathcal{N})$ and $\mathrm{non}(\mathcal{N})$) are indeed not determined by BC (we put □/■ at the respective positions in Figure 3): this is exemplified by Laver's original model of BC together with some of the models of BC discussed in Section 2.7 (namely the

[31] In fact, Y coincides with the branches of such a tree, due to the fact that Y is – as the preimage of the closed set $2^\omega \setminus U$ under the homeomorphism ι – indeed a closed set.

[32] This is easy to verify in a purely combinatorial way. More abstractly, one could argue as follows: Y is – as the preimage of the compact set $2^\omega \setminus U$ under the homeomorphism ι – a compact set, so it is (see also footnote 31) the set of branches of a *finitely splitting* tree (since an infinite split immediately yields a cover without finite subcover, i.e., a counterexample to compactness).

Mathias model and the Laver-random model); see Table 2, in which all the relevant theorems and figures (displaying the respective Cichoń's diagrams) are listed.

TABLE 2. The four $\mathrm{cov}(\mathcal{N})$-$\mathrm{non}(\mathcal{N})$-combinations in models of BC

theorem	Cichoń	name of model	BC	$\mathrm{cov}(\mathcal{N})$	$\mathrm{non}(\mathcal{N})$
Theorem 2.6	Figure 5	Laver	yes	□	□
Theorem 2.23	Figure 6	Mathias	yes	□	■
Theorem 2.31	Figure 7	Laver-random	yes	■	□
Conjecture 2.32	Figure 8	Mathias-random	yes?	■	■
Theorem 2.34	Figure 9	Laver $* \mathbb{B}_{\omega_2}$	yes	■	□
Theorem 2.34	Figure 9	Mathias $* \mathbb{B}_{\omega_2}$	yes	■	□

Note, however, that we are not completely sure whether all four combinations (of $\mathrm{cov}(\mathcal{N})$ and $\mathrm{non}(\mathcal{N})$ being small or large) are consistent with BC (plus continuum equals $\aleph_2$); we only know for sure that (individually) neither the value of $\mathrm{cov}(\mathcal{N})$ nor the value of $\mathrm{non}(\mathcal{N})$ is determined by BC; in fact, we have not checked in detail (see Conjecture 2.32, and Conjecture 2.33) whether BC holds true in the Mathias-random model (in which both $\mathrm{cov}(\mathcal{N})$ and $\mathrm{non}(\mathcal{N})$ are $\aleph_2$).

Further note that we can also take the model "Laver$* \mathbb{B}_{\omega_2}$" (or "Mathias$* \mathbb{B}_{\omega_2}$") instead of the Laver-random model to exemplify that $\mathrm{cov}(\mathcal{N}) = \aleph_2$ is consistent with BC (the proof that BC holds in "Laver $* \mathbb{B}_{\omega_2}$" or "Mathias $* \mathbb{B}_{\omega_2}$" is much more complicated though; see Theorem 2.34).

Before presenting Laver's famous theorem (using *countable support* iteration) that the Borel Conjecture is consistent, let us argue[33] that the two necessary conditions from Corollary 2.4 are in fact not sufficient for the Borel Conjecture (see Figure 4 for the Cichoń's diagram of the respective model):

THEOREM 2.5. *There is a model of ZFC in which BC fails, but* $\mathrm{cov}(\mathcal{M}) = \omega_1$ *and* $\mathfrak{b} > \omega_1$ *hold true.*

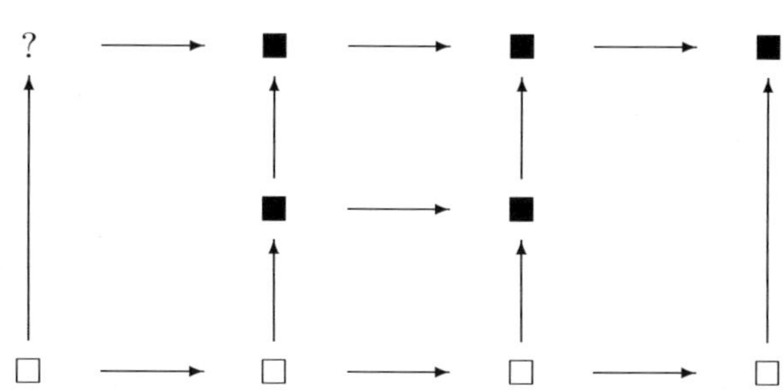

FIGURE 4. (□ $= \aleph_1$, ■ $= \aleph_2$) Cichoń's diagram in the model obtained from the Hechler model by adding $\aleph_1$ many random reals

However, it is not clear to me whether it is possible to get such a model via a (non-trivial) countable support iteration of length ω_2 (see Question 2.22).

[33] I thank Martin Goldstern for coming up with this argument.

PROOF OF THEOREM 2.5. The idea is to go to a model where BC fails and $\mathfrak{b} > \omega_1$, and then to adapt the value of $\mathrm{cov}(\mathcal{M})$ without changing the former two properties. This is accomplished by first taking the *Hechler model*, and then adding ω_1 many *random reals* (side-by-side). More precisely, argue as follows.

Start with a model V satisfying CH, and let $\mathbb{P}_{\omega_2}$ be the finite support iteration of *Hechler forcing* of length ω_2; moreover, in (the Hechler model) $V^{\mathbb{P}_{\omega_2}}$, let $\mathbb{B}_{\omega_1}$ be the forcing for simultaneously adding $\aleph_1$ many random reals (see [**BJ95**, Section 3.1] for a definition[34] of $\mathbb{B}_{\omega_1}$); then the final model is as desired:

$$V^{\mathbb{P}_{\omega_2} * \mathbb{B}_{\omega_1}} \models \neg BC \wedge \mathrm{cov}(\mathcal{M}) = \omega_1 \wedge \mathfrak{b} > \omega_1.$$

This can be seen as follows:

(1) Since Hechler forcing adds *Cohen reals* (or, alternatively, since any finite support iteration adds Cohen reals at limit stages anyway), we know that $\mathrm{cov}(\mathcal{M})$ is large in the Hechler model:

$$V^{\mathbb{P}_{\omega_2}} \models \omega_1 < \mathrm{cov}(\mathcal{M}) = \omega_2 = 2^{\aleph_0};$$

therefore, by Theorem 2.2, the Borel Conjecture fails in $V^{\mathbb{P}_{\omega_2}}$; in other words, we can fix an uncountable strong measure zero set $X \in V^{\mathbb{P}_{\omega_2}}$; it remains to show that X is strong measure zero in the final model $V^{\mathbb{P}_{\omega_2} * \mathbb{B}_{\omega_1}}$ as well.

Since $\mathbb{B}_{\omega_1}$ is ω^ω*-bounding*, for each "challenging sequence" $\langle n_k : k \in \omega \rangle$ in the final model $V^{\mathbb{P}_{\omega_2} * \mathbb{B}_{\omega_1}}$, there is a sequence $\langle n'_k : k \in \omega \rangle$ in the intermediate model $V^{\mathbb{P}_{\omega_2}}$ satisfying $n'_k \geq n_k$ for each $k \in \omega$; since X is strong measure zero in $V^{\mathbb{P}_{\omega_2}}$, there is a sequence $\langle s_k : k \in \omega \rangle \in V^{\mathbb{P}_{\omega_2}} \subseteq V^{\mathbb{P}_{\omega_2} * \mathbb{B}_{\omega_1}}$ with $|s_k| = n'_k \geq n_k$ for all $k \in \omega$ such that $X \subseteq \bigcup_{k \in \omega} [s_k]$, which finishes the argument that (the uncountable set) X is strong measure zero in the final model, i.e., BC fails.

(2) Even though $\mathrm{cov}(\mathcal{M}) > \omega_1$ in the Hechler model $V^{\mathbb{P}_{\omega_2}}$ (see above), it is straightforward to check that $\mathrm{cov}(\mathcal{M}) = \omega_1$ holds true in the final model, as follows.

Note that the set $\{r_\alpha : \alpha < \omega_1\} \in V^{\mathbb{P}_{\omega_2} * \mathbb{B}_{\omega_1}}$ of random reals added by $\mathbb{B}_{\omega_1}$ is (of size $\aleph_1$ and) *not measure zero*: if it were measure zero, then it would be covered by a Borel measure zero set with Borel code already added by a "countably indexed" part of $\mathbb{B}_{\omega_1}$; however, random reals do not belong to measure zero sets coded in the model over which they are random, so there are going to be reals r_α which are actually not covered, a contradiction. Therefore, there is (in the final model) a set of size $\aleph_1$ which is not measure zero; in other words, $\mathrm{non}(\mathcal{N}) = \aleph_1$, yielding $\mathrm{cov}(\mathcal{M}) = \aleph_1$ (due to $\mathrm{cov}(\mathcal{M}) \leq \mathrm{non}(\mathcal{N})$ which is one of the easy ZFC provable relations in Cichoń's diagram; see Corollary 1.14 and Figure 2).

(3) Since Hechler forcing adds *dominating reals*, we have $\mathfrak{b} = \aleph_2$ in the Hechler model $V^{\mathbb{P}_{\omega_2}}$, i.e., the least size of an unbounded family is $\aleph_2$.

But since $\mathbb{B}_{\omega_1}$ is ω^ω-bounding (i.e., for every $f \in \omega^\omega \cap V^{\mathbb{P}_{\omega_2} * \mathbb{B}_{\omega_1}}$ there exists a $g \in \omega^\omega \cap V^{\mathbb{P}_{\omega_2}}$ with $f \leq g$), and $\mathbb{B}_{\omega_1}$ is c.c.c., it is easy to show that the same holds true in the final model $V^{\mathbb{P}_{\omega_2} * \mathbb{B}_{\omega_1}}$. □

[34] In [**BJ95**], the forcing is denoted by $\mathbf{B}(\omega_1)$.

2.2. How to prove the consistency of the Borel Conjecture? In his 1976 paper [**Lav76**], Laver used[35] the method of countable support iteration to get a model satisfying the Borel Conjecture:

THEOREM 2.6 (Laver, 1976). *Assume CH, and let* $\mathbb{P}_{\omega_2}$ *be the countable support iteration of Laver forcing of length* ω_2*.*

Then in a generic extension by $\mathbb{P}_{\omega_2}$ *the Borel Conjecture holds (i.e., there are no uncountable strong measure zero sets).*

This model is called the *Laver model* (see Figure 5 for its Cichoń's diagram).

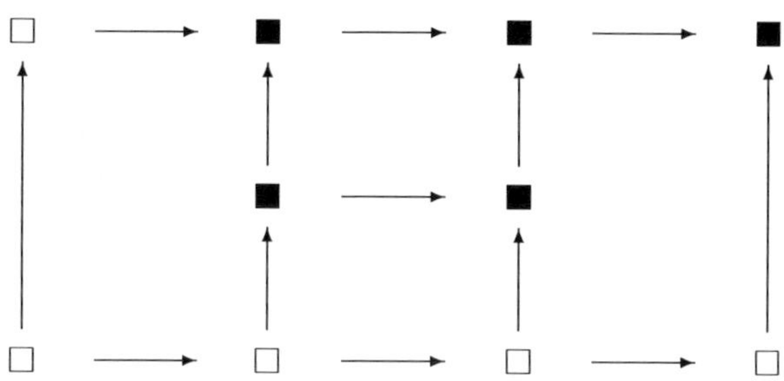

FIGURE 5. ($\square = \aleph_1$, $\blacksquare = \aleph_2$) Cichoń's diagram in the Laver model

In the following (Sections 2.3–2.6), we are going to give a detailed[36] proof of this theorem; the key points of the proof are:

- Each of the iterands $\dot{\mathbb{Q}}_\alpha$ is *Laver forcing* $\mathbb{L}$ (see Definition 2.8), which "kills" all uncountable strong measure zero sets from the ground model. Actually, the following property of Laver forcing is used (see Lemma 2.15):
  ("$\mathbb{L}$ kills smz") If X is an uncountable set of reals and belongs to the ground model, then $\Vdash_{\mathbb{L}}$ "X is not strong measure zero", witnessed by the sequence $\langle \varepsilon_k : k \in \omega \rangle$ with
  $$\varepsilon_k := 2^{-\dot{g}(k)},$$
  where $\dot{g} \in \omega^\omega$ is the "Laver real" (the generic real added by $\mathbb{L}$).
- Once an uncountable set X "has been killed", it "stays dead" (until the end of the iteration). For this, the so-called *Laver property* (see Definition 2.12) is used:
  (1) Laver forcing $\mathbb{L}$ has the *Laver property* (see Proposition 2.13).
  (2) The Laver property is preserved under *countable support* iteration of proper forcings (see Proposition 2.14).
  (3) "X is not strong measure zero" remains true after a forcing satisfying the Laver property (see Lemma 2.20).

[35]This was one of the first proofs (probably the first published one) which used a *countable support* iteration.

[36]However, we concentrate on proving the facts related to the Borel Conjecture and feel free to use well-known facts about Laver forcing (such as properness and the Laver property) as well as standard iteration techniques (see also Remark 2.7) without giving the respective proofs; for these, we refer to the literature.

REMARK 2.7. In fact, the "modern presentation" of the proof proceeds as described above; Laver's original proof precedes concepts such as "properness" and "iteration theorems for proper forcings" (which were introduced by Shelah a couple of years later), so Laver had to accomplish several arguments "by hand" instead of invoking preservation theorems.

Our presentation is mainly based on the treatments of the Borel Conjecture in [**Jec03**, Chapter 28, pp. 564–568] and [**BJ95**, Section 8.3.A].

2.3. Laver forcing, pure decision, and Laver property. Laver forcing $\mathbb{L}$ is a so-called *tree forcing* on ω^ω; we give the definition of $\mathbb{L}$ (see also [**BJ95**, Definition 7.3.26], or [**Jec03**, Definition 28.15]) and present some of the basic properties needed later; for most of the proofs, we refer to the literature (mainly to [**BJ95**, Section 7.3.D] which is devoted to Laver forcing).

Recall that $p \subseteq \omega^{<\omega}$ is a *tree* if it is closed under initial segments; two nodes $\sigma, \tau \in p$ are *comparable* if $\sigma \subseteq \tau$ or $\tau \subseteq \sigma$; moreover, let

$$\mathrm{succ}_p(\tau) := \{i \in \omega : \tau^\frown i \in p\}.$$

Laver forcing consists of trees which are infinitely splitting above the stem:

DEFINITION 2.8. A tree $p \subseteq \omega^{<\omega}$ is called a *Laver tree* if there exists a $\sigma \in p$ (the *stem* of p) such that σ is comparable with every node in p, and $|\mathrm{succ}_p(\tau)| = \aleph_0$ for all $\tau \in p$ with $\sigma \subseteq \tau$.

Laver forcing $\mathbb{L}$ is the collection of Laver trees, ordered by inclusion: $q \leq p$ (i.e., q is stronger than p) if $q \subseteq p$.

For any $p \in \mathbb{L}$, there is a unique node $\sigma \in p$ with the above property, which we denote by $\mathrm{stem}(p)$.

Given a filter $G \subseteq \mathbb{L}$ generic over the ground model V, the generic branch

$$g := \bigcup\{\mathrm{stem}(p) : p \in G\} \in \omega^\omega$$

is called the *Laver real* added by $\mathbb{L}$. Note that any $p \in \mathbb{L}$ forces that $\dot{g}$ is a (new) branch through p (where $\dot{g}$ is an $\mathbb{L}$-name for the Laver real g).

Recall that $\leq^*$ denotes eventual domination on ω^ω, i.e., $f \leq^* h$ if and only if $f(n) \leq h(n)$ for all but finitely many $n \in \omega$. Laver forcing adds a *dominating real*: in fact, an easy density argument shows that the Laver real g is *dominating* over V, i.e., g eventually dominates all ground model reals:

PROPOSITION 2.9. *Let $f \in \omega^\omega \cap V$. Then $f \leq^* g$.*

PROOF. It is enough to show that the set

$$D_f := \{q \in \mathbb{L} : q \Vdash f \leq^* \dot{g}\}$$

(which is in V) is dense in $\mathbb{L}$. Given $p \in \mathbb{L}$, define $q \leq p$ by removing finitely many successors from each node above the stem of p, in such a way that each branch through q eventually (namely above the stem) dominates f; more formally, let

$$q := \{\tau \in p : \tau(n) \geq f(n) \text{ for each } n \in [|\mathrm{stem}(p)|, |\tau|)\};$$

clearly, $q \in \mathbb{L}$ and $q \leq p$; note that[37] $\mathrm{stem}(q) = \mathrm{stem}(p)$; it is easy to check that

$$q \Vdash f(n) \leq \dot{g}(n) \text{ for each } n \geq |\mathrm{stem}(p)|,$$

hence $q \in D_f$, as desired. □

[37] I.e., $q \leq_0 p$ (see below).

As mentioned above (see item (2) in the discussion after Corollary 2.4), it is *necessary* to add dominating reals to get a model of the Borel Conjecture.

For $q, p \in \mathbb{L}$, we say that q is a *pure extension* of p (denoted by $q \leq_0 p$) if $q \leq p$ and $\mathrm{stem}(q) = \mathrm{stem}(p)$.

Laver forcing $\mathbb{L}$ has the *pure decision property*:

PROPOSITION 2.10. *Suppose that A is a finite set, $p \in \mathbb{L}$, and $p \Vdash \dot{a} \in A$. Then there exists a pure extension $q \leq_0 p$ and an element $b \in A$ such that $q \Vdash \dot{a} = b$.*

PROOF. See [**BJ95**, Lemma 7.3.32], or [**Jec03**, Lemma 28.19]. □

The following is essential for Laver forcing being well-behaved in iterations:

PROPOSITION 2.11. *$\mathbb{L}$ satisfies Axiom A and (hence) is proper.*

PROOF. See [**BJ95**, Lemma 7.3.27] (or [**Jec03**, Lemma 28.16 and Lemma 28.17] for similar arguments). □

In particular, $\mathbb{L}$ does not collapse ω_1; moreover, a countable support iteration of proper forcing notions is proper (see [**BJ95**, Lemma 6.1.4]), and hence ω_1 is preserved also by the countable support iteration of Laver forcing. Larger cardinals are preserved as well, due to the general fact that (starting with a model of CH) the countable support iteration of forcings of size continuum (such as Laver forcing) has the $\aleph_2$-c.c. (see, e.g., [**Bau83**]).

The following property (see also[38] [**BJ95**, Definition 6.3.27]) will be crucial for the preservation of being *not* strong measure zero (see Lemma 2.20):

DEFINITION 2.12. A forcing $\mathbb{P}$ has the *Laver property* if for every filter G which is $\mathbb{P}$-generic over V the following holds: for any function $f \in \omega^\omega \cap V[G]$ which is bounded by a ground model function (i.e., there is an $h \in \omega^\omega \cap V$ with $f(n) \leq h(n)$ for each $n \in \omega$), there is an $S \in ([\omega]^{<\omega})^\omega \cap V$ with $|S(n)| \leq n$ for each $n \in \omega \setminus \{0\}$ such that $f(n) \in S(n)$ for each $n \in \omega$.

A sequence S as above[39] is called a *slalom*; so in a Laver property extension, every function bounded by a ground model function can be captured by a ground model slalom.

PROPOSITION 2.13. *$\mathbb{L}$ has the Laver property.*

PROOF. See [**BJ95**, Theorem 7.3.29]. □

Therefore, $\mathbb{L}$ does not add Cohen (or random) reals (see [**BJ95**, Lemma 7.2.3]); this is good because we are not allowed to (cofinally) add Cohen reals when aiming at a model of the Borel Conjecture (compare with item (1) in the discussion after Corollary 2.4).

Moreover, the Laver property is an iterable property:

---

[38]In [**BJ95**, Definition 6.3.27], the growth rate of the slalom is different (instead of $|S(n)| \leq n$, they actually use $|S(n)| \leq 2^n$); however, this does not change the notion of Laver property, see footnote 39 below.

[39]Actually, it doesn't really matter whether $|S(n)| \leq n$ is used as the prescribed growth rate of the slalom, or any other appropriate function from the ground model; more precisely, the following is quite straightforward to check: if $H \in \omega^\omega \cap V$ is any (fixed) function with $\lim_{n\to\infty} H(n) = \infty$, then the Laver property as defined above is equivalent to the property when $|S(n)| \leq n$ is replaced by $|S(n)| \leq H(n)$.

PROPOSITION 2.14. *The Laver property is preserved under countable support iteration of proper forcing notions.*

PROOF. See [**BJ95**, Theorem 6.3.34]. □

Together with the fact that Laver forcing adds dominating reals (see the above Proposition 2.9), this easily yields that[40] $\mathrm{cov}(\mathcal{M}) = \omega_1$ and $\mathfrak{b} > \omega_1$ hold true in the Laver model (see Theorem 2.6 and Figure 5); in other words, the two necessary conditions for the Borel Conjecture from Corollary 2.4 are indeed satisfied.

In the following, we are going to prove that the Borel Conjecture itself holds true in the Laver model.

2.4. Laver forcing kills strong measure zero sets. Let $X \subseteq 2^\omega$, and let $f \in \omega^\omega$. We say that X *is strong measure zero with respect to all tails of* f if for each $m \in \omega$ there exists a sequence $\langle s_k : k \geq m \rangle \subseteq 2^{<\omega}$ with $|s_k| \geq f(k)$ for each $k \geq m$ such that

$$X \subseteq \bigcup_{k \geq m} [s_k].$$

Clearly, a set $X \subseteq 2^\omega$ is strong measure zero (see Definition 1.3) if and only if

$$\forall f \in \omega^\omega \, (X \text{ is strong measure zero with respect to all tails of } f). \tag{2.2}$$

The following key lemma shows that Laver forcing $\mathbb{L}$ "kills" all uncountable strong measure zero sets from the ground model V (witnessed by the Laver real):

LEMMA 2.15. *Let $\dot{g}$ be an $\mathbb{L}$-name for the Laver real (i.e., the generic branch), and let $X \subseteq 2^\omega$ be an uncountable set in V. Then*[41]

$$\Vdash_{\mathbb{L}} X \textit{ is } \text{not} \textit{ strong measure zero with respect to all tails of } \dot{g},$$

and hence $\Vdash_{\mathbb{L}} X \notin \mathcal{SN}$.

---

[40]Since the iteration has the Laver property, no Cohen reals over V are added by the iteration (see [**BJ95**, Lemma 7.2.3]), hence – recall that a real is Cohen over V if and only if it does not belong to any meager set coded in V – all reals in the final model are covered by the $\aleph_1$ many meager sets coded in V, i.e., the final model satisfies $\mathrm{cov}(\mathcal{M}) = \omega_1$.

Analogously, it can be shown that $\mathrm{cov}(\mathcal{N}) = \omega_1$, due to the fact that no random reals are added by forcings with the Laver property (again, see [**BJ95**, Lemma 7.2.3]). Moreover, since Laver forcing preserves random reals (see Theorem 2.27), and this is an iterable property (see Proposition 2.26), which implies preservation of positivity, the ground model reals are positive in the final model, witnessing $\mathrm{non}(\mathcal{N}) = \omega_1$.

[41]Note that "X is strong measure zero with respect to all tails of f" is stronger than "X is strong measure zero with respect to f" (just saying that there exists a sequence $\langle s_k : k \in \omega \rangle \subseteq 2^{<\omega}$ with $|s_k| \geq f(k)$ for each $k \in \omega$ such that $X \subseteq \bigcup_{k \in \omega} [s_k]$); the latter property would be too weak for Lemma 2.15 to work, since a finite initial segment of the Laver real can be as slowly growing as we wish, and therefore any set X could be covered (e.g., let g be a Laver real with $g(0) = 0$, and let $s_0 := \langle\rangle$; note that $|s_0| = g(0)$, and $[s_0] = 2^\omega$; so any X is strong measure zero with respect to g).

However, we could work with the following stronger (and maybe more natural) property (in fact, it is obtained by changing the order of the first two quantifiers in the definition of "strong measure zero with respect to all tails of f"): there exists a sequence $\langle s_k : k \in \omega \rangle \subseteq 2^{<\omega}$ with $|s_k| \geq f(k)$ for each $k \in \omega$ such that

$$X \subseteq \bigcap_{m \in \omega} \bigcup_{k \geq m} [s_k]$$

(i.e., any point is covered by infinitely many of the clopen sets); as in (2.2), X having the above property for each $f \in \omega^\omega$ is equivalent to X being strong measure zero (the proof is similar to

We essentially follow the proof of [**Jec03**, Lemma 28.20]; also compare with the (quite similar proof of the) respective lemma (see [**GKSW14**, Lemma 1.13]) for "ultralaver forcing" used for constructing the model satisfying both the Borel Conjecture and the dual Borel Conjecture (see Theorem 4.1).

Before proving Lemma 2.15, we introduce some more notation and give two technical lemmas. For $p \in \mathbb{L}$, let

$$\operatorname{split}(p) := \{\eta \in p \colon \eta \supseteq \operatorname{stem}(p)\}$$

denote the set of splitting nodes of p (indeed, $\operatorname{split}(p) = \{\eta \in p \colon |\operatorname{succ}_p(\eta)| = \aleph_0\}$); furthermore, for $\eta \in \operatorname{split}(p)$, let

$$p^{[\eta]} := \{\tau \in p \colon \tau \subseteq \eta \vee \eta \subseteq \tau\}$$

denote the "restriction of p to η", i.e., the set of nodes comparable with η. Note that $p^{[\eta]} \in \mathbb{L}$, and $\operatorname{stem}(p^{[\eta]}) = \eta$.

LEMMA 2.16. *Let $p \in \mathbb{L}$ with $\operatorname{stem}(p) =: \sigma \in \omega^{<\omega}$, and let $\dot{x}$ be an $\mathbb{L}$-name for a real in 2^ω. Then there exist a pure extension $q \leq_0 p$ and a real $y \in 2^\omega$ such that for every $n \in \omega$,*

$$q^{[\sigma^\frown i]} \Vdash \dot{x} \restriction n = y \restriction n$$

for all but finitely many $i \in \operatorname{succ}_q(\sigma)$.

PROOF. For each $i \in \operatorname{succ}_p(\sigma)$, apply Proposition 2.10 (i.e., the pure decision property of $\mathbb{L}$) to obtain a $q_i \leq_0 p^{[\sigma^\frown i]}$ such that q_i decides $\dot{x} \restriction i$, i.e., there is a $t_i \in 2^{<\omega}$ (with $|t_i| = i$) such that $q_i \Vdash \dot{x} \restriction i = t_i$.

Since $\{t_i : i \in \operatorname{succ}_p(\sigma)\} \subseteq 2^{<\omega}$ is infinite, König's lemma yields an infinite branch $y \in 2^\omega$ through the tree generated by $\{t_i \colon i \in \operatorname{succ}_p(\sigma)\}$; therefore, we can pick an infinite set $A \subseteq \operatorname{succ}_p(\sigma)$ such that $\{t_i : i \in A\}$ "converges" to y, i.e., for every $n \in \omega$,

$$t_i \restriction n = y \restriction n$$

for all but finitely many $i \in A$.

Finally, let

$$q := \bigcup_{i \in A} q_i;$$

it is easy to check that q and y are as desired. $\square$

LEMMA 2.17. *Let $p \in \mathbb{L}$ with $\operatorname{stem}(p) =: \sigma \in \omega^{<\omega}$, and let $\{\dot{x}_k \colon k \geq |\sigma|\}$ be a sequence of $\mathbb{L}$-names for reals in 2^ω. Then there exist a pure extension $q \leq_0 p$ and a (countable) family of reals $\{y_\eta \colon \eta \in \operatorname{split}(q)\} \subseteq 2^\omega$ such that for each $\eta \in \operatorname{split}(q)$ and every $n \in \omega$,*

$$q^{[\eta^\frown i]} \Vdash \dot{x}_{|\eta|} \restriction n = y_\eta \restriction n \tag{2.3}$$

for all but finitely many $i \in \operatorname{succ}_q(\eta)$.

PROOF. To obtain q, we inductively apply Lemma 2.16 to each splitting node η (i.e., each node above the stem σ); more precisely, we proceed as follows.

showing that the strong measure zero sets form a σ-ideal; see [**BJ95**, Lemma 8.1.2(3)]); nevertheless, we decided to use the weaker property "X is strong measure zero with respect to all tails of f" because this seems to be exactly what is needed in the proof of Lemma 2.15.

We start at σ, and apply Lemma 2.16 to $p =: p_0$ and $\dot{x}_{|\sigma|}$ to get $p_1 \leq_0 p$ and $y_\sigma \in 2^\omega$ such that for every $n \in \omega$,

$$p_1^{[\sigma^\frown i]} \Vdash \dot{x}_{|\sigma|} \restriction n = y_\sigma \restriction n$$

for all but finitely many $i \in \operatorname{succ}_{p_1}(\sigma)$.

Let $S := \{\sigma^\frown i : i \in \operatorname{succ}_{p_1}(\sigma)\}$ be the set of immediate successor nodes of σ within p_1. For every $\eta \in S$, we apply Lemma 2.16 to $p_1^{[\eta]}$ and $\dot{x}_{|\eta|}$ to get $q_\eta \leq_0 p_1^{[\eta]}$ and $y_\eta \in 2^\omega$ with the respective properties; let $p_2 := \bigcup_{\eta \in S} q_\eta$.

We continue in this way, climbing all the way up in the tree; we end up with a sequence $p = p_0 \geq p_1 \geq p_2 \geq \ldots$, and let

$$q := \bigcap_{k \in \omega} p_k.$$

It is easy to check that $q \in \mathbb{L}$, and that q and $\{y_\eta : \eta \in \operatorname{split}(q)\}$ are as desired. □

Proof of Lemma 2.15. Let $X \subseteq 2^\omega$ be an uncountable set in V, and assume towards a contradiction that for some $p \in \mathbb{L}$,

$$p \Vdash_{\mathbb{L}} X \text{ is strong measure zero with respect to all tails of } \dot{g};$$

let $\sigma := \operatorname{stem}(p) \in \omega^{<\omega}$ be the stem of p, and observe that we can fix a sequence $\{\dot{x}_k : k \geq |\sigma|\}$ of $\mathbb{L}$-names for reals in 2^ω such that

$$p \Vdash_{\mathbb{L}} X \subseteq \bigcup_{k \geq |\sigma|} [\dot{x}_k \restriction \dot{g}(k)]. \tag{2.4}$$

By Lemma 2.17, we can fix a pure extension $q \leq_0 p$ and a (countable) family $\{y_\eta : \eta \in \operatorname{split}(q)\} \subseteq 2^\omega$ such that for each $\eta \in \operatorname{split}(q)$ and every $n \in \omega$, condition (2.3) holds for all but finitely many $i \in \operatorname{succ}_q(\eta)$.

Since X is (in V and) uncountable, we can find a real $x^* \in X$ which is different from each real in the countable family $\{y_\eta : \eta \in \operatorname{split}(q)\}$; more specifically, we can pick a family of natural numbers $\{n_\eta : \eta \in \operatorname{split}(q)\}$ such that

$$x^* \restriction n_\eta \neq y_\eta \restriction n_\eta \tag{2.5}$$

for any η.

By removing finitely many successors from each splitting node, we can easily find a condition $r \leq_0 q$ such that for all $\eta \in \operatorname{split}(r)$, and all $i \in \operatorname{succ}_r(\eta)$, we have

(1) $i \geq n_\eta$, and
(2) $r^{[\eta^\frown i]} \Vdash_{\mathbb{L}} \dot{x}_{|\eta|} \restriction n_\eta = y_\eta \restriction n_\eta$.

Note that (1) yields $r^{[\eta^\frown i]} \Vdash_{\mathbb{L}} \dot{g}(|\eta|) = i \geq n_\eta$, and therefore (2) together with (2.5) yield

$$r^{[\eta^\frown i]} \Vdash_{\mathbb{L}} x^* \notin [\dot{x}_{|\eta|} \restriction n_\eta] \supseteq [\dot{x}_{|\eta|} \restriction \dot{g}(|\eta|)].$$

It is straightforward to derive that

$$r \Vdash_{\mathbb{L}} x^* \notin \bigcup_{k \geq |\sigma|} [\dot{x}_k \restriction \dot{g}(k)];$$

since $r \leq p$ and $x^* \in X$, this contradicts (2.4). □

Actually, not only Laver forcing $\mathbb{L}$ (see Lemma 2.15) but also other forcings such as Mathias forcing (see Lemma 2.25 in Section 2.7.1 below) kill all uncountable strong measure zero sets from the ground model; moreover, the proof of Lemma 2.15 generalizes (by generalizing Lemma 2.16 accordingly) to any *Laver forcing with respect to* $\mathcal{I}$, for any ideal $\mathcal{I}$ on ω (i.e., the set of successors is not just required to be infinite, but $\mathcal{I}$-positive); it is even possible to use individual ideals assigned to the nodes in $\omega^{<\omega}$ (instead of a common single ideal for all splittings); in particular, this is true for *ultralaver forcings*, using systems $\bar{D} = (D_\tau)_{\tau\in\omega^{<\omega}}$ of ultrafilters on ω; these forcings are an important ingredient of the proof of Theorem 4.1 (see Section 4.5, as well as [**GKSW14**, Lemma 1.13]).

However, it is not clear to me to which extent Lemma 2.15 can be generalized; in other words, one can ask whether the conclusion of Lemma 2.15 (i.e., killing all uncountable strong measure zero sets from the ground model) holds true for all forcings which add dominating reals but do not add Cohen reals (also compare with Question 2.22):

QUESTION 2.18. Is there a (nicely definable?) forcing notion $\mathbb{P}$ satisfying the Laver property[42] and adding dominating reals such that $\Vdash_{\mathbb{P}} X \in \mathcal{SN}$ for some uncountable ground model set X of reals?

It might be the case (but rather unlikely) that *increasing Sacks amoeba forcing* $\mathbb{A}$ from [**SW**, Definition 3.1] is such a forcing (see Section 2.7.7 for further discussion and more context). Note that a negative answer to Question 2.42 would be necessary for this.

REMARK 2.19. Let us remark that there seem to be three *typical* behaviors of (tree) forcing notions with respect to how they act on the smz status of uncountable sets *from the ground model*:

(1) turning all sets of reals from the ground model into strong measure zero sets (examples include[43] *Cohen forcing*, as well as other forcings which add Cohen reals such as *Hechler forcing*),
(2) making all (uncountable) sets of reals from the ground model *not* strong measure zero (examples include[44] *Laver forcing* as well as *Mathias forcing*), and
(3) preserving the status of being strong measure zero or not (examples include[45] *Sacks forcing*, *Silver forcing*, as well as *random forcing*).

[42]One could also ask for forcings preserving random reals (instead of the Laver property); after all, there is an alternative argument that BC holds in the Laver model using preservation of random reals (see Section 2.7.2, in particular, Theorem 2.27 and Lemma 2.29).

[43]This is due to the well-known fact (for a proof, see, e.g., [**Gol**]; we are not aware of any published proof) that a single Cohen real turns the set of all reals from the ground model into a strong measure zero set.

[44]Due to Lemma 2.15 and Lemma 2.25, respectively.

[45]This can be seen as follows: first, it is easy to see that any forcing which is ω^ω-bounding preserves being strong measure zero, basically due to the fact that no new ε_n-sequences (i.e., no new "challenges" for being strong measure zero) are added (and the forcings mentioned are all ω^ω-bounding); second, a forcing having the Laver property (see Lemma 2.20) or preserving positivity (see Lemma 2.29) preserve being *not* strong measure zero (and Sacks forcing as well as Silver forcing have the Laver property, and random forcing preserves positivity).

Note that the fact that random forcing falls into this category (which has also been explicitly proved in [**BRS96**, Theorem 2.2]) does not mean that the Borel Conjecture is necessarily preserved by adding a single random real (which is an open question; also compare with Section 2.7.3 and,

One exception (of probably several), i.e., an example of a forcing which perhaps does not fall into one of these categories, is *Miller forcing*: due to the fact that it has the Laver property, being *not* strong measure zero is preserved; on the other hand, we have not investigated whether it also preserves being strong measure zero (i.e., whether it falls into category (3)); however, it cannot fall into category (2) – at least not in ZFC, due to the following: if it would, then the "usual" proof would go through and the Borel Conjecture were true in the Miller model, which is not the case because $\mathfrak{b} = \omega_1$ (see Theorem 2.3).

2.5. No resurrection after forcings with the Laver property. We now prove that being *not* strong measure zero is preserved by forcings which have the Laver property:

LEMMA 2.20. *Suppose $\mathbb{P}$ is a forcing which has the Laver property, and G is a $\mathbb{P}$-generic filter over V. Let $X \in V \cap \mathcal{P}(2^\omega)$. Then*

$$V \models X \notin \mathcal{SN} \implies V[G] \models X \notin \mathcal{SN}.$$

PROOF. Let $X \in V$ be a set of reals; suppose that X is strong measure zero in $V[G]$; we are going to show that X is strong measure zero in V. So fix a strictly increasing function $f \in V \cap \omega^\omega$; it is enough to find a sequence $\langle s_k : k \in \omega \rangle$ satisfying $|s_k| \geq f(k)$ for each $k \in \omega$ such that

$$X \subseteq \bigcup_{k \in \omega} [s_k].$$

Define $g \in \omega^\omega \cap V$ such that $g(l) = f(l^2)$ for each $l \in \omega$. Since

$$V[G] \models X \in \mathcal{SN},$$

and $g \in V \subseteq V[G]$, we can fix a sequence $\langle t_l : l \in \omega \rangle \in V[G]$ such that $t_l \in 2^{g(l)}$ for each $l \in \omega$ and

$$X \subseteq \bigcup_{l \in \omega} [t_l]. \tag{2.6}$$

Since $\mathbb{P}$ has the Laver property (see Definition 2.12), we can find a ground model slalom capturing $\langle t_l : l \in \omega \rangle$, i.e., there is a sequence $\langle J_l : l \in \omega \rangle \in V$ with $J_l \subseteq 2^{g(l)}$ for each $l \in \omega$ such that

(1) $t_l \in J_l$ for each $l \in \omega$,
(2) $|J_l| \leq l$ for each $l \in \omega \setminus \{0\}$ (and, say, $|J_0| = 1$).

Let $\langle s_k : k \in \omega \rangle$ be the list of all elements of $\bigcup_{l \in \omega} J_l$ ordered with respect to the length; in other words, $\langle s_k : k \in \omega \rangle$ starts with the element(s) of J_0, then continues with the ones of J_1, then the ones of J_2, etc.

Using (2.6) and item (1) above, we clearly obtain

$$X \subseteq \bigcup_{l \in \omega} [t_l] \subseteq \bigcup_{l \in \omega} \bigcup_{s \in J_l} [s] = \bigcup_{k \in \omega} [s_k].$$

To finish the proof of the lemma, it remains to show that $|s_k| \geq f(k)$ for each $k \in \omega$. So let $k \in \omega$, and fix $l \in \omega$ such that $s_k \in J_l$. Since $J_l \subseteq 2^{g(l)}$, we have

$$|s_k| = g(l). \tag{2.7}$$

---

in particular, Question 2.35): the Borel Conjecture only holds true in the extension if being not strong measure zero also applies to all *new* uncountable sets.

Using the definition of the list $\langle s_k : k \in \omega \rangle$, the fact that $s_k \in J_l$, and the bounds for the sizes of the sets J_l (see item (2) above), it is easy to estimate k:

$$k + 1 = |\{s_i : i \leq k\}| \leq \sum_{j=0}^{l} |J_j| \leq 1 + \sum_{j=1}^{l} j \leq 1 + l^2,$$

hence $k \leq l^2$. Therefore, using (2.7), the definition of the function g, and the fact that f is increasing, we get

$$|s_k| = g(l) = f(l^2) \geq f(k),$$

as desired. □

In Section 2.7.2, we are going to present an alternative way to preserve being *not* strong measure zero, using preservation of positivity (see Lemma 2.29).

2.6. No uncountable strong measure zero sets in the Laver model. To prove Theorem 2.6, i.e., to show that the Borel Conjecture holds true in the Laver model, we are actually going to prove the following slightly more general theorem:

THEOREM 2.21. *Assume CH, and let $\{\mathbb{P}_\alpha, \dot{\mathbb{Q}}_\alpha : \alpha < \omega_2\}$ be a countable support iteration such that for every $\alpha < \omega_2$,*

$$\Vdash_\alpha \dot{\mathbb{Q}}_\alpha \textit{ is proper and has the Laver property,}$$

and

$$\{\alpha < \omega_2 : \Vdash_\alpha \dot{\mathbb{Q}}_\alpha \textit{ is Laver forcing}\}$$

is cofinal in ω_2. Then

$$V^{\mathbb{P}_{\omega_2}} \models BC.$$

Recall that Laver forcing $\mathbb{L}$ is proper (see Proposition 2.11) and has the Laver property (see Proposition 2.13), so Theorem 2.6 immediately follows from the above theorem.

PROOF OF THEOREM 2.21. We have to prove that there is no uncountable strong measure zero set in the final model $V^{\mathbb{P}_{\omega_2}}$. This follows quite easily from the fact that the *Laver property* is preserved under *countable support* iteration of proper forcings, together with Lemma 2.15 and Lemma 2.20. More precisely, we can argue as follows.

Assume towards a contradiction that $X \in V^{\mathbb{P}_{\omega_2}}$ is an uncountable set which is strong measure zero (in the final model $V^{\mathbb{P}_{\omega_2}}$). Since the collection of strong measure zero sets is closed under subsets, we can assume without loss of generality that X is actually of size $\aleph_1$. Recall that $\mathbb{P}_{\omega_2}$ has the $\aleph_2$-c.c. (see the discussion after Proposition 2.11), so standard arguments show that there is an $\alpha < \omega_2$ such that $X \in V^{\mathbb{P}_\alpha}$. Since cofinally many iterands are actually Laver forcing, we can also assume (by making α larger if necessary) that $\Vdash_\alpha \dot{\mathbb{Q}}_\alpha$ is Laver forcing.

But then $V^{\mathbb{P}_{\alpha+1}}$ is an extension of $V^{\mathbb{P}_\alpha}$ by Laver forcing $\mathbb{L}$, and so we can apply Lemma 2.15 to conclude that

$$V^{\mathbb{P}_{\alpha+1}} \models X \notin \mathcal{SN}.$$

Moreover, Proposition 2.14 yields that (in $V^{\mathbb{P}_{\alpha+1}}$) the quotient forcing $\mathbb{P}_{\omega_2} : \mathbb{P}_{\alpha+1}$ has the Laver property (as a countable support iteration of proper forcings having the Laver property); hence we can apply Lemma 2.20 to derive that

$$V^{\mathbb{P}_{\omega_2}} \models X \notin \mathcal{SN},$$

i.e., X is not strong measure zero in the final model, which is a contradiction. □

It is not clear to me to which extent Theorem 2.21 above can be generalized; in other words, one can ask whether any model obtained via a countable support iteration happens to satisfy the Borel Conjecture as long as it satisfies the two necessary conditions from Corollary 2.4 (also compare with Theorem 2.5, as well as Question 2.18 and Question 2.44):

QUESTION 2.22. Assume CH, and let $\{\mathbb{P}_\alpha, \dot{\mathbb{Q}}_\alpha : \alpha < \omega_2\}$ be any non-trivial[46] proper countable support iteration. Is it possible that

$$V^{\mathbb{P}_{\omega_2}} \models \neg BC \wedge \operatorname{cov}(\mathcal{M}) = \omega_1 \wedge \mathfrak{b} > \omega_1 \,? \tag{2.8}$$

Moreover, one could ask the above question while additionally demanding that all the involved forcing iterands are nicely definable (in this case, their definition *must not* be based on non-principal ultrafilters on ω).

2.7. Other models satisfying the Borel Conjecture. We now present several models of BC other than the Laver model, and discuss some of the ideas involved: in Section 2.7.1, we explain why Laver forcing can be replaced by Mathias forcing to obtain a model of BC (the Mathias model); in Section 2.7.2, we provide an alternative way (which is also relevant for the proof of Theorem 4.1) to preserve being *not* strong measure zero (using preservation of positivity instead of the Laver property), which yields both an alternative proof that BC holds in the Laver model, and a proof that BC holds in the Laver-random model; in Section 2.7.3, we present the theorem that BC is consistent with continuum larger than $\aleph_2$.

Moreover, we are going to discuss a weak version of the Borel Conjecture, which holds true in the random model and the Corazza model (see Section 2.7.6); finally, Section 2.7.7 is concerned with the question whether it is easy to modify suitable forcing constructions to additionally (i.e., while keeping the desired properties of the resulting model) ensure the Borel Conjecture, or additionally ensure the negation of the Borel Conjecture, using a recent model from [**SW**] as an example.

2.7.1. *Using Mathias forcing instead of Laver forcing.* Even though Laver's original approach to obtain a model satisfying the Borel Conjecture used a countable support iteration of Laver forcing (see Theorem 2.6), it was realized soon afterwards that not just Laver forcing, but also Mathias forcing (see [**BJ95**, Section 7.4.A]) does the job of killing strong measure zero sets; this yields the Borel Conjecture in the *Mathias model* (see Figure 6 for its Cichoń's diagram):

THEOREM 2.23. *Assume CH, and let $\mathbb{P}_{\omega_2}$ be the countable support iteration of Mathias forcing of length ω_2.*

Then in a generic extension by $\mathbb{P}_{\omega_2}$ the Borel Conjecture holds.

---

[46] To make the question non-trivial, we should at least assume something like

$$V^{\mathbb{P}_\alpha} \models CH \text{ for each } \alpha < \omega_2 \tag{2.9}$$

for the iteration in question (note that all models of this paper which are obtained by a countable support iteration fulfill this standard requirement anyway); otherwise, a model satisfying (2.8) could be obtained by a "trivial" countable support iteration as follows: let $\mathbb{Q}_0$ be the forcing (iteration) from Theorem 2.5 (i.e., Hechler iteration, followed by $\aleph_1$ many random reals), and for each $\alpha > 0$, let $\dot{\mathbb{Q}}_\alpha$ be the trivial forcing; then $V^{\mathbb{P}_{\omega_2}}$ equals $V^{\mathbb{Q}_0} = V^{\mathbb{P}_1}$ and is therefore the desired model, but already $V^{\mathbb{P}_1} \models 2^{\aleph_0} = \aleph_2$, violating (2.9).

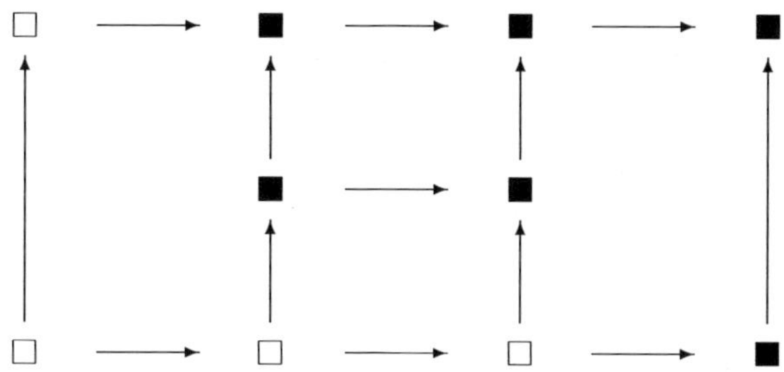

FIGURE 6. ($\square = \aleph_1$, $\blacksquare = \aleph_2$) Cichoń's diagram in the Mathias model

The (structure of its) proof is pretty similar to the one of Theorem 2.6 involving Laver forcing. We provide the new ingredients.

Recall the definition of *Mathias forcing* (see also [**BJ95**, Definition 7.4.1]): the conditions are of the form (s, A), where

- s is a finite subset of ω,
- A is an infinite subset of ω, and
- $\max(s) < \min(A)$;

the ordering is as follows: $(t, B) \leq (s, A)$ – i.e., (t, B) is stronger than (s, A) – if

- $t \supseteq s$,
- $B \subseteq A$, and
- $t \setminus s \subseteq A$.

For a filter G which is Mathias-generic over the ground model, let

$$\bigcup\{s\colon (s, A) \in G \text{ for some } A\} \subseteq \omega$$

be the *Mathias real*.

An easy density argument shows (also compare with Proposition 2.9 where the same is proved for Laver forcing) that the (increasing enumeration of the) Mathias real is dominating over the ground model (see also [**BJ95**, Lemma 7.4.4]); i.e., Mathias forcing adds dominating reals.

We say that (t, B) is a *pure extension* of (s, A) if $(t, B) \leq (s, A)$ and $t = s$; as Laver forcing, Mathias forcing has the pure decision property (see [**BJ95**, Theorem 7.4.6]) and satisfies Axiom A (and, hence, is proper).

We also have the following (see Proposition 2.13 for the analogous fact about Laver forcing):

PROPOSITION 2.24. *Mathias forcing has the Laver property.*

PROOF. See [**BJ95**, Corollary 7.4.7, Definition 7.2.1, and Lemma 7.2.2]. $\square$

The main (new) ingredient for the proof of Theorem 2.23 is the following lemma, which is the analogue of Lemma 2.15:

LEMMA 2.25. *Let $X \subseteq 2^\omega$ be an uncountable set in V. Then $\Vdash_{\text{Mathias}} X \notin \mathcal{SN}$.*

PROOF. See [**BJ95**, Theorem 8.3.2]. $\square$

In fact, X being not strong measure zero in the extension is witnessed by the (increasing enumeration of the) Mathias real.

The rest of the proof of Theorem 2.23 is exactly the same as the one of Theorem 2.6: if there were an uncountable strong measure zero set in the final model (the Mathias model), this set would already appear in the course of the iteration; but then the next Mathias real kills it (due to the above Lemma 2.25), and there is no resurrection possible, due to Proposition 2.24, Proposition 2.14, and Lemma 2.20.

2.7.2. *Preservation of positivity (and interleaving with random reals).* So far, we always used the Laver property to prevent resurrection of strong measure zero sets; more precisely, our strategy was as follows: after making sure that uncountable strong measure zero sets are killed (by either Laver or Mathias forcing), we used the fact that the tail of the iteration has the Laver property, and so we could exploit Lemma 2.20 to finish the proof.

In fact, there is the following alternative way: instead of the Laver property, one can use *preservation of positivity*, which is the following property of a forcing notion: sets from the ground model which are *not* measure zero (i.e., "positive") in the ground model remain *not* measure zero in the forcing extension; this property might not be iterable, so we use the technical strengthening *preservation of random reals* (see[47] [**BJ95**, Section 6.3.B] for the definition, and [**BJ95**, Lemma 6.3.11] for the proof that it implies preservation of positivity) which is indeed an iterable property:

PROPOSITION 2.26. *Preservation of random reals is preserved under countable support iteration of proper forcing notions.*

PROOF. See [**BJ95**, Theorem 6.3.13] (or rather [**BJ95**, Theorem 6.1.13]). □

Laver forcing[48] indeed preserves positivity (and also *outer measure*; see [**BJ95**, Theorem 7.3.37]); this is an unpublished theorem of Hugh Woodin (and later it was independently proved in [**JS90**]). In fact, it even enjoys the stronger iterable property:

THEOREM 2.27. $\mathbb{L}$ *preserves random reals.*

PROOF. See [**BJ95**, Theorem 7.3.39], or Pawlikowski's paper [**Paw96c**]. □

The following characterization of strong measure zero sets (by Pawlikowski) is the key ingredient for replacing the Laver property by preservation of positivity:

THEOREM 2.28. *Let $X \subseteq 2^\omega$ be a set of reals. Then X is strong measure zero if and only if $X + E$ is measure zero for each closed measure zero set E.*

PROOF. See [**BJ95**, Theorem 8.1.18], or [**Paw96a**] for the original paper. □

So the property "X not strong measure zero" is translated into the property "not measure zero" (for the associated set $X + E$, for a certain closed measure zero set E provided by the above theorem).

Therefore, it is straightforward to check that Theorem 2.28 yields the following analogue of Lemma 2.20:

---

[47]In [**BJ95**, Section 6.1.B], one can find a general framework for preservation theorems for countable support iteration: in particular, see [**BJ95**, Definition 6.1.6] for the relation $\sqsubseteq$, [**BJ95**, Definition 6.1.10] for "$\mathbb{P}$ preserves $\sqsubseteq$", as well as [**BJ95**, Theorem 6.1.13] for the general preservation theorem for $\sqsubseteq$.

[**BJ95**, Section 6.3.B] deals with $\sqsubseteq^{\text{random}}$, a special instance of $\sqsubseteq$. What we call "preserving random reals" is called "preserving $\sqsubseteq^{\text{random}}$" there.

[48]In contrast to Mathias forcing; see the discussion after Conjecture 2.32.

LEMMA 2.29. *Suppose $\mathbb{P}$ is a forcing which preserves positivity, and G is a $\mathbb{P}$-generic filter over V. Let $X \in V \cap \mathcal{P}(2^\omega)$. Then*

$$V \models X \notin \mathcal{SN} \implies V[G] \models X \notin \mathcal{SN}.$$

PROOF. Since X is not strong measure zero in V, Theorem 2.28 provides a closed measure zero set $E \in V$ such that $X + E$ is not measure zero (in V), and this property is preserved in the extension by $\mathbb{P}$, i.e., (the very set) $X + E$ is not measure zero in $V[G]$; in particular, this is true when E is replaced by the Borel code for E reinterpreted in $V[G]$ (yielding a superset of E), so there is a closed measure zero set $E' \in V[G]$ (namely the evaluation of the Borel code of E within $V[G]$) such that $X + E'$ is not measure zero, and hence (applying the other direction of Theorem 2.28 within $V[G]$)) X is not strong measure zero in $V[G]$. □

Combining the above facts, it is possible to give the following *alternative argument that BC holds in the Laver model*: like in the usual proof (see the proof of Theorem 2.21), assume towards a contradiction that there is an uncountable (without loss of generality of size $\aleph_1$) strong measure zero set X in the final model; this set would already appear in the course of the iteration, and the next Laver real kills it, i.e., there is $\alpha < \omega_2$ such that $V^{\mathbb{P}_{\alpha+1}} \models X \notin \mathcal{SN}$ (see Lemma 2.15); but due to Theorem 2.27 and Proposition 2.26, the tail $\mathbb{P}_{\omega_2} : \mathbb{P}_{\alpha+1}$ of the iteration preserves random reals; so, in particular, $\mathbb{P}_{\omega_2} : \mathbb{P}_{\alpha+1}$ preserves positivity, hence Lemma 2.29 guarantees that there is no resurrection possible, i.e., X is not strong measure zero in the final model, a contradiction.

This alternative approach (using preservation of positivity) for proving BC in the Laver model has the advantage that the same argument can be used for showing BC in the model obtained by interleaving Laver forcing with random forcing along the iteration (see Theorem 2.31 below).

A similar approach is used for proving BC in the model of BC+dBC (see Theorem 4.1, and the discussion in Section 4.5 and Section 4.6 about the respective properties of ultralaver forcing and Janus forcing, as well as Section 4.11 for the proof of BC in the final model).

In fact, random forcing (even though it does not have the Laver property) shares the alternative iterable property with Laver forcing (compare with Theorem 2.27):

THEOREM 2.30. *Random forcing $\mathbb{B}$ preserves random reals.*

PROOF. See [**BJ95**, Lemma 6.3.12(2)]. □

Arguing like above, Lemma 2.15, Theorem 2.27, Theorem 2.30, Proposition 2.26, and Lemma 2.29 yield the Borel Conjecture in the *Laver-random model* (see Figure 7 for its Cichoń's diagram):

THEOREM 2.31. *Assume CH, and let $\{\mathbb{P}_\alpha, \dot{\mathbb{Q}}_\alpha : \alpha < \omega_2\}$ be a countable support iteration such that for every* even $\alpha < \omega_2$,

$$\Vdash_\alpha \dot{\mathbb{Q}}_\alpha \textit{ is Laver forcing } \mathbb{L},$$

and for every odd $\alpha < \omega_2$,

$$\Vdash_\alpha \dot{\mathbb{Q}}_\alpha \textit{ is random forcing } \mathbb{B}.$$

Then

$$V^{\mathbb{P}_{\omega_2}} \models BC.$$

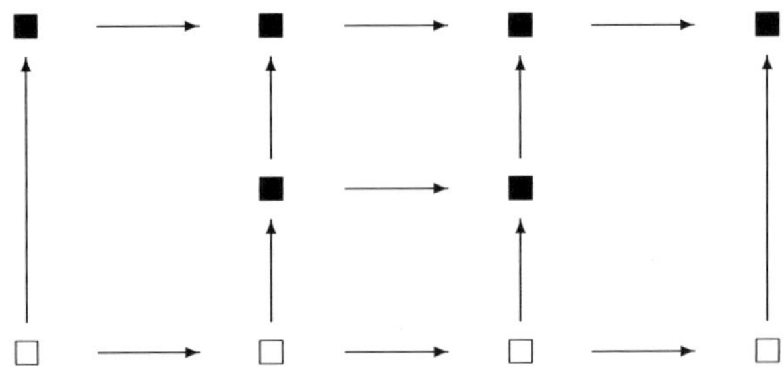

FIGURE 7. ($\square = \aleph_1$, $\blacksquare = \aleph_2$) Cichoń's diagram in the Laver-random model

We believe that the Borel Conjecture also holds true in the *Mathias-random model* (see Figure 8 for its Cichoń's diagram), i.e., when Laver forcing is replaced by Mathias forcing in Theorem 2.31:

CONJECTURE 2.32. *Assume CH, and let* $\{\mathbb{P}_\alpha, \dot{\mathbb{Q}}_\alpha : \alpha < \omega_2\}$ *be a countable support iteration such that every even iterand is Mathias forcing and every odd iterand is random forcing. Then the final model* $V^{\mathbb{P}_{\omega_2}}$ *satisfies BC.*

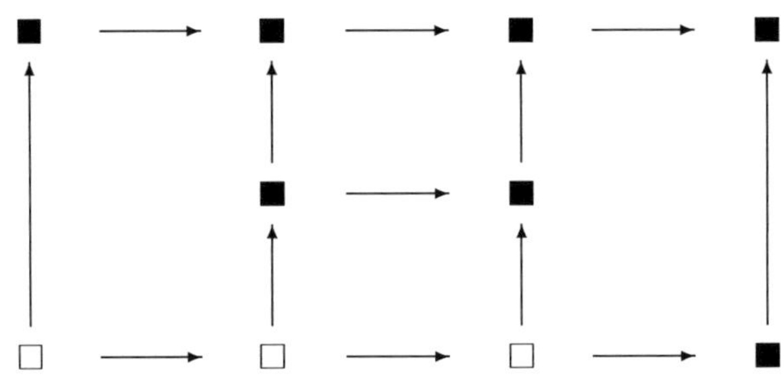

FIGURE 8. ($\square = \aleph_1$, $\blacksquare = \aleph_2$) Cichoń's diagram in the Mathias-random model

Note that Mathias forcing does *not* preserve positivity (in fact, Mathias forcing turns the old reals $2^\omega \cap V$ into a measure zero set; see [**BJ95**, Lemma 7.4.2]); moreover, random forcing does *not* have the Laver property (in fact, a forcing with the Laver property does not add random reals; see [**BJ95**, Lemma 7.2.3]); so – in order to prove Conjecture 2.32 – we can neither use the Laver property (and Lemma 2.20) nor preservation of positivity (and Lemma 2.29) for preventing resurrection of strong measure zero sets; however, the following seems plausible (and would yield a proof of Conjecture 2.32):

CONJECTURE 2.33. *Assume CH, and let* $\{\mathbb{P}_\alpha, \dot{\mathbb{Q}}_\alpha : \alpha < \omega_2\}$ *be a countable support iteration such that for every* $\alpha < \omega_2$,

$$\Vdash_\alpha \dot{\mathbb{Q}}_\alpha \text{ is proper and either has the Laver property or preserves random reals,}$$

and

$$\{\alpha < \omega_2 \colon \Vdash_\alpha \dot{\mathbb{Q}}_\alpha \text{ kills smz}\}$$

is cofinal in ω_2. *Then*

$$V^{\mathbb{P}_{\omega_2}} \models BC.$$

The proof is supposed to be somewhat as follows:[49] to prove that a set X which has been killed (i.e., is not smz in some intermediate model) remains not smz, Theorem 2.28 is applied, yielding (a Borel code for) a closed measure zero set E such that $X + E$ is not measure zero; it has to be shown that

$$\text{“}X + E \text{ is not measure zero”} \tag{2.10}$$

remains valid for the rest of the iteration (where E is the respective re-evaluation of the very same Borel code); for iterands which preserve random reals (and hence preserve positivity) – such as random forcing –, this is clear anyway (compare with Lemma 2.29), whereas for iterands with the Laver property (but without preserving positivity) – such as Mathias forcing –, this can probably be shown using Lemma 2.20 together with *the proof of* Theorem 2.28; moreover, one has to show that (2.10) is also preserved at limit stages.

2.7.3. *Borel Conjecture and large continuum (by adding many random reals).* Since the consistency of the Borel Conjecture is usually proved by a countable support iteration of length ω_2 over a model of CH, the typical model for BC satisfies $2^{\aleph_0} = \aleph_2$. So the question arises, whether BC is consistent with larger continuum, i.e., $2^{\aleph_0} > \aleph_2$.

Recall that finite support iterations cannot be used to settle this question because they always add Cohen reals at limit stages, and this inevitably destroys BC (see Theorem 2.2 and the discussion after Corollary 2.4).

However, it is possible to proceed as follows to get a model of BC with large continuum: start with one of the two standard models for BC, i.e., the Laver model (see Theorem 2.6) or the Mathias model (see Theorem 2.23); then increase the continuum by simultaneously adding many random reals:

THEOREM 2.34 (Judah-Shelah-Woodin, 1990). *Assume GCH, and let $\mathbb{P}_{\omega_2}$ be the countable support iteration of Laver forcing of length ω_2; moreover, in the Laver model $V^{\mathbb{P}_{\omega_2}}$, let $\mathbb{B}_\lambda$ be the forcing for simultaneously adding λ many random reals. Then*

$$V^{\mathbb{P}_{\omega_2} * \mathbb{B}_\lambda} \models BC \wedge 2^{\aleph_0} = \lambda.$$

(Actually, the same holds when Laver forcing is replaced by Mathias forcing.)

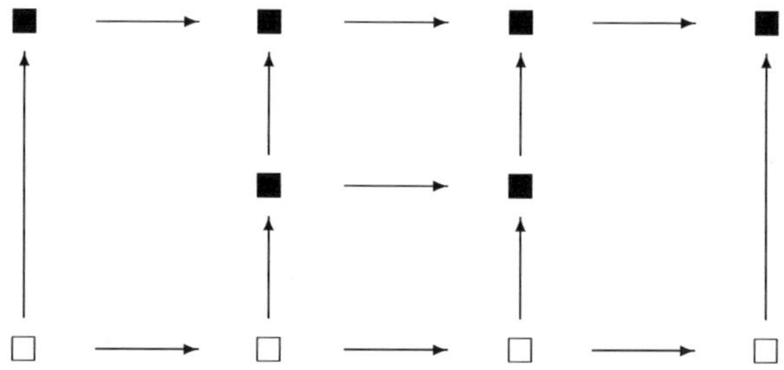

FIGURE 9. ($\square = \aleph_1$, $\blacksquare = \aleph_2$) Cichoń's diagram in the model obtained from the Laver model (or the Mathias model, respectively) by adding $\aleph_2$ many random reals

[49] I thank Tomek Bartoszyński for pointing out this proof idea to me.

PROOF. For the (quite long and technical) proof, see [**BJ95**, Theorem 8.3.7] (or [**JSW90**] for the original[50] paper). □

One (comparably easy) ingredient of the proof is the fact that the two statements

(1) forcing with single random forcing $\mathbb{B}$ preserves BC
(2) forcing with $\mathbb{B}_\lambda$ preserves BC

are equivalent (see [**BJ95**, Theorem 8.3.6]). Proving Theorem 2.34 then amounts to proving that (which is the difficult part) forcing with single random forcing $\mathbb{B}$ does not destroy[51] the Borel Conjecture (of the Laver or Mathias model). Whether this is true in general, seems to be open (compare with Remark 2.19 and footnote 45):

QUESTION 2.35. Does adding random reals (always) preserve the Borel Conjecture?

2.7.4. *Joint consistency with the dual Borel Conjecture.* There is a model of the Borel Conjecture in which also the dual Borel Conjecture holds true (see Theorem 4.1); an overview of the very involved forcing construction is given in Section 4.

2.7.5. *Joint consistency with a projective well-order of the reals.* There is a model of the Borel Conjecture in which there exists a projective well-order of the reals (see Theorem 6.1); an outline of the proof is given in Section 6.2.

2.7.6. *The weak Borel Conjecture (in the random model and the Corazza model).* Let us say that the *weak Borel Conjecture* (weak BC) is the assertion that there is *no* strong measure zero set *of size continuum.*

The following very general theorem essentially says that one obtains a model of weak BC as long as no unbounded reals are added along the (countable support) iteration:

THEOREM 2.36. *Assume CH, and let $\{\mathbb{P}_\alpha, \dot{\mathbb{Q}}_\alpha : \alpha < \omega_2\}$ be a countable support iteration of strongly*[52] *ω^ω-bounding proper forcing notions.*

Then $V^{\mathbb{P}_{\omega_2}}$ satisfies the weak Borel Conjecture (i.e., there are no strong measure zero sets of size continuum).

PROOF. See [**BJ95**, Theorem 8.2.14]. □

Since random forcing is strongly ω^ω-bounding, the above theorem immediately yields that weak BC holds in the *random model* (see Figure 10 for its Cichoń's diagram):

COROLLARY 2.37. *Assume CH, and let $\{\mathbb{P}_\alpha, \dot{\mathbb{Q}}_\alpha : \alpha < \omega_2\}$ be a countable support iteration of random forcing.*

Then $V^{\mathbb{P}_{\omega_2}}$ satisfies the weak Borel Conjecture.

There is actually an easier and more direct argument which shows that weak BC holds in the extension by adding (at least) $\aleph_2$ many random reals (side-by-side) to

[50]However, there is a mistake in the original paper [**JSW90**]; the corrected version can be found in the book [**BJ95**, Theorem 8.3.7].

Moreover, according to Carlson (see [**Car93**, page 580]), Woodin has already shown this result in unpublished notes.

[51]After all, BC is not destroyed by random forcing (at least when we start with the Laver or Mathias model); however, random forcing cannot produce BC: see the discussion after Corollary 2.37.

[52]For the definition of *strongly ω^ω-bounding*, see, e.g., [**BS03**, Definition 15].

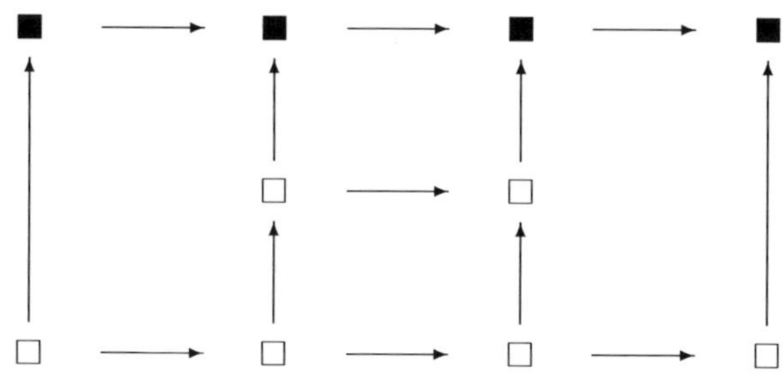

FIGURE 10. ($\square = \aleph_1$, $\blacksquare = \aleph_2$) Cichoń's diagram in the random model

a model of CH. It also works for adding more than $\aleph_2$ many random reals; in this case, the conclusion is slightly stronger than weak BC:

THEOREM 2.38. *Assume CH. Let $\lambda \geq \aleph_2$ be a regular cardinal, and let $\mathbb{B}_\lambda$ be the forcing for simultaneously adding λ many random reals.*

Then in $V^{\mathbb{B}_\lambda}$, each strong measure zero set is of size at most $\aleph_1$.

PROOF. See [**BJ95**, Theorem 8.2.11]. □

Carlson remarks (see [**Car93**, page 580]) that due to the fact that dBC holds in the Cohen model (see Theorem 3.10), and BC and dBC are dual to each other (recall that BC is the $\mathcal{M}$-BC, whereas dBC is the $\mathcal{N}$-BC; see Definition 1.9), one might expect that (full) BC holds in the random model; however, this is not the case: in the random model, $\mathfrak{b} = \omega_1$ holds true (see Figure 10), so one of the necessary conditions for the Borel Conjecture fails (see Theorem 2.3).

Note that there are sets of size $\aleph_1$ in the random model which are *not* strong measure zero (even though BC fails, i.e., there are also strong measure zero sets of size $\aleph_1$): this is due to the fact that $\text{non}(\mathcal{N}) = \aleph_1$ (see Figure 10), i.e., there are even sets of size $\aleph_1$ which are not (Lebesgue) measure zero.

The following theorem (see also [**BJ95**, Theorem 8.2.17]) yields a model of the weak Borel Conjecture in which being strong measure zero is only a matter of size (the *Corazza*[53] *model*; see Figure 11 for its Cichoń's diagram). Let $\mathbb{EE}$ be *infinitely equal forcing* (see [**BJ95**, Definition 7.4.11] for its definition).

THEOREM 2.39. *Assume CH, and let $\{\mathbb{P}_\alpha, \dot{\mathbb{Q}}_\alpha : \alpha < \omega_2\}$ be a countable support iteration of infinitely equal forcing $\mathbb{EE}$.*

Then

$$V^{\mathbb{P}_{\omega_2}} \models \mathcal{SN} = [2^\omega]^{\leq \aleph_1} \wedge 2^{\aleph_0} = \aleph_2,$$

i.e., a set of reals is strong measure zero (in the final model) if and only if it is of cardinality strictly less than the continuum. (In particular, the weak Borel Conjecture holds.)

PROOF. Since the forcing $\mathbb{EE}$ turns the set of reals from the ground model into a strong measure zero set (see [**BJ95**, Lemma 7.4.13]), it is straightforward to check that (in $V^{\mathbb{P}_{\omega_2}}$) any set of size (at most) $\aleph_1$ is strong measure zero.

[53] Actually, Corazza (see [**Cor89**]) provided a slightly different model than the one presented here: in Theorem 2.39 (following [**GJS93**]) only $\mathbb{EE}$ forcing is iterated, whereas Corazza's model is obtained by an alternating iteration of $\mathbb{EE}$ forcing and Sacks forcing; the two models, however, agree on their essential properties.

On the other hand, the forcing $\mathbb{EE}$ is Axiom A (see [**BJ95**, Lemma 7.4.12]) and hence proper, and strongly ω^ω-bounding (see [**BJ95**, Lemma 7.4.14]); therefore, Theorem 2.36 yields the weak Borel Conjecture (in $V^{\mathbb{P}_{\omega_2}}$). □

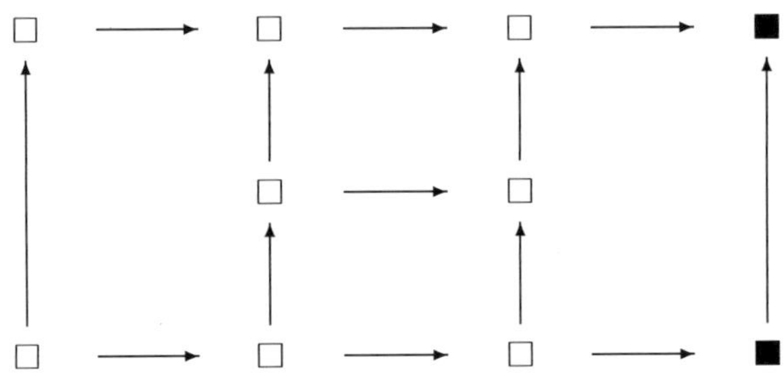

FIGURE 11. ($\square = \aleph_1$, $\blacksquare = \aleph_2$) Cichoń's diagram in the Corazza model

2.7.7. *Modifying an iteration to additionally obtain the Borel Conjecture.* As an example of the issue of modifying a suitable forcing construction to (additionally) ensure the (negation of the) Borel Conjecture, we discuss a (modification of a) recent result from [**SW**] (of Otmar Spinas and the author of this paper) in the context of the Borel Conjecture.

Let $\mathbb{A}$ be[54] *increasing Sacks amoeba forcing* (see [**SW**, Definition 3.1] for its definition); it is a variant of the "usual" Sacks amoeba forcing (which adds Cohen reals); the forcing $\mathbb{A}$, however, has the Laver property (see [**SW**, Theorem 3.13]).

The following theorem[55] is the main result of [**SW**] (see Figure 12 for Cichoń's diagram in the resulting model):

THEOREM 2.40. *Assume CH, and let* $\{\mathbb{P}_\alpha, \dot{\mathbb{Q}}_\alpha : \alpha < \omega_2\}$ *be a countable support iteration of increasing Sacks amoeba forcing* $\mathbb{A}$.

Then in the final model $V^{\mathbb{P}_{\omega_2}}$ *we have*

$$\omega_1 = \operatorname{cov}(\mathcal{N}) = \operatorname{cov}(\mathcal{M}) = \mathfrak{h} < \operatorname{add}(s_0) = \mathfrak{s} = \mathfrak{b} = \mathfrak{c} = \omega_2. \tag{2.11}$$

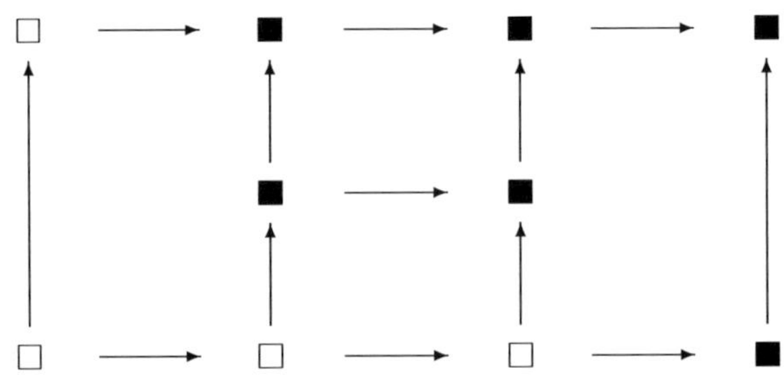

FIGURE 12. ($\square = \aleph_1$, $\blacksquare = \aleph_2$) Cichoń's diagram in the model for $\mathfrak{h} < \operatorname{add}(s_0)$

[54]The forcing $\mathbb{A}$ was implicitly introduced by Louveau, Shelah, and Veličković in their paper [**LSV93**].

[55]The proof of the theorem has been completed in 2019.

The essential part of (2.11) (which the theorem aims at) is $\mathfrak{h} < \text{add}(s_0)$; the rest is a by-product of the way the model is constructed.

We do not know the status of BC in the model of Theorem 2.40:

QUESTION 2.41. Does the Borel Conjecture hold in the model of [**SW**]?

In fact, we do not know whether increasing Sacks amoeba forcing $\mathbb{A}$ adds[56] a Laver real; in other words, the following seems to be open:

QUESTION 2.42. Is Laver forcing $\mathbb{L}$ completely embeddable into increasing Sacks amoeba forcing $\mathbb{A}$?

Observe that an affirmative answer to Question 2.42 would entail an affirmative answer to Question 2.41, essentially due to Theorem 2.21.

Alternatively, the following approach is imaginable: it could be that increasing Sacks amoeba forcing $\mathbb{A}$ kills all uncountable strong measure zero sets from the ground model, i.e., that there is a lemma for $\mathbb{A}$ which is analogous to Lemma 2.15 for Laver forcing (or Lemma 2.25 for Mathias forcing); this would yield an affirmative answer to Question 2.41 (without the need of solving Question 2.42).

By (cofinally) inserting Laver forcing, it is actually possible to sightly modify the construction of Theorem 2.40 to make BC hold *for sure* (while preserving all the other relevant properties of the original model; in particular, $\mathfrak{h} < \text{add}(s_0)$ remains true, and Cichoń's diagram is the same, i.e., as displayed in Figure 12):

THEOREM 2.43. *Assume CH, and let* $\{\mathbb{P}_\alpha, \dot{\mathbb{Q}}_\alpha : \alpha < \omega_2\}$ *be a countable support iteration such that for every* even $\alpha < \omega_2$,

$$\Vdash_\alpha \dot{\mathbb{Q}}_\alpha \textit{ is increasing Sacks amoeba forcing } \mathbb{A},$$

and for every odd $\alpha < \omega_2$,

$$\Vdash_\alpha \dot{\mathbb{Q}}_\alpha \textit{ is Laver}^{57} \textit{ forcing } \mathbb{L}.$$

Then the final model $V^{\mathbb{P}_{\omega_2}}$ *satisfies* (2.11) *as well as the Borel Conjecture.*

The arguments in [**SW**] to show (2.11) still work when Laver forcing is inserted at odd[58] stages; moreover, the Borel Conjecture immediately follows from Theorem 2.21.

---

[56] Note that $\mathbb{A}$ is definitely not forcing equivalent to Laver forcing since otherwise the resulting model would be the Laver model: this is not the case because non($\mathcal{N}$) is small in the Laver model (see Figure 5), whereas it is large in the model of Theorem 2.40 (see Figure 12).

Moreover, $\mathbb{A}$ is not forcing equivalent to Mathias forcing since otherwise the resulting model would be the Mathias model: even though Cichoń's diagram in the Mathias model is the same (see Figure 6), this is not the case because $\mathfrak{h} = \omega_2$ holds true in the Mathias model (see [**Bla10**]).

In fact, an easy density argument shows that a Mathias real "intersects" any system of mad families (by "selecting" one element from each mad family); so the forcing $\mathbb{A}$ cannot even add a Mathias real: otherwise, no (reflected version of a) system of mad families of size ω_1 (supposedly witnessing $\mathfrak{h} = \omega_1$ in the final model) would survive along the iteration.

[57] Surprisingly, Laver forcing can even be replaced by Mathias forcing here: this seems contradictory since $\mathfrak{h}$ is large in the Mathias model, due to the fact that Mathias forcing destroys (reflected) witnesses of $\mathfrak{h} = \omega_1$ (also compare with footnote 56); however, this is not the case when Mathias forcing (as it is done) is only applied at (the non-stationary set of all) *odd* stages: the maximality of the mad families in the system (living in the final model) witnessing small $\mathfrak{h}$ does not reflect to odd stages, therefore the Mathias real is not able to select from them (actually, in the construction of Theorem 2.40, the madness is only established at limit stages of cofinality ω_1).

[58] It is not possible to interchange the roles of even and odd stages because increasing Sacks amoeba forcing $\mathbb{A}$ has to be applied (according to the method used in the proof of Theorem 2.40) at every limit stage of cofinality ω_1.

On the other hand, it is not clear how to modify the construction to get a model satisfying (2.11) – or just $\mathfrak{h} < \operatorname{add}(s_0)$ – together with the negation of BC (also compare with Question 2.22):

QUESTION 2.44. Is $\mathfrak{h} < \operatorname{add}(s_0)$ (or even (2.11)) consistent with the failure of the Borel Conjecture?

Note that it is *not* possible to solve the above question just by adding a single Cohen real to the model of Theorem 2.40: it is true that BC is going to fail (due to the well-known fact that a Cohen real turns[59] the set of reals from the ground model into a strong measure zero set), but $\mathfrak{h} < \operatorname{add}(s_0)$ is going to fail as well; this can be seen as follows: since the model from Theorem 2.40 satisfies $\operatorname{add}(\mathcal{M}) = \omega_1$ (see Figure 12), we can apply [**BJ95**, Theorem 3.3.23] to derive $\mathfrak{b} = \omega_1$ in its single Cohen extension; but then $\operatorname{add}(s_0) = \omega_1$ holds there as well because $\operatorname{add}(s_0) \leq \mathfrak{b}$ is a theorem of ZFC (see [**HS20**, Corollary 2.8]).

3. Dual Borel Conjecture (dBC)

This section is concerned with strongly meager sets and models of the dual Borel Conjecture. First, we discuss necessary conditions for the dual Borel Conjecture and argue why they are not sufficient (see Section 3.1). Subsequently, most of the section is dedicated to a detailed proof of Carlson's theorem [**Car93**] that the dual Borel Conjecture holds true in the Cohen model and is therefore consistent (see Theorem 3.10); at the end (see Section 3.7), we discuss alternative models of the dual Borel Conjecture.

3.1. Necessary conditions for the dual Borel Conjecture. Before discussing two necessary conditions (involving cardinal characteristics from Cichoń's diagram) for the dual Borel Conjecture to hold (see Corollary 3.8), we first recall why $2^{\aleph_0} > \aleph_1$ is a necessary requirement; in other words, the dual Borel Conjecture necessarily fails under the continuum hypothesis:

THEOREM 3.1. *Assume CH. Then there exists an uncountable strongly meager set.*

PROOF. Since $\mathcal{N}$ is a translation-invariant σ-ideal and has a basis consisting of Borel sets, Corollary 1.11 yields an uncountable set which is $\mathcal{N}$-shiftable (see Definition 1.8); in other words, this set is strongly meager (see Definition 1.6). □

We now show that the dual Borel Conjecture necessarily fails under certain assumptions about cardinal characteristics from Cichoń's diagram (see Section 1.6, and, in particular, Figure 2).

The following theorem is the "dual" version of Theorem 2.2; it says that the dual Borel Conjecture necessarily fails if the covering number $\operatorname{cov}(\mathcal{N})$ of the measure zero ideal $\mathcal{N}$ is large, i.e., if the reals cannot be covered by $\aleph_1$ many measure zero sets:

THEOREM 3.2. *Assume* $\operatorname{cov}(\mathcal{N}) > \omega_1$*; then there exists an uncountable strongly meager set. (In fact, each set of size strictly less than* $\operatorname{cov}(\mathcal{N})$ *is strongly meager.)*

PROOF. Let $X \subseteq 2^\omega$ with $|X| < \operatorname{cov}(\mathcal{N})$; since $\mathcal{N}$ is translation-invariant, we can apply Proposition 1.12 to derive that X is $\mathcal{N}$-shiftable (see Definition 1.8); in other words, X is strongly meager (see Definition 1.6). □

[59]For a proof, see, e.g., [**Gol**] (we are not aware of any published proof of this fact).

Note that the above result can be rephrased as

$$\mathrm{cov}(\mathcal{N}) \leq \mathrm{non}(\mathcal{SM});$$

together with[60] $\mathcal{SM} \subseteq \mathcal{M}$ (which trivially yields $\mathrm{non}(\mathcal{SM}) \leq \mathrm{non}(\mathcal{M})$), this reproves the second inequality of Corollary 1.14.

The following theorem shows that the dual Borel Conjecture necessarily fails if the dominating number $\mathfrak{d}$ is small, i.e., if there is a dominating family (of functions in ω^ω) of size $\aleph_1$:

THEOREM 3.3. *Assume $\mathfrak{d} = \omega_1$; then there exists an uncountable strongly meager[61] set.*

Note that this gives another proof of[62] Theorem 3.1, i.e., the fact that there exists an uncountable strongly meager set under CH (since CH implies $\mathfrak{d} = \omega_1$).

PROOF. This result was first shown by Bartoszyński in [**Bar03**]. The author of this paper has independently discovered another proof, which is part of his PhD thesis (see [**Woh13**, Section 6.3]); here, we essentially follow the proof from [**Woh13**].

We start with a few definitions: for $X \subseteq 2^\omega$ and $k \in \omega$, let

$$X \restriction k := \{x \restriction k : x \in X\} \subseteq 2^k.$$

For $f \in \omega^\omega$, a set $X \subseteq 2^\omega$ is *f-tiny* if $|X \restriction f(k)| \leq k$ for almost all $k \in \omega$. Let $\mathcal{I}_f$ be the σ-ideal generated by the f-tiny sets. A set is *completely tiny* if it belongs to $\mathcal{I}_f$ for each $f \in \omega^\omega$.

The *structure of the proof* is going to be as follows: we first show how to construct a dominating tower of length ω_1 under our assumption that $\mathfrak{d} = \omega_1$ (see Claim 3.4), then show that each such tower forms a completely tiny set of reals (see Claim 3.5), and finally prove (see Claim 3.7) that each completely tiny set is null-additive (a notion stronger than strongly meager).

Before going into more details, let us recall that $\mathbb{Q}$ denotes the set of "rationals" in the Cantor space 2^ω (i.e., $x \in \mathbb{Q}$ if and only if $x(n) = 0$ for all but finitely many $n \in \omega$). Let $x \in 2^\omega \setminus \mathbb{Q}$; then x is the characteristic function of the *infinite* set $\{n \in \omega \colon x(n) = 1\} \subseteq \omega$, and $\mathrm{enum}(x) \in \omega^\omega$ shall denote its *enumerating function* (in other words, let $\mathrm{enum}(x)$ be the unique strictly increasing function whose range equals $\{n \in \omega \colon x(n) = 1\}$).

Further recall (see Section 1.6.3) the relation $\leq^*$ of eventual domination on ω^ω, and that the dominating number $\mathfrak{d}$ denotes the smallest size of a dominating family. For $x, y \in 2^\omega \setminus \mathbb{Q}$, we write $x \leq^* y$ to denote that $x(n) \leq y(n)$ for almost all $n \in \omega$; note that $x \leq^* y$ if and only if $x \subseteq^* y$ when x and y are viewed as subsets of ω (where $\subseteq^*$ denotes *almost inclusion*), and $x \leq y$ if and only if $x \subseteq y$.

---

[60]For the proof of $\mathcal{SM} \subseteq \mathcal{M}$, see the discussion after Definition 1.6.

[61]In fact, the constructed set will even be null-additive.

[62]Observe that this also gives yet another proof of Theorem 2.1: this is because our construction actually yields an uncountable completely tiny set of reals (see Claim 3.4 and Claim 3.5), and each completely tiny set can easily seen to be strong measure zero (see [**Woh13**, Section 6.3, discussion after Definition 6.19]); so, in fact, we obtain a weaker version of Theorem 2.3 (namely – instead of $\mathfrak{b} = \omega_1$ – with the stronger assumption that $\mathfrak{d} = \omega_1$).

Alternatively, one could argue as follows: a completely tiny set of reals is null-additive (see Claim 3.7), and every null-additive set is meager-additive (due to a theorem of Shelah; see [**BJ95**, Theorem 2.7.20], or the original paper [**She95**]), and hence strong measure zero (invoking this theorem is an overkill though because showing that completely tiny sets are strong measure zero is much easier than Shelah's theorem).

A sequence $\langle x_\alpha \colon \alpha < \omega_1 \rangle \subseteq 2^\omega \setminus \mathbb{Q}$ is a *dominating tower of length* ω_1 if

- $x_\beta \leq^* x_\alpha$ for each $\alpha < \beta$, and
- $\{\mathrm{enum}(x_\alpha) \colon \alpha < \omega_1\} \subseteq \omega^\omega$ is a dominating family, i.e.,

$$\forall g \in \omega^\omega \; \exists \alpha < \omega_1 \quad g \leq^* \mathrm{enum}(x_\alpha).$$

Note that any $\subseteq^*$-decreasing sequence of infinite subsets of ω of countable (ordinal) length has an infinite *pseudointersection*; in other words, given $\gamma < \omega_1$ and $\langle x_\alpha \colon \alpha < \gamma \rangle \subseteq 2^\omega \setminus \mathbb{Q}$ with $x_\beta \leq^* x_\alpha$ for each $\alpha < \beta$, there exists[63] an $x_\gamma \in 2^\omega \setminus \mathbb{Q}$ such that $x_\gamma \leq^* x_\alpha$ for each $\alpha < \gamma$.

Using this, it is easy to show the following:

Claim 3.4. Let $\mathfrak{d} = \omega_1$. Then there exists a dominating tower of length ω_1.

Proof. By assumption, we can fix a dominating family $\{g_\alpha : \alpha < \omega_1\} \subseteq \omega^\omega$. We construct a dominating tower $\langle x_\alpha \colon \alpha < \omega_1 \rangle \subseteq 2^\omega \setminus \mathbb{Q}$ of length ω_1 by induction on $\alpha < \omega_1$: at step α, let $y \in 2^\omega \setminus \mathbb{Q}$ be a pseudointersection of the x_β constructed so far (i.e., $y \leq^* x_\beta$ for every $\beta < \alpha$), and let $x_\alpha \in 2^\omega \setminus \mathbb{Q}$ be such that $x_\alpha \leq y$ and $g_\alpha \leq \mathrm{enum}(x_\alpha)$ (in particular, we have $x_\alpha \leq^* x_\beta$ for every $\beta < \alpha$). It is easy to verify that $\langle x_\alpha \colon \alpha < \omega_1 \rangle$ is a dominating tower of length ω_1. □

We now show that each such tower forms a completely tiny set of reals:

Claim 3.5. Let $\langle x_\alpha \colon \alpha < \omega_1 \rangle$ be a dominating tower of length ω_1. Then

$$X := \{x_\alpha \colon \alpha < \omega_1\}$$

is completely tiny, i.e., $X \in \mathcal{I}_f$ for each $f \in \omega^\omega$.

Proof. Fix a (without loss of generality strictly increasing) function $f \in \omega^\omega$; we will show that $X \in \mathcal{I}_f$. Let $g \in \omega^\omega$ be defined by $g(n) := f(2^{n+1})$ for each $n \in \omega$. Due to the fact that $\{\mathrm{enum}(x_\alpha) \colon \alpha < \omega_1\}$ is a dominating family, we can fix $\alpha < \omega_1$ such that $g \leq^* \mathrm{enum}(x_\alpha)$.

It is *rather straightforward to check* that[64] $Y := \{y \in 2^\omega : y \leq x_\alpha\}$ is f-tiny. More precisely, we can argue as follows. Let $\mathrm{enum}(x_\alpha) =: h \in \omega^\omega$; since $g \leq^* h$, we can fix $N \in \omega$ such that $g(n) \leq h(n)$ for each $n \geq N$; we will show that for each $k \geq 2^N$, we have $|Y \restriction f(k)| \leq k$. So fix $k \geq 2^N$. Fix $n \in \omega$ such that $k \in [2^n, 2^{n+1})$. Note that $n \geq N$, so $g(n) \leq h(n)$ by our assumption on N. Moreover, we have $f(k) < f(2^{n+1}) = g(n) \leq h(n)$. Observe that

$$|Y \restriction f(k)| = |\{s \in 2^{f(k)} \colon s \leq x_\alpha \restriction f(k)\}| = 2^{|x_\alpha \cap f(k)|},$$

where $x_\alpha \cap f(k)$ actually denotes the set $\{j < f(k) : x_\alpha(j) = 1\}$ (i.e., we view x_α as a subset of ω here). But $|x_\alpha \cap f(k)| \leq |x_\alpha \cap h(n)| = n$ (since $h = \mathrm{enum}(x_\alpha)$ enumerates x_α), therefore $|Y \restriction f(k)| \leq 2^n \leq k$, as desired. This *finishes the argument* that Y is f-tiny.

To show that $X \in \mathcal{I}_f$, note that $\mathcal{I}_f$ is a σ-ideal containing all countable sets of reals (since every singleton is trivially f-tiny). Since $\alpha < \omega_1$, and

$$X = \{x_\beta \colon \beta < \alpha\} \cup \{x_\beta \colon \beta \geq \alpha\},$$

it is enough to show that $\{x_\beta \colon \beta \geq \alpha\} \in \mathcal{I}_f$.

---

[63]This can be seen by an easy diagonalization argument, and corresponds to the fact that the *tower number* $\mathfrak{t}$ (i.e., the smallest length of a $\subseteq^*$-decreasing sequence without pseudointersection) is at least ω_1.

[64]This set is (when the elements are viewed as subsets of ω) the set of all subsets of x_α.

For each $\beta \geq \alpha$, we have $x_\beta \leq^* x_\alpha$, hence there exist $y \leq x_\alpha$ (i.e., $y \in Y$) and $q \in \mathbb{Q}$ such that $x_\beta = q + y$. Therefore,

$$\{x_\beta : \beta \geq \alpha\} \subseteq \mathbb{Q} + Y = \bigcup_{q \in \mathbb{Q}} (q + Y) \in \mathcal{I}_f$$

(due to the fact that Y is f-tiny, and being f-tiny is clearly a translation-invariant property); this finishes the proof of the claim. □

Recall that a set $X \subseteq 2^\omega$ is strongly meager if $X + H \neq 2^\omega$ for each $H \in \mathcal{N}$. The following notion is a strengthening: $X \subseteq 2^\omega$ is *null-additive* if $X + H \in \mathcal{N}$ for each $H \in \mathcal{N}$; clearly, each null-additive set is strongly meager (due to the fact that $2^\omega \notin \mathcal{N}$).

We are now going to prove (see Claim 3.7 below) that each completely tiny set is null-additive (and hence strongly meager).

The following fact is the key part of the proof:

CLAIM 3.6. Let $H \in \mathcal{N}$. Then there exists a function $f \in \omega^\omega$ such that for each $X \subseteq 2^\omega$, the following holds:

$$X \text{ is } f\text{-tiny} \Rightarrow X + H \in \mathcal{N}.$$

PROOF. It is easy to show (see, e.g., [**Gol93**, Fact 6.7(2)]) that there is a sequence of clopen[65] sets $(C_n)_{n\in\omega}$ with measure $\mu(C_n) \leq 2^{-n}$ for each $n \in \omega$ such that $H \subseteq \bigcap_{m\in\omega} \bigcup_{n\geq m} C_n$; each C_n is the union of finitely many basic clopen sets: for each $n \in \omega$, let $l_n \in \omega$ and $s_{n,i} \in 2^{<\omega}$ (for each $i < l_n$) such that $C_n = \bigcup_{i<l_n} [s_{n,i}]$; without loss of generality, we can assume that for each $n \in \omega$, all the $(s_{n,i})_{i<l_n}$ are (distinct and) of the same length; let $f(n)$ be this common length, i.e., let $f \in \omega^\omega$ be such that $f(n) = |s_{n,0}|$ for each $n \in \omega$.[66]

We claim that this f works: so fix a set X which is f-tiny; we have to show that $X + H \in \mathcal{N}$. Since X is f-tiny, we can fix an $N \in \omega$ such that $|X \restriction f(n)| \leq n$ for every $n \geq N$.

For the moment, fix $n \geq N$. Note that[67]

$$X + C_n = X + \bigcup_{i<l_n} [s_{n,i}] = X \restriction f(n) + \bigcup_{i<l_n} [s_{n,i}], \tag{3.1}$$

hence "adding X to C_n" increases the measure of C_n by (at most) a factor of $|X \restriction f(n)| \leq n$, i.e.,[68]

$$\mu(X + C_n) \leq |X \restriction f(n)| \cdot \mu(C_n) \leq n \cdot 2^{-n}.$$

Now for each $m \in \omega$, $H \subseteq \bigcup_{n\geq m} C_n$, hence $X+H \subseteq \bigcup_{n\geq m}(X+C_n)$. Therefore, for each $m \geq N$,

---

[65] Consult the beginning of Section 3.3 for the definition of clopen.

[66] Note that $\mu(C_n) \leq 2^{-n}$ implies $l_n \leq 2^{-n} \cdot 2^{f(n)}$.

[67] Actually, $X \restriction f(n)$ is not a set of reals, but of elements of $2^{f(n)}$ (of size at most n); so what we really mean with the right-most expression of (3.1) is the following: since the question whether (or not) a real belongs to $\bigcup_{i<l_n} [s_{n,i}]$ only depends on its restriction to $f(n)$, it is sufficient to take any set of (at most n) "representative" reals for $X \restriction f(n)$ (i.e., its restrictions to $f(n)$ belong to $X \restriction f(n)$), yielding the same sum as X would yield.

[68] More formally: $z \in X + C_n \iff z \restriction f(n) \in X \restriction f(n) + \{s_{n,i} : i < l_n\} \subseteq 2^{f(n)}$; but $|X \restriction f(n) + \{s_{n,i} : i < l_n\}| \leq |X \restriction f(n)| \cdot |\{s_{n,i} : i < l_n\}| \leq n \cdot l_n \leq n \cdot 2^{-n} \cdot 2^{f(n)}$ (for the last inequality, see footnote 66), hence $\mu(X + C_n) \leq n \cdot 2^{-n}$.

$$\mu(X+H) \leq \sum_{n\geq m} \mu(X+C_n) \leq \sum_{n\geq m} n\cdot 2^{-n} \quad\longrightarrow\quad 0 \quad (\text{for } m\to\infty),$$

so $\mu(X+H)=0$, i.e., $X+H\in\mathcal{N}$. □

It is easy to conclude that every completely tiny set is null-additive:

CLAIM 3.7. Let $X\subseteq 2^\omega$ be completely tiny (i.e., $X\in\mathcal{I}_f$ for each $f\in\omega^\omega$). Then X is null-additive (and hence strongly meager).

PROOF. Fix $H\in\mathcal{N}$; by the above Claim 3.6, we can fix $f\in\omega^\omega$ such that $Y+H\in\mathcal{N}$ whenever $Y\subseteq 2^\omega$ is f-tiny.

By assumption, $X\in\mathcal{I}_f$, i.e., there are f-tiny sets $(X_k)_{k\in\omega}$ with $X\subseteq\bigcup_{k\in\omega}X_k$. So $X_k+H\in\mathcal{N}$ for each $k\in\omega$, hence

$$X+H\subseteq\bigcup_{k\in\omega}(X_k+H)\in\mathcal{N}$$

(due to $\mathcal{N}$ being a σ-ideal), i.e., $X+H\in\mathcal{N}$, as desired. □

By putting together the above claims, we can complete the proof of the theorem: since $\mathfrak{d}=\omega_1$, Claim 3.4 yields a dominating tower of length ω_1, which is (clearly uncountable and) completely tiny by Claim 3.5; finally, Claim 3.7 shows that this uncountable set is indeed strongly meager, as desired. □

Let us summarize the two necessary conditions involving the cardinal characteristics discussed above:

COROLLARY 3.8. *Assume dBC. Then both* $\mathrm{cov}(\mathcal{N})=\omega_1$ *and* $\mathfrak{d}>\omega_1$ *hold true.*

PROOF. Immediately follows from Theorem 3.2 and Theorem 3.3 above. □

In particular, a forcing iteration (starting with a model of CH) yielding a model of the dual Borel Conjecture *must add unbounded reals*, but *must not (cofinally) add random reals* (since random reals increase $\mathrm{cov}(\mathcal{N})$); however (in contrast to the situation when aiming for a model of the Borel Conjecture; see the discussion after Corollary 2.4), this does not impose any restrictions on the kind of iteration used since neither finite nor countable iterations necessarily add random reals; indeed, Carlson's theorem (see Theorem 3.10) says that the dual Borel Conjecture is valid in the standard Cohen model.

It is straightforward to check that there are nine possible configurations of Cichoń's diagram consistent with the two necessary conditions for dBC from Corollary 3.8 (see[69] Figure 13, and recall from (1.2) that $\mathrm{add}(\mathcal{M})=\min(\mathfrak{b},\mathrm{cov}(\mathcal{M}))$).

For only three of these nine configurations, we know models witnessing their consistency with dBC (see Table 3); for the remaining six configurations, we are not sure;[70] observe, however, that the first three models listed in Table 3 are enough to guarantee that (individually) none of the values of the five cardinal characteristics in question (marked by □/■ in Figure 13) is determined by dBC: in the Hechler model, all of them are large, whereas $\mathfrak{b}$ and $\mathrm{non}(\mathcal{M})$ are small in the Cohen model,

[69] For the notation using □ (standing for $\aleph_1$) and ■ (standing for $\aleph_2$), see also the last paragraph of Section 1.6, as well as Figure 3 and the discussion there (i.e., after Corollary 2.4).

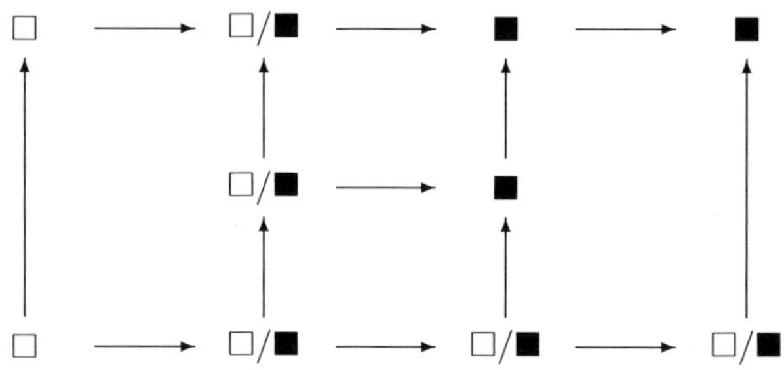

FIGURE 13. ($\square = \aleph_1$, $\blacksquare = \aleph_2$) Cichoń's diagram in models of dBC

and $\mathrm{cov}(\mathcal{M})$ and $\mathrm{non}(\mathcal{N})$ are small in the BC+dBC model (and hence $\mathrm{add}(\mathcal{M})$ is small as well in either of the latter two models).

TABLE 3. Some of the nine configurations realized in models of dBC

theorem	Cichoń	name of model	$\mathfrak{b}$	$\mathrm{non}(\mathcal{M})$	$\mathrm{cov}(\mathcal{M})$	$\mathrm{non}(\mathcal{N})$
Theorem 3.10	Figure 15	Cohen	□	□	■	■
Theorem 3.28	Figure 16	Hechler	■	■	■	■
Theorem 4.1	Figure 19	BC+dBC	■	■	□	□
Theorem 3.30	Figure 17	dBC w/o Cohens	□?	?	□	□

Before presenting Carlson's theorem that the dual Borel Conjecture holds true in the Cohen model and is therefore consistent, let us argue that the two necessary conditions from Corollary 3.8 are in fact not sufficient for the dual Borel Conjecture (see Figure 14 for the Cichoń's diagram of the respective model):

THEOREM 3.9. *There is a model of ZFC in which dBC fails, but* $\mathrm{cov}(\mathcal{N}) = \omega_1$ *and* $\mathfrak{d} > \omega_1$ *hold true.*

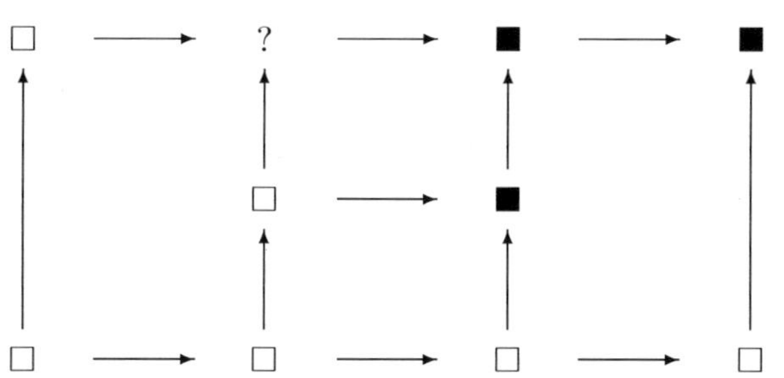

FIGURE 14. ($\square = \aleph_1$, $\blacksquare = \aleph_2$) Cichoń's diagram in the model obtained from the Cohen model by adding $\aleph_1$ many random reals

PROOF. The result follows from [**BS03**, Theorem 30] and its proof: start with a model V_0 satisfying CH, and let V_1 be obtained by adding $\aleph_2$ many Cohen reals to

[70] It is very likely though that the Cichoń's diagram of the model from [**BS10**] for dBC without Cohen reals (see Theorem 3.30) belongs to the remaining six (see the last line of Table 3).

V_0 (in other words, V_1 is the Cohen model; see Figure 15 for its Cichoń's diagram); finally, let V_2 be obtained by adding $\aleph_1$ many random reals (side-by-side) to V_1; then the final model V_2 is as desired, i.e.,

$$V_2 \models \neg dBC \wedge \operatorname{cov}(\mathcal{N}) = \omega_1 \wedge \mathfrak{d} > \omega_1;$$

the failure of dBC as well as $\operatorname{cov}(\mathcal{N}) = \omega_1$ follows from [**BS03**, Theorem 30], whereas $\mathfrak{d} > \omega_1$ easily follows from the fact that the same holds true in the intermediate model V_1, which is not changed by adding random reals since random forcing is ω^ω-bounding. □

3.2. How to prove the consistency of the dual Borel Conjecture? In his 1993 paper [**Car93**], Carlson showed[71] that one can obtain a model of the dual Borel Conjecture by adding many Cohen reals:

THEOREM 3.10 (Carlson, 1993). *Let $\mathbb{P}_{\omega_2}$ be the finite support iteration*[72] *of Cohen forcing of length*[73] *ω_2.*

Then in a generic extension by $\mathbb{P}_{\omega_2}$ the dual Borel Conjecture holds (i.e., there are no uncountable strongly meager sets).

The above model is the well-known *Cohen model* (see Figure 15 for its Cichoń's diagram).

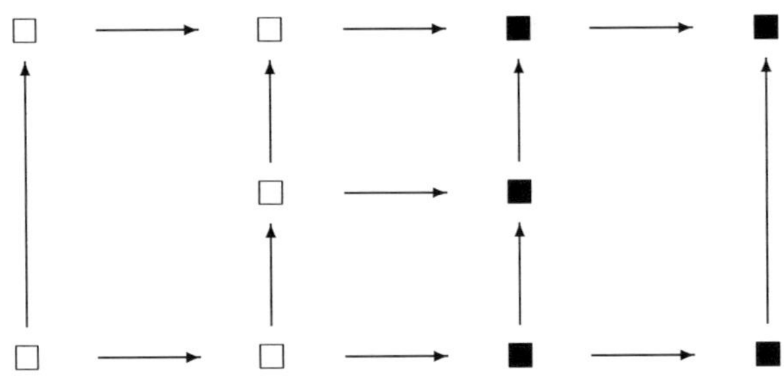

FIGURE 15. (□ $= \aleph_1$, ■ $= \aleph_2$) Cichoń's diagram in the Cohen model

In the following (Sections 3.3–3.6), we are going to give a detailed proof of this theorem; the key points of the proof are:

- Each of the iterands $\dot{\mathbb{Q}}_\alpha$ is *(single) Cohen forcing* $\mathbb{C}$, which "kills" all uncountable strongly meager sets from the ground model. Actually, the following property of Cohen forcing is used (see Corollary 3.19, as well as Lemma 3.18):

[71] Actually, Carlson proved the theorem several years earlier, but the result was published only in 1993; it was already build upon in, e.g., the 1989 paper [**JS89**], where it is proved that dBC is consistent with MA(σ-centered) (see Theorem 3.29).

[72] It is well-known that this is the same as adding ω_2 many Cohen reals "simultaneously", i.e., by the poset of finite partial functions from $\omega_2 \times \omega$ to 2.

[73] In fact, the same holds when more than ω_2 many Cohen reals are added (see also Theorem 3.26); therefore, the dual Borel Conjecture is also consistent with arbitrarily large continuum.

Recall that the respective claim for the Borel Conjecture is true as well, but much more complicated to prove (see Theorem 2.34).

("$\mathbb{C}$ kills sm") If X is an uncountable set of reals and belongs to the ground model, then $\Vdash_{\mathbb{C}}$ "X is not strongly meager", witnessed by the "generic G_δ measure zero set"

$$H := \bigcap_{n\in\omega} \bigcup_{m\geq n} C^m$$

(i.e., $\Vdash_{\mathbb{C}} X+\dot{H} = 2^\omega$), where the sequence $\langle C^m : m \in \omega\rangle$ consists of "Cohenly chosen" clopen sets (see Definition 3.15 and the notation introduced before) with measures $\mu(C^m) \leq 2^{-m-1}$. (For the proof, a *finite combinatorial fact* is used; see Theorem 3.11.)

- Once an uncountable set X "has been killed", it "stays dead" (until the end of the iteration). For this, the property *precaliber* $\aleph_1$ (see Definition 3.20) is used:
  (1) Cohen forcing $\mathbb{C}$ is (countable, hence trivially) *precaliber* $\aleph_1$.
  (2) Precaliber $\aleph_1$ is preserved under *finite support* iteration (see Proposition 3.21).
  (3) "X is not strongly meager" remains true[74] after a forcing which is precaliber $\aleph_1$ (see Lemma 3.22).

Our presentation is mainly based on the treatment of the dual Borel Conjecture in [**BJ95**, Section 8.5.B] and the respective lemmas referred to[75] there (which can be found in other parts of the book [**BJ95**]).

3.3. Finite combinatorics. We start with a combinatorial fact, which is (essentially) only concerned with finite sets (see Theorem 3.11).

The reader can skip the proof of this theorem in case he or she is just interested in the forcing part of the construction, and not so much in the details of the finite combinatorics; at least, the details of the proof of Lemma 3.13 can be omitted.

We say that a set $C \subseteq 2^\omega$ is *clopen* if it is both closed and open; it is well-known and easy to check that this is the case if and only if C can be written as a *finite* union of basic clopen sets $([s_i])_{i<n}$. Recall that μ denotes the (usual product) measure on 2^ω (see Section 1.2).

THEOREM 3.11. *Let $X \subseteq 2^\omega$ be infinite, and let $\varepsilon > 0$. Then there exists a clopen set $C \subseteq 2^\omega$ such that $\mu(C) < \varepsilon$ and $X + C = 2^\omega$.*

In fact, we prove the following stronger statement (see Lemma 3.12), from which one can easily derive the above theorem; for that, just observe[76] that

$$\lim_{n\to\infty} \frac{1}{n} \sum_{i=1}^{n} \frac{1}{i} = 0. \tag{3.2}$$

[74]In fact, it might be the case that this is not literally true (see Question 3.24); we rather mean the following: "X is not strongly meager" remains true provided this is witnessed by the canonical measure zero set H derived from the Cohen real (see Definition 3.15).

[75]More specifically, see [**BJ95**, Section 8.5.B (Theorem 8.5.22 and Theorem 8.5.23), Section 3.3.B (Theorem 3.3.9–Lemma 3.3.13), and Section 8.4.A (Definition 8.4.3 and Theorem 8.4.4)].

[76]Actually, $\sum_{i=1}^{n} \frac{1}{i} \leq 1 + \int_1^n \frac{1}{x}dx = 1 + \ln(n)$, and $\frac{1+\ln(n)}{n} \to 0$ for $n \to \infty$.

LEMMA 3.12. *Let $F \subseteq 2^\omega$ be a finite set. Then there exists a clopen set $C \subseteq 2^\omega$ such that*

$$\mu(C) \leq \frac{1}{|F|} \sum_{i=1}^{|F|} \frac{1}{i}$$

and $F + C = 2^\omega$.

PROOF. We start with the following technical lemma whose proof is the major part of the proof of Lemma 3.12.

For a set $E \subseteq A \times A$, and $x \in A$, let $(E)_x := \{y \in A \colon (x,y) \in E\}$ denote the vertical section of E at column x, and, for $y \in A$, let $(E)^y := \{x \in A \colon (x,y) \in E\}$ denote the horizontal section of E at row y.

LEMMA 3.13. *Let A be a finite set. Suppose $E \subseteq A \times A$ with the property that $\bigcup_{x \in A} (E)_x = A$. Let $K := \max\{|(E)_x| : x \in A\}$ and $L := \min\{|(E)^y| : y \in A\}$. Then there exists a $B \subseteq A$ such that*

$$|B| \leq \frac{|A|}{L} \cdot \sum_{i=1}^{K} \frac{1}{i}$$

and $A = \bigcup_{x \in B} (E)_x$.

PROOF. The idea of the proof is the following: the set B is constructed by performing a *greedy algorithm* which tries to collect all the elements of A (from vertical sections of E) "as fast as possible"; in each step, a column of E is selected in such a way that the number of elements which belong to the respective vertical section and have not been collected yet is maximized; then the required number $\hat{n}$ of steps (to collect all elements of A) is estimated.

More precisely, we proceed as follows. By induction, choose a sequence $x_n \in A$ for $n < |A|$ such that $|(E)_{x_0}| = K$ and $|(E)_{x_n} \setminus \bigcup_{i<n} (E)_{x_i}|$ has the maximal possible value on the n-th step of the induction. Let $\hat{n}$ be the smallest number such that $|(E)_{x_{\hat{n}}} \setminus \bigcup_{i<\hat{n}} (E)_{x_i}| = 0$. Let $B := \{x_i \colon i < \hat{n}\}$. Clearly,

$$A = \bigcup_{i<\hat{n}} (E)_{x_i} = \bigcup_{x \in B} (E)_x.$$

To estimate the size of B, we have to estimate $\hat{n}$.

Let $I_k := \{n < \hat{n} \colon |(E)_{x_n} \setminus \bigcup_{i<n} (E)_{x_i}| = k\}$ for $k \leq K$. Define for $k \leq K$,

$$Y_k := \bigcup_{i \leq k} \bigcup_{n \in I_i} (E)_{x_n} \setminus \bigcup_{i>k} \bigcup_{n \in I_i} (E)_{x_n}.$$

It is clear that $\emptyset = Y_0 \subseteq Y_1 \subseteq \cdots \subseteq Y_K = A$. Observe that $|Y_1 \setminus Y_0| = |Y_1| = |I_1|$, $|Y_2 \setminus Y_1| = |Y_2| - |Y_1| = 2 \cdot |I_2|$ and, in general,

$$|Y_k| - |Y_{k-1}| = k \cdot |I_k|. \tag{3.3}$$

On the other hand, for every $x \in A$ and $k \leq K$, we have $|(E)_x \cap Y_k| \leq k$ by the choice of the x_n (according to the greedy algorithm). Therefore, for all $k \leq K$,

$$L \cdot |Y_k| \leq |E \cap (A \times Y_k)| \leq k \cdot |A|,$$

and hence

$$|Y_k| \leq k \frac{|A|}{L}. \tag{3.4}$$

The following easy computation finally yields the desired upper bound for $\hat{n}$; note that the second equality is due to (3.3), whereas the only occurrence of $\leq$ in this series of (in)equalities is due to (3.4):

$$\hat{n} = \sum_{i=1}^{K} |I_i| = \sum_{i=1}^{K} \frac{|Y_i| - |Y_{i-1}|}{i} = \sum_{i=1}^{K} \frac{|Y_i|}{i} - \sum_{i=0}^{K-1} \frac{|Y_i|}{i+1} =$$

$$\frac{1}{K} \cdot |Y_K| + \sum_{i=1}^{K-1} \frac{1}{i(i+1)} \cdot |Y_i| \leq \frac{1}{K} \cdot K\frac{|A|}{L} + \sum_{i=1}^{K-1} \frac{1}{i(i+1)} \cdot i\frac{|A|}{L} =$$

$$\frac{|A|}{L}\left(1 + \sum_{i=1}^{K-1} \frac{1}{i+1}\right) = \frac{|A|}{L} \sum_{i=1}^{K} \frac{1}{i}.$$

This finishes the proof of the lemma. □

Suppose $F = \{z_0, \ldots, z_{n-1}\} \subseteq 2^\omega$ is a finite set of size n. Fix $m \in \omega$ large enough such that $z_0 \restriction m, \ldots, z_{n-1} \restriction m$ are all different; let

$$F' := \{z_i \restriction m : i < n\} \subseteq 2^m.$$

Let $E \subseteq 2^m \times 2^m$ be a matrix containing all the shifts of F' as its columns; more precisely, define

$$E := \bigcup_{x \in 2^m} \{x\} \times (F' + x) \subseteq 2^m \times 2^m;$$

equivalently, E is the following binary relation on 2^m: a pair $\langle x, y \rangle$ belongs to E if and only if there is an $i < n$ such that $y = z_i \restriction m + x$.

We want to apply Lemma 3.13 to $A := 2^m$ and E: it is straightforward to check that $\bigcup_{x \in 2^m} (E)_x = 2^m$, and that $K = L = |F| = n$; so the lemma is applicable and yields a set $B \subseteq 2^m$ such that

$$|B| \leq \frac{2^m}{|F|} \cdot \sum_{i=1}^{|F|} \frac{1}{i}$$

and

$$2^m = \bigcup_{x \in B} (E)_x = \bigcup_{x \in B} (F' + x) = F' + B.$$

Finally, we can derive the following clopen set C from the set B:

$$C := \bigcup_{s \in B} [s];$$

it is easy to check that $F + C = 2^\omega$; moreover, $\mu([s]) = \frac{1}{2^m}$ for every $s \in 2^m$, and so

$$\mu(C) = \frac{1}{2^m} \cdot |B| \leq \frac{1}{2^m} \cdot \frac{2^m}{|F|} \cdot \sum_{i=1}^{|F|} \frac{1}{i} = \frac{1}{|F|} \sum_{i=1}^{|F|} \frac{1}{i},$$

as desired. □

Straightforward computations suggest that the *upper bound*

$$\frac{1}{|F|} \sum_{i=1}^{|F|} \frac{1}{i}$$

for $\mu(C)$ given in Lemma 3.12 is in general not optimal. On the other hand,

$$\frac{1}{|F|}$$

is a *trivial lower bound* for the measure of a clopen set C satisfying $F + C = 2^\omega$ (due to the fact that $\mu(F + C) \leq |F| \cdot \mu(C)$ and $\mu(2^\omega) = 1$).

It is natural to ask for the optimal (or at least a better) upper bound in Lemma 3.12:

PROBLEM 3.14. Given $n \in \omega \setminus \{0\}$, determine the minimal μ_n such that for each set $F \subseteq 2^\omega$ with $|F| = n$ there is a clopen set $C \subseteq 2^\omega$ with $\mu(C) \leq \mu_n$ and $F + C = 2^\omega$.

By the above, we know (only) that

$$\frac{1}{n} \leq \mu_n \leq \frac{1}{n} \sum_{i=1}^{n} \frac{1}{i}$$

for each $n \in \omega \setminus \{0\}$.

Note that the upper bound provided by Lemma 3.12 has been obtained by estimating possible outcomes of the greedy algorithm used in the proof of Lemma 3.13; however, this doesn't mean that the estimate is optimal, even for results produced by the greedy algorithm; therefore one could also ask for the optimal (or at least a better) upper bound μ_n^{greedy} for results provided by the greedy algorithm (which might be strictly larger than the "true" upper bound μ_n).

Of course, the exact nature of the upper bound is of no relevance whatsoever (for the infinitary issues): indeed, the only important feature to prove Theorem 3.11 is the fact that it converges to zero (see (3.2)).

3.4. Cohen reals kill strongly meager sets. Let $(\mathbb{C}, \leq) := (\omega^{<\omega}, \supseteq)$ be *Cohen forcing.* Given a generic filter $G \subseteq \mathbb{C}$, let

$$c := \bigcup G \in \omega^\omega$$

be the generic real (the *Cohen real* added by $\mathbb{C}$). Moreover, let $\dot{c}$ be a $\mathbb{C}$-name for the Cohen real c.

Recall the notion of clopen set (see the beginning of Section 3.3). Note that there are only countably many clopen subsets of 2^ω. For $m \in \omega$, let $\{C_k^m : k \in \omega\}$ be an enumeration of all clopen subsets of 2^ω with measure at most 2^{-m-1}; in other words, let $\{C_k^m : m, k \in \omega\}$ be so that for each $m, k \in \omega$, we have $\mu(C_k^m) \leq 2^{-m-1}$, and for each $m \in \omega$ and each clopen set $C \subseteq 2^\omega$ with $\mu(C) \leq 2^{-m-1}$ there is a $k \in \omega$ such that $C = C_k^m$.

DEFINITION 3.15. For $n \in \omega$, define

$$H_n := \bigcup_{m \geq n} C_{c(m)}^m; \tag{3.5}$$

moreover, let

$$H := \bigcap_{n \in \omega} H_n, \tag{3.6}$$

the *canonical measure zero set derived from the Cohen real* c.

The following is rather obvious:

PROPOSITION 3.16. *For each $n \in \omega$, the set H_n is an open subset of 2^ω of measure at most 2^{-n}.*

In particular, $\mu(H) = 0$, i.e., H is a G_δ measure zero set.[77]

In the following, let $\dot{H}_n$ be a $\mathbb{C}$-name for the set H_n defined in (3.5) (for $n \in \omega$), and let $\dot{H}$ be a $\mathbb{C}$-name for the set H defined in (3.6).

LEMMA 3.17. *Let $X \subseteq 2^\omega$ be an infinite set in V, and let $n \in \omega$. Then*

$$\Vdash_{\mathbb{C}} X + \dot{H}_n = 2^\omega.$$

PROOF. Assume towards a contradiction that the conclusion of the lemma fails; then we can fix $s \in \omega^{<\omega}$ such that $s \Vdash_{\mathbb{C}} X + \dot{H}_n \neq 2^\omega$; without loss of generality we can assume that $|s| \geq n$. Since $X \in V$ is infinite, we can apply Theorem 3.11 to obtain $k \in \omega$ such that

$$X + C_k^{|s|} = 2^\omega.$$

Let $s' := s^\frown\langle k \rangle$. But then $s' \leq s$ and

$$s' \Vdash_{\mathbb{C}} X + \dot{H}_n \supseteq X + C_k^{|s|} = 2^\omega,$$

which gives a contradiction. □

We are now going to prove that adding a single Cohen real "kills all uncountable strongly meager sets from the ground model"; in fact, we show (and are going to need) slightly more (see Lemma 3.18 below): in the extension by $\mathbb{C}$, there is no uncountable strongly meager set consisting of ground model reals, and this is witnessed by H (see (3.6)).

LEMMA 3.18. *Let $\dot{H}$ be the $\mathbb{C}$-name for the canonical measure zero set H derived from the Cohen real. Then*

$$\Vdash_{\mathbb{C}} \forall X \in [2^\omega \cap V]^{\aleph_1}\ X + \dot{H} = 2^\omega.$$

PROOF. Assume towards a contradiction that the conclusion of the lemma fails; so we can fix a $\mathbb{C}$-name $\dot{X}$ for a subset of $2^\omega \cap V$ of size $\aleph_1$, a $\mathbb{C}$-name $\dot{\eta}$ for an element of 2^ω, and a condition $s \in \mathbb{C}$ such that

$$s \Vdash_{\mathbb{C}} \dot{\eta} \notin \dot{X} + \dot{H}.$$

Let $\dot{X} = \{\dot{x}_\xi : \xi < \omega_1\}$. For every $\xi < \omega_1$, let $s_\xi \leq s$, $x_\xi \in V$ and $n_\xi \in \omega$ such that

$$s_\xi \Vdash_{\mathbb{C}} \dot{x}_\xi = x_\xi \text{ and } s_\xi \Vdash_{\mathbb{C}} \dot{\eta} \notin \dot{x}_\xi + \dot{H}_{n_\xi}.$$

Since $\mathbb{C}$ is countable, there exists $s \in \mathbb{C}$, $n \in \omega$, and an uncountable set $A \subseteq \omega_1$ such that $s_\xi = s$ and $n_\xi = n$ for each $\xi \in A$. Let $Y := \{x_\xi : \xi \in A\}$. But then

$$s \Vdash_{\mathbb{C}} \dot{\eta} \notin Y + \dot{H}_n;$$

on the other hand, Y is an uncountable (hence in particular infinite) set in V, so Lemma 3.17 yields

$$\Vdash_{\mathbb{C}} Y + \dot{H}_n = 2^\omega,$$

which is a contradiction. □

In particular, uncountable sets from the ground model cannot be strongly meager in the Cohen extension:

[77] Furthermore, an easy genericity argument shows that each H_n is indeed open dense, hence H is dense G_δ (i.e., co-meager); so H is actually a measure zero set whose complement is meager.

COROLLARY 3.19. *Let $X \subseteq 2^\omega$ be an uncountable set in V. Then*

$$\Vdash_{\mathbb{C}} X + \dot{H} = 2^\omega,$$

and hence $\Vdash_{\mathbb{C}} X \notin \mathcal{SM}$.

However, we will use the stronger version above (i.e., Lemma 3.18) to show that being not strongly meager (when witnessed by the canonical measure zero set H) is preserved by forcings having precaliber $\aleph_1$.

3.5. No resurrection after forcings with precaliber $\aleph_1$. Let $\mathbb{P}$ be a forcing notion. Recall that $A \subseteq \mathbb{P}$ is *centered* (or: directed) if any finitely many conditions from A have a common lower bound (i.e., for any finite set $F \subseteq A$ there is $q \in \mathbb{P}$ with $q \leq p$ for each $p \in F$).

DEFINITION 3.20. We say that $\mathbb{P}$ has *precaliber* $\aleph_1$ if each set of conditions of size $\aleph_1$ has a centered subset of size $\aleph_1$.

We say that $\mathbb{P}$ is *σ-centered* if $\mathbb{P} = \bigcup_{n\in\omega} \mathbb{P}_n$ where each $\mathbb{P}_n$ is centered.

Note that a forcing notion $\mathbb{P}$ which is σ-centered is trivially precaliber $\aleph_1$.

PROPOSITION 3.21. *Precaliber $\aleph_1$ is preserved under finite support iteration.*

PROOF. See [**BJ95**, Theorem 1.5.13]. □

LEMMA 3.22. *Let $X \subseteq 2^\omega$ be an uncountable set in V. Suppose c is a Cohen real over V, and H the canonical measure zero set derived from it (see Definition 3.15).*

Then the following holds in $V[c]$: for any partial order $\mathbb{P}$ having precaliber $\aleph_1$,

$$\Vdash_{\mathbb{P}} X + H = 2^\omega \tag{3.7}$$

(and hence $\Vdash_{\mathbb{P}} X \notin \mathcal{SM}$).

In (3.7), H is of course understood to be the evaluation (in the respective model) of the *Borel code*[78] defined in $V[c]$, not the very set H as evaluated in $V[c]$ (in which case (3.7) would necessarily be false, provided $\mathbb{P}$ adds new reals).

PROOF. Work in $V[c]$. Assume towards a contradiction that the partial order $\mathbb{P}$ has precaliber $\aleph_1$, and $\dot\eta$ is a $\mathbb{P}$-name for an element of 2^ω, with $p \in \mathbb{P}$ being a condition such that

$$p \Vdash_{\mathbb{P}} \dot\eta \notin X + H.$$

For every $x \in X$, let $p_x \leq p$ and $n_x \in \omega$ such that

$$p_x \Vdash_{\mathbb{P}} \dot\eta \notin x + H_{n_x}.$$

Since $\mathbb{P}$ has precaliber $\aleph_1$, we can find $Y \subseteq X$ of size $\aleph_1$ such that $\{p_x : x \in Y\}$ is centered.

CLAIM 3.23. Let $Z \subseteq Y$ be finite. Then

$$\bigcup_{x\in Z} x + H_{n_x} \neq 2^\omega. \tag{3.8}$$

[78] For an introduction to Borel codes, see, e.g., [**Jec03**, Chapter 25, pp. 504–507].

PROOF. Since $\{p_x : x \in Y\}$ is centered, we can fix $q \in \mathbb{P}$ such that $q \leq p_x$ for every $x \in Z$. But then $q \Vdash_{\mathbb{P}} \dot{\eta} \notin x + H_{n_x}$ for each $x \in Z$; therefore, for a filter G which is $\mathbb{P}$-generic over $V[c]$ and contains q, we have

$$V[c][G] \models \bigcup_{x \in Z} x + H_{n_x} \neq 2^\omega,$$

i.e., (3.8) holds in $V[c][G]$, and hence, by absoluteness of the property

"(the evaluation of the Borel code of) $\bigcup_{x \in Z} x + H_{n_x}$ is the whole space"

between the models $V[c]$ and $V[c][G]$ (see [**Jec03**, Lemma 25.45], and the discussion afterwards), (3.8) holds in $V[c]$ as well. □

Note that $x + H_{n_x}$ is an open subset of 2^ω (for every $x \in Y$); so by compactness of 2^ω (i.e., the fact that any open cover of 2^ω has a finite subcover), Claim 3.23 yields that

$$\bigcup_{x \in Y} x + H_{n_x} \neq 2^\omega,$$

so the same holds for its subset $\bigcup_{x \in Y} x + H$, i.e.,

$$Y + H \neq 2^\omega;$$

but since $Y \subseteq X$ is a set (in $V[c]$) of size $\aleph_1$ consisting of reals in V, Lemma 3.18 actually yields $Y + H = 2^\omega$, a contradiction. □

Note that Lemma 3.22 guarantees preservation (by precaliber $\aleph_1$ forcings) of the property "X is not strongly meager" only in case this property has been accomplished by Cohen forcing; we do not know the following:

QUESTION 3.24. Is it true in general that the property "X is not strongly meager" is preserved by any forcing which is precaliber $\aleph_1$?

REMARK 3.25. In this section, we are concerned with (killing) strongly meager sets (and avoiding their resurrection) in the context of *finite support iteration*; in [**BS03**, Section 2], a respective theory for *countable support iteration* is established: *being not very meager* (which is slightly stronger than being not strongly meager; see[79] Definition 5.4) is actually preserved by proper forcings which are *nice*[80] (e.g., all forcings with the Laver property are nice), and being nice is preserved under countable support iteration.

3.6. No uncountable strongly meager sets in the Cohen model. To prove Theorem 3.10, i.e., to show that the dual Borel Conjecture holds true in the Cohen model, we are actually going to prove the following slightly more general theorem:

THEOREM 3.26. *Assume that* $\mathrm{cf}(\delta) > \aleph_1$. *Let* $\{\mathbb{P}_\alpha, \dot{\mathbb{Q}}_\alpha : \alpha < \delta\}$ *be a finite support iteration such that for every* $\alpha < \delta$,

$$\Vdash_\alpha \dot{\mathbb{Q}}_\alpha \textit{ has precaliber } \aleph_1,$$

and[81]

$$\{\alpha < \delta \colon \Vdash_\alpha \dot{\mathbb{Q}}_\alpha \textit{ is Cohen forcing}\}$$

[79] Compare with the equivalent [**BS03**, Definition 4].
[80] See [**BS03**, Definition 6].

is cofinal in δ*. Then*

$$V^{\mathbb{P}_\delta} \models dBC.$$

Since Cohen forcing $\mathbb{C}$ is countable, it trivially has precaliber $\aleph_1$, so Theorem 3.10 immediately follows from the above theorem.

REMARK 3.27. Demanding that Cohen reals are explicitly added cofinally often is actually not necessary (see also [**BJ95**, Theorem 8.5.22]): since the iteration in Theorem 3.26 is a finite support iteration, Cohen reals are added at limits anyway; moreover, the respective quotient forcings (i.e., the remaining parts of the forcing after the single Cohen real added at the limit) has precaliber $\aleph_1$ as well; therefore, the proof still goes through.

PROOF OF THEOREM 3.26. We have to prove that there is no uncountable strongly meager set in the final model $V^{\mathbb{P}_\delta}$. This follows quite easily from the fact that having *precaliber* $\aleph_1$ is preserved under *finite support* iteration, together with Lemma 3.22. More precisely, we can argue as follows.

Assume towards a contradiction that $X \in V^{\mathbb{P}_\delta}$ is an uncountable set which is strongly meager (in the final model $V^{\mathbb{P}_\delta}$). Since the collection of strongly meager sets is closed under subsets, we can assume without loss of generality that X is actually of size $\aleph_1$. Since $\mathrm{cf}(\delta) > \aleph_1$, and the forcing is c.c.c., standard arguments show that there is an $\alpha < \delta$ such that $X \in V^{\mathbb{P}_\alpha}$. Since cofinally many iterands are Cohen forcing, we can also assume (by making α larger if necessary) that $\Vdash_\alpha \dot{\mathbb{Q}}_\alpha$ is Cohen forcing.

But then $V^{\mathbb{P}_{\alpha+1}} = V^{\mathbb{P}_\alpha}[c]$ for a Cohen real c over $V^{\mathbb{P}_\alpha}$, and Proposition 3.21 yields that (in $V^{\mathbb{P}_{\alpha+1}}$) the quotient forcing $\mathbb{P}_\delta : \mathbb{P}_{\alpha+1}$ has precaliber $\aleph_1$ (as a finite support iteration of forcings having precaliber $\aleph_1$); hence we can apply Lemma 3.22 (with $V^{\mathbb{P}_\alpha}$ as our "new ground model") to derive that

$$V^{\mathbb{P}_\delta} \models X \notin \mathcal{SM},$$

i.e., X is not strongly meager in the final model, which is a contradiction. □

3.7. Other models satisfying the dual Borel Conjecture. We now present several models of dBC other than the Cohen model: in particular, we are going to discuss the Hechler model (see Section 3.7.1), a model satisfying MA(σ-centered) (see Section 3.7.3), and a model of dBC obtained without adding Cohen reals (see Section 3.7.4); moreover, we discuss a weak version of the dual Borel Conjecture (see Section 3.7.7), which holds true in the Laver model; we do not know, however, whether the Laver model satisfies the (usual) dual Borel Conjecture (see Section 3.7.8).

3.7.1. *Hechler model.* Recall *Hechler forcing* (also called "dominating forcing"; see [**BJ95**, Definition 3.1.9] and the discussion afterwards); it is σ-centered (and therefore c.c.c.), and adds a dominating real as the generic real.

We show that dBC holds in the *Hechler model* (see Figure 16 for its Cichoń's diagram):

THEOREM 3.28. *Assume CH, and let* $\mathbb{P}_{\omega_2}$ *be the finite support iteration of Hechler forcing of length* ω_2*.*

Then in a generic extension by $\mathbb{P}_{\omega_2}$ *the dual Borel Conjecture holds.*

[81]Demanding that Cohen reals are explicitly added is not necessary, see [**BJ95**, Theorem 8.5.22], or Remark 3.27.

PROOF. This immediately follows from Theorem 3.26 (and Remark 3.27, saying that we do not have to worry about ensuring that cofinally many iterands are Cohen forcing): just observe that Hechler forcing is σ-centered and therefore precaliber $\aleph_1$ (see Definition 3.20). □

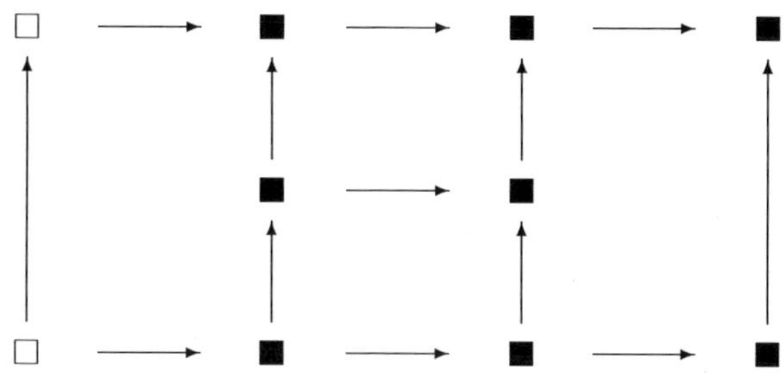

FIGURE 16. ($\square = \aleph_1$, $\blacksquare = \aleph_2$) Cichoń's diagram in the Hechler model (and under dBC + MA(σ-centered))

3.7.2. *No random reals added by σ-centered forcings.* It is well known that a forcing which is σ-centered cannot add random reals; so, in particular, random forcing $\mathbb{B}$ is not σ-centered (see [**BJ95**, Theorem 3.2.4] for an elementary proof).

Observe that this can also be derived[82] from our machinery to produce models of dBC: assume towards a contradiction that $\mathbb{Q}$ is a σ-centered forcing adding a random real; perform a finite support iteration of length ω_2 whose iterands are $\mathbb{Q}$ (interleaving them with Cohen forcing; or, once more, have Remark 3.27 in mind); all the iterands are σ-centered and hence precaliber $\aleph_1$, so Theorem 3.26 guarantees that the final model satisfies dBC; moreover, $\mathrm{cov}(\mathcal{N})$ is large in the final model because random reals had been added cofinally along the iteration (due to the assumption that $\mathbb{Q}$ adds random reals); however, $\mathrm{cov}(\mathcal{N})$ being large is inconsistent with dBC (see Theorem 3.2), which is a contradiction.

3.7.3. *MA(σ-centered).* We will now argue (see also [**BJ95**, Theorem 8.5.23], or [**JS89**] for the original paper) that the dual Borel Conjecture is consistent with Martin's axiom for σ-centered forcings; note, however, that dBC is not consistent with (full) MA (see also Theorem 4.8), due to the fact that (random forcing is[83] c.c.c. and hence) $\mathrm{cov}(\mathcal{N})$ is large under MA, but $\mathrm{cov}(\mathcal{N})$ being small is necessary for dBC (see Theorem 3.2).

THEOREM 3.29. *There is a model of ZFC satisfying*

$$dBC \wedge MA(\sigma\text{-centered}).$$

[82]Of course, this is an unnecessarily complicated way to prove this fact; the elementary proof that random forcing is not σ-centered is the much easier argument.

[83]In fact, random forcing $\mathbb{B}$ is σ-linked, or, even stronger, *σ-n-linked* for every $n \in \omega$, i.e., given any fixed $n \in \omega$ in advance, it is possible to partition $\mathbb{B}$ into countably many parts each of which is *n-linked*: any set of conditions of size n has a common lower bound.

However, even though being σ-centered (see Definition 3.20) seems to be only a slight strengthening of being σ-n-linked for every $n \in \omega$ (namely a change in the order of the quantifiers concerning the natural number n and the partition of the forcing), $\mathbb{B}$ *fails* to satisfy this stronger property (see also Section 3.7.2 above); this is an immediate corollary of Theorem 3.29 as well: if $\mathbb{B}$ were σ-centered, then $\mathrm{cov}(\mathcal{N})$ would be large under MA(σ-centered), contradicting the fact that dBC implies $\mathrm{cov}(\mathcal{N})$ being small (see Theorem 3.2).

PROOF. Again, this is a straightforward instance of Theorem 3.26 (keeping Remark 3.27 in mind): by letting each $\dot{\mathbb{Q}}_\alpha$ be σ-centered, and careful bookkeeping, we can ensure that the final model satisfies MA(σ-centered); moreover, our final model satisfies dBC due to Theorem 3.26. $\square$

Note that every model of dBC+MA(σ-centered) has the same Cichoń's diagram as the Hechler model (see Figure 16): Hechler forcing is σ-centered, and adds dominating reals as well as Cohen reals, so MA(σ-centered) implies that both $\mathfrak{b}$ and $\text{cov}(\mathcal{M})$ (and hence also $\text{add}(\mathcal{M})$) are large; on the other hand, dBC implies that $\text{cov}(\mathcal{N})$ is small.

Furthermore, (the easy direction of) Bell's theorem says that $\mathfrak{p} = 2^{\aleph_0}$ follows from MA(σ-centered); therefore, Theorem 3.29 in particular yields that dBC is consistent with $\mathfrak{p}$ being large.

3.7.4. *Dual Borel Conjecture without Cohen reals.* For a long time, all known models of dBC had been obtained by forcing constructions adding Cohen reals. In [**BS10**], Bartoszyński and Shelah demonstrated that it is possible to get a model of *dBC without adding Cohen reals* (i.e., a model of dBC in which $\text{cov}(\mathcal{M})$ is small).

Therefore, their theorem is a precursor of [**GKSW14**] (the joint consistency of BC and dBC; see Theorem 4.1) because $\text{cov}(\mathcal{M}) = \omega_1$ is one of the two necessary conditions for BC concerning Cichoń's diagram (see Theorem 2.2, or Figure 3):

THEOREM 3.30. *There is a model of ZFC (with $2^{\aleph_0} = \aleph_2$) satisfying*

$$dBC \wedge \text{cov}(\mathcal{M}) = \text{non}(\mathcal{N}) = \omega_1.$$

In fact, $\text{cov}(\mathcal{M})$ is guaranteed to be small by showing that the set of reals from the ground model is not measure zero (see [**BS10**, Theorem 29(2)]); in other words, $\text{non}(\mathcal{N}) = \omega_1$, hence $\text{cov}(\mathcal{M}) = \omega_1$ (due to $\text{cov}(\mathcal{M}) \leq \text{non}(\mathcal{N})$; see Corollary 1.14).

The construction of the model in Theorem 3.30 (see Figure 17 for its Cichoń's diagram) is rather complicated; in particular, the iteration technique is hard to understand, also because it is based on the non-Cohen oracle c.c.c. method from Shelah's paper [**She06**] (which is not well understood).

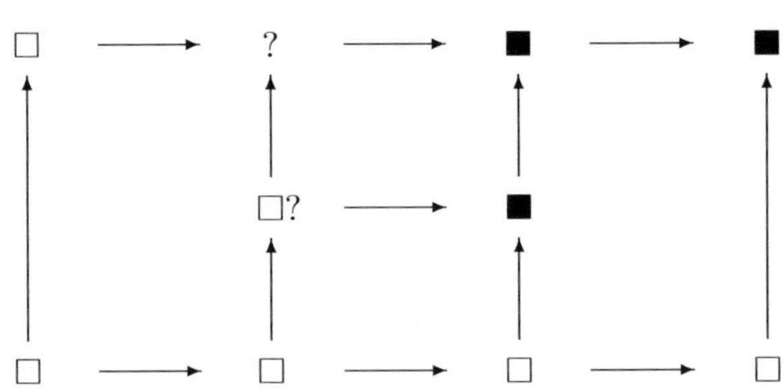

FIGURE 17. ($\square = \aleph_1$, $\blacksquare = \aleph_2$) Cichoń's diagram in the [**BS10**] model for dBC without Cohen reals

3.7.5. *Joint consistency with the Borel Conjecture.* There is a model of the dual Borel Conjecture in which also the Borel Conjecture holds true (see Theorem 4.1); an overview of the very involved forcing construction is given in Section 4.

In fact, this construction supersedes the method used in [**BS10**] for the model of dBC without Cohen reals (see Theorem 3.30 above).

3.7.6. *Joint consistency with a projective well-order of the reals.* There is a model of the dual Borel Conjecture in which there exists a projective well-order of the reals (see Theorem 6.3); an outline of the proof is given in Section 6.3.

3.7.7. *The weak dual Borel Conjecture (in the Laver model and the Miller model).* Let us say that the *weak dual Borel Conjecture* (weak dBC) is the assertion that there is *no* strongly meager set *of size continuum.*

It has been proved by Bartoszyński and Shelah that both in the Laver model (recall Theorem 2.6 and Figure 5) and in the *Miller model* (see Figure 18 for its Cichoń's diagram) weak dBC holds true:

THEOREM 3.31. *The Laver model and the Miller model satisfy the weak dual Borel Conjecture (i.e., there are no strongly meager sets of size continuum).*

PROOF. See [**BS03**, Theorem 24 and Theorem 26]. □

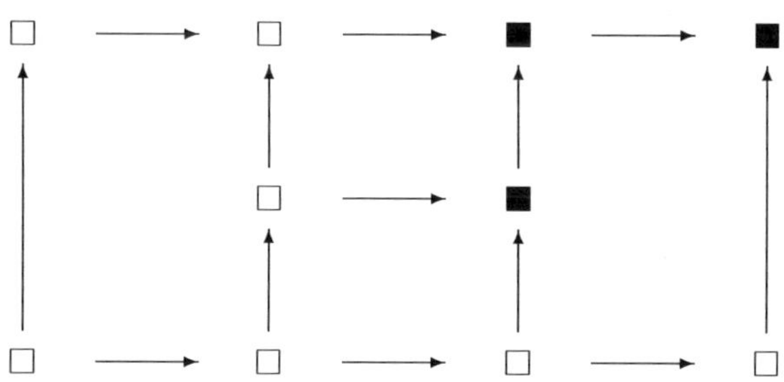

FIGURE 18. (□ = $\aleph_1$, ■ = $\aleph_2$) Cichoń's diagram in the Miller model

Moreover, it can be shown[84] that weak dBC holds true in the *Sacks*[85] *model* (compare with Theorem 4.4 below, and see Figure 21 for its Cichoń's diagram):

THEOREM 3.32. *The Sacks model satisfies the weak dual Borel Conjecture (i.e., there are no strongly meager sets of size continuum).*

PROOF. This can be derived, e.g., as follows: by [**CP04**, Chapter 7], the Sacks model satisfies the *Covering Property Axiom* (CPA); but [**CP04**, Theorem 1.1.4] shows that CPA_{cube} (a weaker version of CPA) implies the non-existence of *perfectly meager* sets of size continuum; the well-known fact that every strongly meager set is perfectly meager finishes the proof. □

Since the Sacks model does *not* satisfy the dual Borel Conjecture (see Theorem 4.4), it follows that weak dBC does not imply dBC.

[84]The first proof was given by Arnold Miller (see [**Mil83**]).

[85]We conjecture that the same is true for the Silver model; indeed, we believe (but have not checked) that both Sacks forcing and Silver forcing are *good* in the sense of [**BS03**, Definition 22]; this would yield the weak dBC in both the Sacks model and the Silver model, analogously to the proof of Theorem 3.31 (and give yet another proof of Theorem 3.32).

3.7.8. *Does the dual Borel Conjecture hold true in the Laver model?* Observe that the Laver model satisfies all the necessary conditions for dBC, as far as Cichoń's diagram is concerned (see Corollary 3.8 and Figure 5).

It is surprising that the following question seems to be open:

QUESTION 3.33. Does the Laver model satisfy the *dual* Borel Conjecture?

The above question is tightly related to the following (see also Remark 3.25):

QUESTION 3.34. Does (single) Laver forcing $\mathbb{L}$ make[86] all uncountable sets from the ground model *not* strongly meager?

Note that a positive answer to Question 3.33 would yield another (potentially much easier) proof of the joint consistency (see Theorem 4.1) of the Borel Conjecture and the dual Borel Conjecture (recall that the Laver model satisfies the Borel Conjecture anyway).

We rather believe that the dual Borel Conjecture fails in the Laver model; in this case, the Laver model would be an alternative model (compare with Theorem 3.9 and Figure 14) to exemplify that the necessary conditions for dBC are not sufficient.

4. Joint consistency of Borel Conjecture and dual Borel Conjecture

Most of this section is dedicated to a (rather informal) overview of the proof from **[GKSW14]** which shows the joint consistency of BC and dBC (see Theorem 4.1); at the end, we also discuss all combinations of BC or dBC being true or false (see Section 4.12), as well as stronger versions of BC+dBC which we do not know to be consistent (see Section 4.13).

Recall that the *Borel Conjecture* (BC) is the statement that there are no uncountable *strong measure zero* (smz) sets, whereas the *dual Borel Conjecture* (dBC) is the statement that there are no uncountable *strongly meager* (sm) sets.

The following theorem[87] states the existence of a model of ZFC satisfying both BC and dBC (see Figure 19 for its Cichoń's diagram):

THEOREM 4.1 (Goldstern-Kellner-Shelah-Wohofsky, 2014). *It is consistent that the Borel Conjecture and the dual Borel Conjecture hold simultaneously.*

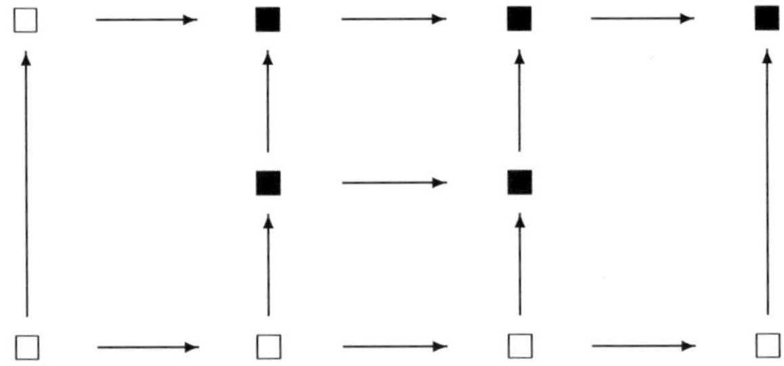

FIGURE 19. ($\square = \aleph_1$, $\blacksquare = \aleph_2$) Cichoń's diagram in our BC+dBC model

In the following (see Section 4.1–Section 4.11), we try to explain some of the major ideas of the proof of Theorem 4.1 (with brief summaries of the proofs of

[86]... as Cohen forcing $\mathbb{C}$ does (see Lemma 3.18 and Corollary 3.19).

[87]The proof of the theorem has actually been completed in 2011.

BC in the Laver model – for the detailed proof, see Section 2 – and dBC in the Cohen model – for the detailed proof, see Section 3 – as a starting point); in our presentation, we provide several footnotes pointing to the respective details given in the original paper [**GKSW14**].

For an alternative (and presumably more technical) overview of the proof of Theorem 4.1, see [**GKSW**].

4.1. The usual proof of Con(BC). Recall the key points of the proof of *BC in the Laver model* (for more details, see Section 2.2):

- each iterand is *Laver forcing* $\mathbb{L}$
  $\mathbb{L}$ kills smz (witnessed by the Laver real)
- being *not* smz is preserved:
  (1) $\mathbb{L}$ has the *Laver property*
  (2) Laver property is preserved under *countable support* iteration
  (3) "X not smz" remains true after forcings with the Laver property

To prove that the Borel Conjecture holds in our model of BC+dBC from Theorem 4.1, we are going to use "preservation of positivity" (and the the stronger iterable property "preserving random reals", respectively) instead of the "Laver property". For more details, see Section 2.7.2, where we gave an alternative proof of BC in the Laver model (as well as a proof of BC in the Laver-random model); in particular, see Theorem 2.27 (replacing item (1) above), Proposition 2.26 (replacing item (2) above), and Lemma 2.29 (replacing item (3) above).

Two types of iterands will be used for the construction of the BC+dBC model: ultralaver forcings (see Section 4.5) and Janus forcings (see Section 4.6); in fact, both types will be (capable of potentially) preserving (certain) random reals.

4.2. The usual proof of Con(dBC). Recall the key points of the proof of *dBC in the Cohen model* (for more details, see Section 3.2):

- each iterand is *Cohen forcing* $\mathbb{C}$
  $\mathbb{C}$ kills sm (witnessed by the generic null set $H = \bigcap_{n\in\omega} \bigcup_{m\geq n} C^m$)
- being *not* sm is preserved:
  (1) $\mathbb{C}$ is *precaliber* $\aleph_1$
  (2) Precaliber $\aleph_1$ is preserved under *finite support* iteration
  (3) "X not sm" remains true after forcings with precaliber $\aleph_1$

To prove that the dual Borel Conjecture holds in our model of BC+dBC from Theorem 4.1, we are going to replace "precaliber $\aleph_1$" by "σ-centered" (see Definition 3.20), which is a stronger property, and is preserved under finite support iteration of countable[88] length.

Our two types of iterands used for the construction of the BC+dBC model behave well: ultralaver forcings (see Section 4.5) will be σ-centered anyway (this will easily follow from the definition), and Janus forcings (see Section 4.6) will be (capable of potentially being countable and hence) σ-centered as well.

4.3. Obstacles in combining the proofs of Con(BC) and Con(dBC). To get a model of ZFC satisfying both BC and dBC, one might attempt to combine the proof that BC holds in the Laver model and the proof that dBC holds in the Cohen model, by just mixing the two sorts of iterands, i.e., alternately forcing with Laver and Cohen forcing. This would "kill" all smz and sm sets which belong to

[88] Indeed, it is even preserved under finite support iteration of length strictly less than $(2^{\aleph_0})^+$.

the respective intermediate model. However, there are severe problems with this approach.

The methods to preserve "X is not smz" and "X is not sm", respectively, do not fit to each other (keep item (1) and item (2) of the respective lists of Section 4.1 and Section 4.2 in mind): Cohen forcing $\mathbb{C}$ does not have the Laver property (and does not preserve random reals either), whereas Laver forcing $\mathbb{L}$ is not precaliber $\aleph_1$ (in fact not even c.c.c.); moreover, countable support iteration is essential for the preservation of the Laver property (as well as the preservation of "preserving random reals"), whereas finite support iteration is essential to preserve precaliber $\aleph_1$.

Note that it is the *proofs* of BC and dBC (in the Laver model and the Cohen model, respectively) that are seemingly irreconcilable, not necessarily the models; in case of the Cohen model, however, the situation is worse, since adding (cofinally many) Cohen reals (which is not only because of the Cohen iterands, but also due to the finite support iteration which adds Cohen reals at limits) inevitably destroys the Borel Conjecture: it makes $\mathrm{cov}(\mathcal{M})$ large, but $\mathrm{cov}(\mathcal{M}) = \omega_1$ is necessary for BC (recall[89] Theorem 2.2).

On the other hand, it is not clear whether maybe the Laver model could already satisfy the dual Borel Conjecture: this is surprisingly open (see Question 3.33).

4.4. The idea of the proof of Con(BC+dBC). So how to reconcile the two strategies?

We have to come up with an iteration which works with respect to both tasks (i.e., killing both smz and sm, and preserving both "X not smz" and "X not sm"). This is accomplished by using a "generic" iteration (with "generic iterands" and "generic support", so to speak).

For this purpose, we start with a "preparatory" forcing $\mathbb{R}$ (which is σ-closed and $\aleph_2$-c.c.); any generic filter $G \subseteq \mathbb{R}$ will yield a c.c.c.(!) iteration $\bar{\mathbf{P}}$ of length ω_2, whose "generic iterands" are (alternately) forcings "killing smz" and forcings "killing sm". Additionally, both types of iterands are capable of preserving both "X not smz" and "X not sm" (either behavior is "prepared" within the preparatory forcing $\mathbb{R}$, hence both at the same time is possible); also, both preserving "X not smz" and preserving "X not sm" is preserved at limits (since both finite and countable support is "prepared" within the preparatory forcing $\mathbb{R}$). Finally, the forcing $\mathbb{R} * \mathbf{P}_{\omega_2}$ yields BC+dBC.

In the following, we are going to describe the construction in more detail. We start with defining two (classes of) forcing notions: Ultralaver forcing and Janus forcing.

4.5. Ultralaver forcings (the iterands for killing smz). Instead of standard Laver forcing $\mathbb{L}$ (as for the Laver model), we use "ultralaver" forcings $\mathbb{L}_{\bar{D}}$, which are "filtered" versions of $\mathbb{L}$. An[90] *ultralaver forcing* $\mathbb{L}_{\bar{D}}$ is based on a system $\bar{D} = (D_s)_{s\in\omega^{<\omega}}$ of (non-principal) ultrafilters on ω and consists of all Laver trees $p \subseteq \omega^{<\omega}$ with the property that for any node $s \in p$ (above the stem) the set

$$\{n \in \omega \colon s^\frown n \in p\}$$

(is not only infinite but) belongs to the ultrafilter D_s.

[89]In fact, $\mathrm{cov}(\mathcal{M}) > \omega_1$ even yields $\mathrm{non}(\mathcal{SN}) > \omega_1$, i.e., the following strong failure of BC: not only *there is*, but *all* sets of size $\aleph_1$ are smz.

[90]See also [**GKSW14**, Definition 1.1] for the definition of ultralaver forcing.

Like Laver forcing, it "kills" uncountable smz sets from the ground model (the proof is actually pretty similar[91] to the one of Lemma 2.15). Moreover, it is clearly[92] σ-centered (which was the incentive to replace Laver forcing by ultralaver forcings); this is towards preserving "X not sm". By choosing appropriate ultrafilters, it can also be made to "preserve (certain[93]) random reals" (in place of the Laver property; compare with Section 2.7.2); this is towards preserving "X not smz".

For more details on ultralaver forcing, see [**GKSW14**, Section 1].

4.6. Janus forcings (the iterands for killing sm). "Janus" forcing has "two faces" (therefore the name[94]). It will always be preceded by an ultralaver forcing (the definition is actually dependent[95] on the ultralaver real which has been added by this ultralaver forcing).

First of all, a *Janus forcing* consists[96] of a (countable) core which is (essentially) Cohen forcing; as in the proof of dBC in the Cohen model, the core provides a measure zero set[97]

$$H := \bigcap_{n\in\omega} \bigcup_{m\geq n} C^m$$

which will kill[98] any old sm set X (and there is no resurrection after σ-centered forcings).

Second, there are (countably or uncountably many) additional conditions which are "wrapped around" the core. On the one hand,

a Janus forcing can[99] be countable, and hence σ-centered;

this is towards preserving "X not sm". On the other hand,

a Janus forcing can[100] be (equivalent to) random forcing;

this is towards preserving "X not smz", since random forcing "preserves random reals" (see Theorem 2.30) which is the iterable property in place of the Laver property (compare with Section 2.7.2).

In both cases, however, a Janus forcing has to satisfy a certain combinatorial property which makes sure that its core yields a null set $\dot{H}$ which indeed fulfills

$$\Vdash_{\text{Janus}} X + \dot{H} = 2^\omega$$

[91] See [**GKSW14**, Lemma 1.13] for the proof.

[92] Just observe that any finite set of conditions sharing the same stem has a common lower bound; see also [**GKSW14**, Lemma 1.2].

[93] In fact, given a system of ultrafilters (in a model), and given a (single) random real (over that model), we can extend the (system of) ultrafilters in such a way that the random real is preserved (in an iterable way); see [**GKSW14**, Section 1.E], in particular, [**GKSW14**, Lemma 1.30].

[94] *Janus* – the ancient Roman god of beginnings, gates, transitions, time, duality, doorways, passages, and endings – is usually depicted as having *two faces* (since he looks to the future and to the past).

[95] For the exact way the definition of Janus forcing depends on the preceding ultralaver real, see [**GKSW14**, Assumption 2.3 (based on Lemma 1.23 and Corollary 2.2)].

[96] See [**GKSW14**, Definition 2.5] for the (quite complicated) details of the definition of Janus forcing; the core is defined in [**GKSW14**, Definition 2.4].

[97] See also [**GKSW14**, Definition 2.6].

[98] See [**GKSW14**, Lemma 2.9 (or Corollary 2.15)] for the proof that strongly meager sets are killed, and there is no resurrection after σ-centered forcings.

[99] This follows from [**GKSW14**, Fact 2.8].

[100] See [**GKSW14**, Lemma 2.16].

for old X (i.e., kills old sm sets X). The property essentially says[101] that there is no condition outside the core which excludes too many potential clopen sets from being selected as one of the C^m; in other words, it is true that the sequence $(C^m)_{m\in\omega}$ of clopen sets is *not* completely "Cohenly chosen" (as it is done in the usual proof, where the C^m's are derived from the Cohen real straight away), but still the C^m's are chosen in a sufficiently "free" way in order to make the arguments work. (In fact, a quite involved *combinatorial lemma*[102] *from* [**BS10**] is used, thereby replacing Theorem 3.11 – the finite combinatorial fact on which the usual proof is based.)

For more details on Janus forcing, see [**GKSW14**, Section 2].

4.7. Alternating (partial countable support) iterations. An *alternating (partial countable support) iteration*[103] is a (proper) iteration $\bar{P} = (P_\alpha, \dot{Q}_\alpha)_{\alpha<\omega_2}$ of length ω_2 with the following properties:

- For even[104] α, the iterand $\dot{Q}_\alpha$ is an ultralaver forcing.
- For odd α, the iterand $\dot{Q}_\alpha$ is a Janus forcing.
- For limit δ, the forcing P_δ is a *partial countable support limit*[105] of the $(P_\alpha)_{\alpha<\delta}$; this is a subset of the countable support limit containing the direct limit.

The concept "partial countable support iteration" gives the freedom to use either an iteration which is (close to) a finite support iteration or an iteration which is (close to) a countable support iteration. Using (almost) finite support iterations is towards preserving "X not sm", whereas using (almost) countable support iterations is towards preserving "X not smz".

For more details on the notion of partial countable support, see [**GKSW14**, Section 3.B]; for more on alternating iterations, see [**GKSW14**, Section 4.A].

4.8. The preparatory forcing $\mathbb{R}$ and M-complete embeddings. The *preparatory forcing* $\mathbb{R}$ consists of alternating iterations which approximate the desired (generic) alternating iteration $\bar{\mathbf{P}}$. More precisely, $\mathbb{R}$ consists[106] of conditions $x = (M^x, \bar{P}^x)$, where M^x is a countable ord-transitive model of (some fragment of) ZFC, and $\bar{P}^x$ is an alternating iteration in M^x.

A (countable) *ord-transitive*[107] *model* M is somewhere in between a transitive model and an elementary (sub)model (of some $H(\chi)$); it essentially equals a transitive model, but $M \cap \mathrm{Ord}$ is not a countable ordinal (as in the case of a real countable transitive model), but a non-transitive set of ordinals (as in the case of an elementary model). One gets an ord-transitive model when "ord-collapsing"[108] an elementary model, i.e., collapsing everything except for ordinals (which are "treated as urelements"). The reason for using ord-transitive models instead of transitive

[101]The crucial combinatorial property of Janus forcing (describing how the additional conditions are "wrapped around" the core) is [**GKSW14**, Definition 2.5(3)] (which is called "fatness" there).

[102]See [**GKSW14**, Lemma 2.1 (and Corollary 2.2)].

[103]See [**GKSW14**, Definition 4.1]; also see footnote 104.

[104]In fact, it is also allowed that the iterand $\dot{Q}_\alpha$ (and in this case also the succeeding iterand $\dot{Q}_{\alpha+1}$) is the trivial forcing.

[105]See [**GKSW14**, Section 3.B].

[106]See [**GKSW14**, Definition 4.3].

[107]Actually, so-called *(nice) candidates* are used, see [**GKSW14**, Definition 3.1].

[108]See [**GKSW14**, Fact 3.2(1)].

ones is the following: Our conditions $x = (M^x, \bar{P}^x) \in \mathbb{R}$ are supposed to approximate a generic iteration of length ω_2, so our approximating iterations $\bar{P}^x$ have to allow for non-trivial iterands $\dot{Q}_\alpha$ for coordinates α arbitrarily large in ω_2.

The *order on* $\mathbb{R}$ is defined as follows: a condition $y = (M^y, \bar{P}^y)$ is stronger than $x = (M^x, \bar{P}^x)$, i.e., $y \leq x$, if

- $M^x \in M^y$ (and M^y already "knows" that M^x is in fact countable),
- $\bar{P}^x$ is "canonically M^x-completely embeddable" into $\bar{P}^y$.

"*Canonically M^x-completely embeddable*"[109] means the following: One can find a (canonical) "coherent" sequence $(i_\alpha)_{\alpha\leq\omega_2}$ of M^x-complete embeddings

$$i_\alpha : P^x_\alpha \to P^y_\alpha.$$

(An embedding $i : P \to Q$ is *M-complete* if each maximal antichain $A \subseteq P$ *which belongs to* M is mapped to a maximal antichain $i[A] \subseteq Q$.)

For $\bar{P}^x$ being canonically M^x-completely embedded into $\bar{P}^y$, it is in particular necessary[110] that $\dot{Q}^x_\alpha$ is (forced to be) an M^x-complete subforcing of $\dot{Q}^y_\alpha$. In case of ultralaver forcing (i.e., even α), this will be true[111] whenever the ultrafilter system $\bar{D}^y$ (defining $\dot{Q}^y_\alpha = \mathbb{L}_{\bar{D}^y}$) "extends" the ultrafilter system $\bar{D}^x$, i.e., for each node $s \in \omega^{<\omega}$, $D^y_s \supseteq D^x_s$.

For more details on the preparatory forcing $\mathbb{R}$ and M-complete embeddings, see [**GKSW14**, Section 4.A (and Section 3.B)]; for more on ord-transitive models, see [**GKSW14**, Section 3.A].

4.9. Almost finite support and almost countable support. At limit ordinals δ, we have the following situation (still assuming that $\bar{P}^x$ is canonically M^x-completely embedded into $\bar{P}^y$).

The forcing P^y_δ is a partial countable support limit of the $(P^y_\alpha)_{\alpha<\delta}$ such that the "canonically[112] defined embedding" $i_\delta : P^x_\delta \to P^y_\delta$ (satisfies $i_\delta[P^x_\delta] \subseteq P^y_\delta$ and) is M^x-complete. Note that we would like to define P^y_δ as the finite (or countable) support limit of the $(P^y_\alpha)_{\alpha<\delta}$. However, this is not quite possible; instead, we will take the "almost finite support limit over x" and the "almost countable support limit over x", respectively, which will make sure that i_δ is an M^x-complete embedding.

More precisely, the *almost finite support limit over x* is the "minimal" partial countable support limit; it essentially equals[113] the direct limit of $(P^y_\alpha)_{\alpha<\delta}$ together with the (countable) set $i_\delta[P^x_\delta]$. The *almost countable support limit over x* is kind of "maximal": it essentially consists[114] of all those conditions in the countable support limit of $(P^y_\alpha)_{\alpha<\delta}$ which do not outright force that i_δ is not M^x-complete.

They are sufficiently close to the finite/countable support limit to make the respective arguments work: The almost finite/countable support limit will preserve σ-centeredness[115]/"preserving random reals"[116], which is towards preserving "X not sm"/"X not smz".

---

[109]See also [**GKSW14**, Definition 3.10], and the discussion after [**GKSW14**, Definition 4.1].

[110]See [**GKSW14**, Fact 4.2].

[111]See the first item of [**GKSW14**, Fact 4.2], or, more explicitly, [**GKSW14**, Lemma 1.5].

[112]See also [**GKSW14**, Definition 3.9] for the definition of *canonical extension*.

[113]See [**GKSW14**, Definition 3.13], and [**GKSW14**, Lemma 3.14] for the proof that it "works", i.e., that it makes i_δ an M^x-complete embedding.

[114]See [**GKSW14**, Definition and Claim 3.21], and [**GKSW14**, Lemma 3.26] for the proof that it "works", i.e., that it makes i_δ an M^x-complete embedding.

[115]See [**GKSW14**, Lemma 3.17].

[116]See [**GKSW14**, Lemma 3.33] and the lemmas afterwards.

For more details on the notions of almost finite support and almost countable support, see [**GKSW14**, Section 3.C] and [**GKSW14**, Section 3.D], respectively.

4.10. The generic alternating iteration $\bar{\mathbf{P}}$. Fix a generic filter $G \subseteq \mathbb{R}$. The filter G yields a directed[117] system of alternating iterations together with (not quite) complete embeddings; we can derive[118] a *"generic" alternating iteration* $\bar{\mathbf{P}}$ (with limit $\mathbf{P}_{\omega_2}$) from the (direct) limit of this directed system.

The generic iteration $\bar{\mathbf{P}}$ is approximated by all the $\bar{P}^x$ for which $x = (M^x, \bar{P}^x)$ belongs to the generic filter G. More precisely,[119] each such $\bar{P}^x$ can be canonically M^x-completely embedded into $\bar{\mathbf{P}}$, via embeddings $i^x_\alpha : P^x_\alpha \to \mathbf{P}_\alpha$; in other words, whenever $H \subseteq \mathbf{P}_{\omega_2}$ is a generic filter over V[G], then the preimage of H under $i^x_{\omega_2}$ (which is a subset of $P^x_{\omega_2}$) is a generic filter over M^x.

The iterands $\dot{\mathbf{Q}}_\alpha$ of the generic iteration $\bar{\mathbf{P}} = (\mathbf{P}_\alpha, \dot{\mathbf{Q}}_\alpha)_{\alpha<\omega_2}$ are "generic" ultralaver forcings[120] (in case of α even) and "generic" Janus forcings (in case of α odd). Surprisingly, $\bar{\mathbf{P}}$ is a c.c.c. iteration[121] (which is essential for the proof of BC and dBC): The generic ultralaver forcings are σ-centered (hence c.c.c.), and the generic Janus forcings can also be shown to be c.c.c. (by constructing a model M which "catches" a countable part of the maximal antichain, and using M-completeness to preserve maximality); moreover, the c.c.c. is preserved at all limits (which can be shown by a similar argument).

For more details on the generic alternating iteration, see [**GKSW14**, Section 4.B and Section 4.D] (as well as [**GKSW14**, Section 4.C] for the inductive proof of c.c.c.).

4.11. The proof of BC and dBC in the final model. We can now prove that *both BC and dBC hold in the final model* $\mathbb{R} * \mathbf{P}_{\omega_2}$. This is due to the fact that the (generic) ultralaver/Janus forcings kill smz/sm sets, and the fact that the set of all $x = (M^x, \bar{P}^x) \in \mathbb{R}$ with either of the following two properties is dense in $\mathbb{R}$:

- $\bar{P}^x$ is build towards preserving "X not smz".
- $\bar{P}^x$ is build towards preserving "X not sm".

In the following, we will sketch the basic strategy for the proof of BC (as well as dBC).

For more details, see [**GKSW14**, Section 5]: first, in [**GKSW14**, Section 5.A], a simplified version is presented (only dealing with uncountable sets from the ground model); finally, in [**GKSW14**, Section 5.B], a "factor lemma" is introduced, allowing to prove the statements (that there are no uncountable smz or sm sets in the final model) in full generality.

4.11.1. *Showing BC.* The proof that BC holds[122] in the final model is (roughly speaking) as follows.

[117] See [**GKSW14**, Definition 4.10] and the discussion before, for how to build the direct limit of the alternating iterations which belong to the generic filter G.

[118] See [**GKSW14**, Lemma 4.17(1) (and Corollary 4.26)].

[119] See [**GKSW14**, Lemma 4.17(1) (and Lemma 4.27 for more)].

[120] In particular, all the involved ultrafilters assigned to the nodes in $\omega^{<\omega}$ are generic objects provided by the preparatory forcing $\mathbb{R}$.

[121] See [**GKSW14**, Lemma 4.17(4)].

[122] See [**GKSW14**, Lemma 5.1] for the simplified version (showing that there is no uncountable set from the ground model which is smz in the final model). To show that BC really holds (i.e., that there is no uncountable smz set in the final model), consult [**GKSW14**, Section 5.B] (the "factor lemma").

Assume towards a contradiction that (in the final model) the uncountable set X is smz. We can assume that X is of size $\aleph_1$, and that X already appears at some stage $\alpha_0 < \omega_2$. Consider an ultralaver forcing $\dot{\mathbf{Q}}_\alpha$ at any even stage $\alpha > \alpha_0$; it will kill X, by adding a witness E (a Borel code for a closed null set) such that $X + E$ is not null (compare with Section 2.7.2, in particular, Theorem 2.28 and Lemma 2.29). Since X is smz in the final model, we know that $X + E$ will become null at some later stage, i.e., we can assume that (a Borel code for) a null set containing $X + E$ appears at some $\beta < \omega_2$ with $\beta > \alpha$.

Since $\bar{\mathbf{P}}$ is c.c.c., names for reals (such as Borel codes) are countable objects, so they can be "seen" by countable models. The whole situation can be "reflected down" to a condition x in the preparatory forcing. This condition x can be strengthened to a condition $y = (M^y, \bar{P}^y)$ which belongs to the first of the two dense sets mentioned above, i.e., $\bar{P}^y$ is build towards preserving "X not smz": The iterands are ultralaver forcings "preserving (certain) random reals", and Janus forcings which are equivalent to random forcing (hence "preserving random reals"), respectively; at limits we take almost countable support limits over x (which preserve "preserving random reals"). This shows that "$X + E$ not null" is preserved, which can be shown to lead to a contradiction (by using absoluteness of names for reals etc.).

4.11.2. *Showing dBC.* The proof that dBC holds[123] in the final model is similar: we again get a condition x "capturing" the whole situation; we choose a stronger condition y which belongs to the second of the two dense sets, i.e., $\bar{P}^y$ is build towards preserving "X not sm": The iterands are ultralaver forcings which are σ-centered anyway, and Janus forcings which are countable (hence σ-centered), respectively; at limits we now take almost finite support limits over x (which preserve σ-centeredness). Again, this leads to a contradiction.

4.12. All four combinations of truth values for BC and dBC. We are now going to discuss all (four) combinations of BC or dBC being true or false, and argue that they are consistent (see Table 4).

TABLE 4. All four combinations of BC/dBC

BC	dBC	canonical model	theorem	Cichoń
no	no	under MA (or CH)	Theorem 4.8	Figure 23
yes	no	Laver-random	Theorem 2.31	Figure 7
no	yes	Cohen	Theorem 3.10	Figure 15
yes	yes	BC+dBC	Theorem 4.1	Figure 19

We will also give a brief overview (see Section 4.12.5 and Table 5) about what is known concerning the (nine) combinations when the weak version of the (dual) Borel Conjecture (see Section 2.7.6, and Section 3.7.7, respectively) is taken into consideration.

[123]See [**GKSW14**, Lemma 5.2] for the simplified version (showing that there is no uncountable set from the ground model which is sm in the final model). To show that dBC really holds (i.e., that there is no uncountable sm set in the final model), again use [**GKSW14**, Section 5.B] (the "factor lemma").

4.12.1. *Models for* BC $\wedge$ dBC. The only known model for BC and dBC is the one from [**GKSW14**] discussed in this section (see Theorem 4.1).

On the other hand, one can consider natural strengthenings of BC+dBC – such as $(\mathcal{M} \cap \mathcal{N})$-BC, or $\mathcal{E}$-BC – and ask whether they are consistent (see Section 4.13).

Furthermore, one can ask whether BC+dBC is consistent with a projective well-order of the reals (see Section 6, and, more specifically, Question 6.5).

Since all cardinal characteristics of Cichoń's diagram (in models of $2^{\aleph_0} = \aleph_2$) except for $\mathrm{cov}(\mathcal{N})$ and $\mathrm{non}(\mathcal{N})$ are determined by BC (see Corollary 2.4 and Figure 3), and (among others) $\mathrm{cov}(\mathcal{N})$ is determined by dBC (see Corollary 3.8 and Figure 13), all characteristics except $\mathrm{non}(\mathcal{N})$ are determined by the conjunction of BC and dBC for sure (see Figure 20).

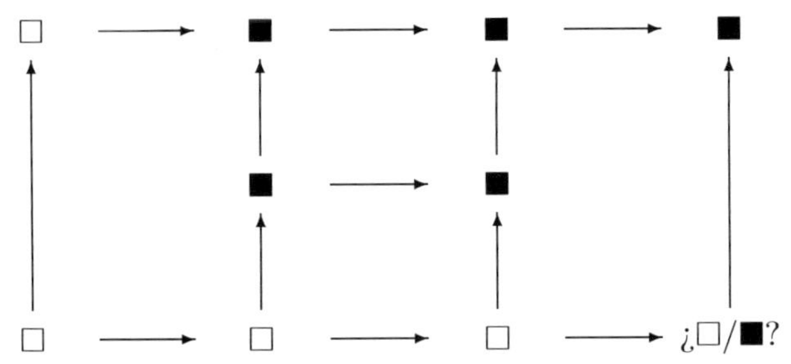

FIGURE 20. ($\square = \aleph_1$, $\blacksquare = \aleph_2$) Cichoń's diagram in models of BC+dBC

However, we do not know whether $\mathrm{non}(\mathcal{N})$ is determined by BC+dBC; since our method from [**GKSW14**] is heavily based on *preservation of positivity* (keeping $\mathrm{non}(\mathcal{N})$ small; see also Figure 19), it is not clear at all whether it can be adapted to allow for $\mathrm{non}(\mathcal{N})$ being large:

QUESTION 4.2. Is there a model of ZFC satisfying both BC and dBC as well as $\mathrm{non}(\mathcal{N}) = \aleph_2$?

4.12.2. *Models for* $\neg$BC$\wedge$dBC. The canonical way to obtain a model satisfying $\neg$BC and dBC is to aim for a model of the dual Borel Conjecture in which some of the necessary conditions for the Borel Conjecture (namely $\mathrm{cov}(\mathcal{M}) = \omega_1$ and $\mathfrak{b} > \omega_1$; see Corollary 2.4) fail.

This is indeed easy: recall that the usual model for the dual Borel Conjecture is the Cohen model (see Theorem 3.10 and Figure 15) in which $\mathrm{cov}(\mathcal{M})$ is large (see also the discussion in Section 4.3), and hence the Borel Conjecture fails. (In fact, also the other necessary condition for BC fails in the Cohen model: since no dominating reals are added, $\mathfrak{b}$ is small.)

4.12.3. *Models for* BC$\wedge\neg$dBC. Like above, the canonical way to obtain a model satisfying BC and $\neg$dBC is to aim for a model of the Borel Conjecture in which some of the necessary conditions for the dual Borel Conjecture (namely $\mathrm{cov}(\mathcal{N}) = \omega_1$ and $\mathfrak{d} > \omega_1$; see Corollary 3.8) fail.

First of all, observe that $\mathfrak{d} > \omega_1$ holds true in every model of the Borel Conjecture (see Figure 3), due to $\mathfrak{b} > \omega_1$ being necessary for BC, and $\mathfrak{b} \leq \mathfrak{d}$ in ZFC; so we have to aim at violating the other necessary condition for dBC, i.e., making $\mathrm{cov}(\mathcal{N})$ large.

The two standard models for the Borel Conjecture are not suitable for this approach: both the Laver model (see Theorem 2.6 and Figure 5) and the Mathias model (see Theorem 2.23 and Figure 6) are obtained by countable support iterations with the Laver property, so $\mathrm{cov}(\mathcal{N})$ is small in these models.

However, we can take the Laver-random model (see Theorem 2.31 and Figure 7): since random reals are cofinally added along the iteration, $\mathrm{cov}(\mathcal{N}) = \omega_2$ holds true in this model, hence the dual Borel Conjecture fails.

On the other hand, even the Laver model might be an example of a model for $\mathrm{BC} \wedge \neg\mathrm{dBC}$: it could be the case – and in fact is not too unlikely – that the Laver model (even though it satisfies both of the necessary conditions for dBC) does *not* satisfy dBC (see Question 3.33).

4.12.4. *Models for* $\neg\mathrm{BC} \wedge \neg\mathrm{dBC}$. There are several models in which both BC and dBC fail, i.e., both uncountable strong measure zero and uncountable strongly meager sets exist; actually, there are various reasons for the existence of these sets.

First of all, let us recall that this is the case under the continuum hypothesis:

COROLLARY 4.3. *Assume CH. Then both BC and dBC fail.*

PROOF. Immediately follows from Theorem 2.1 and Theorem 3.1. □

Also, both BC and dBC fail in the *Sacks model* as well as in the *Silver model* (see Figure 21 for their Cichoń's diagram); this is because the respective cardinal characteristics are *small*:

THEOREM 4.4. *Assume CH, and let $\mathbb{P}_{\omega_2}$ be the countable support iteration of Sacks forcing (or, alternatively, Silver forcing) of length ω_2.*

Then in a generic extension by $\mathbb{P}_{\omega_2}$ both BC and dBC fail.

PROOF. It is well-known that Sacks forcing (as well as Silver forcing) has the Sacks property, and therefore all cardinal characteristics of Cichoń's diagram are small in the Sacks model (as well as the Silver model); see also [**BJ95**, Model 7.6.2].

In particular, $\mathfrak{b} = \mathfrak{d} = \omega_1$, so the failure of BC (and the failure of dBC) immediately follows from Theorem 2.3 (and Theorem 3.3, respectively).

(Alternatively, one could argue with Proposition 1.10: since all cardinal characteristics of Cichoń's diagram are small, we have $\mathrm{cov}(\mathcal{M}) = \mathrm{cof}(\mathcal{M}) = \omega_1$ as well as $\mathrm{cov}(\mathcal{N}) = \mathrm{cof}(\mathcal{N}) = \omega_1$, yielding both strong measure zero and strongly meager sets of size ω_1.) □

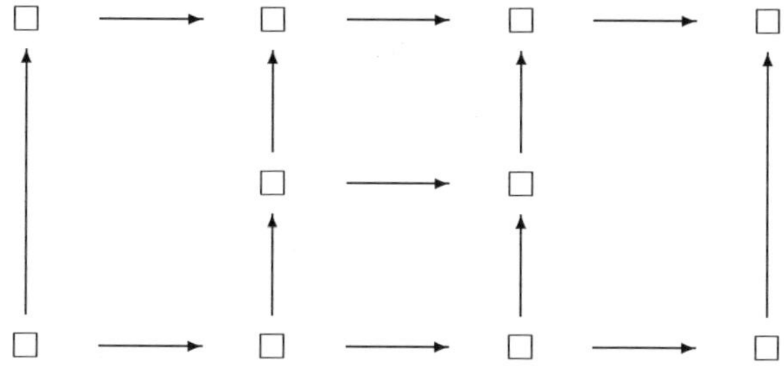

FIGURE 21. ($\square = \aleph_1$, $\blacksquare = \aleph_2$) Cichoń's diagram in the Sacks model (and the Silver model)

The same applies to the *random model* (see Figure 10 for its Cichoń's diagram); sufficiently many (even though not all) cardinal characteristics of Cichoń's diagram happen to be *small*:

THEOREM 4.5. *Assume CH, and let $\mathbb{P}_{\omega_2}$ be the countable support iteration of random forcing of length ω_2.*

Then in a generic extension by $\mathbb{P}_{\omega_2}$ both BC and dBC fail.

PROOF. Since random forcing is ω^ω-bounding, we have $\mathfrak{b} = \mathfrak{d} = \omega_1$ in the final model, so – again – the failure of BC (and the failure of dBC) immediately follows from Theorem 2.3 (and Theorem 3.3, respectively). □

Note that the random model satisfies the weak Borel Conjecture (see Corollary 2.37), i.e., – even though there are uncountable strong measure zero (and strongly meager) sets – there is no strong measure zero set of size continuum.

The same is true for the Sacks model (and the Silver model): this follows from Theorem 2.36, due to the fact that Sacks forcing (and Silver forcing) is strongly ω^ω-bounding.

Moreover, the Sacks[124] model satisfies the weak dual Borel Conjecture (see Theorem 3.32), i.e., there is no strongly meager set of size continuum.

Another model for the failure of both BC and dBC is the *Cohen-random model* (see Figure 22 for its Cichoń's diagram), i.e., the model obtained by cofinally adding both Cohen and random reals; this time, the uncountable strong measure zero and strongly meager sets exist for a different reason, namely the fact that the respective covering numbers are *large*:

THEOREM 4.6. *Assume CH, and let $\{\mathbb{P}_\alpha, \dot{\mathbb{Q}}_\alpha : \alpha < \omega_2\}$ be a countable*[125] *support iteration such that every even iterand is Cohen forcing and every odd iterand is random forcing.*

Then in a generic extension by $\mathbb{P}_{\omega_2}$ both BC and dBC fail (and even weak BC and weak dBC fail).

PROOF. Since Cohen forcing increases $\mathrm{cov}(\mathcal{M})$, and random forcing increases $\mathrm{cov}(\mathcal{N})$, both $\mathrm{cov}(\mathcal{M})$ and $\mathrm{cov}(\mathcal{N})$ are large in the final model, so the failure of BC (and the failure of dBC) immediately follows from Theorem 2.2 (and Theorem 3.2, respectively): in fact, it is even the case that *all* sets of size $\aleph_1$ are both strong measure zero and strongly meager.

Furthermore, there exist both strong measure zero sets and strongly meager sets of size continuum (i.e., weak BC and weak dBC fail) in the final model: this follows from Proposition 1.10, due to the fact that both $\mathrm{cov}(\mathcal{M}) = \mathrm{cof}(\mathcal{M}) = 2^{\aleph_0}$ and $\mathrm{cov}(\mathcal{N}) = \mathrm{cof}(\mathcal{N}) = 2^{\aleph_0}$ hold true. □

Moreover, the failure of both BC and dBC can be obtained by making $\mathrm{add}(\mathcal{N})$ (and hence all cardinal characteristics in Cichoń's diagram) *large*; this is the case, e.g., in the *amoeba model* (see Figure 23 for its Cichoń's diagram):

[124] We believe that the same is true for the Silver model (see footnote 85).

[125] In fact, we can alternatively use a finite support iteration, and obtain the same result (as well as the same Cichoń's diagram); in this case, we can even omit the Cohen iterands, since Cohen reals are added at limit stages of finite support iterations anyway (in other words, iterating only random forcing with finite support does the job).

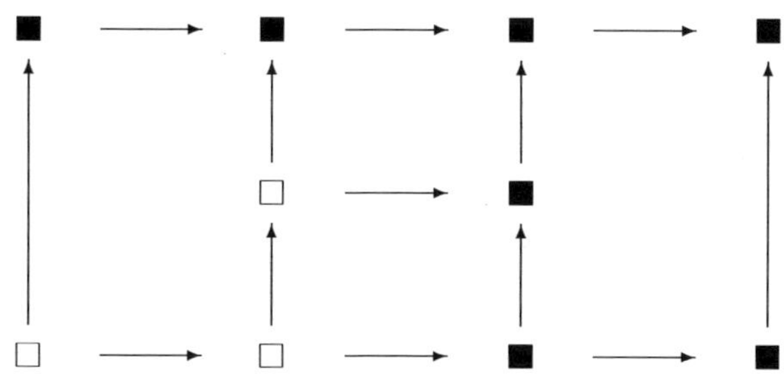

FIGURE 22. ($\square = \aleph_1$, $\blacksquare = \aleph_2$) Cichoń's diagram in the Cohen-random model

THEOREM 4.7. *Assume CH, and let $\mathbb{P}_{\omega_2}$ be the countable*[126] *support iteration of amoeba forcing of length ω_2.*

Then in a generic extension by $\mathbb{P}_{\omega_2}$ both BC and dBC fail (and even weak BC and weak dBC fail).

PROOF. Note that amoeba forcing increases $\text{add}(\mathcal{N})$ (see the discussion after [**BJ95**, Definition 3.1.10], and [**BJ95**, Model 7.6.1]); therefore $\text{add}(\mathcal{N})$ – and hence also $\text{cov}(\mathcal{M})$ and $\text{cov}(\mathcal{N})$ – are large in the final model, and the conclusion follows as in the proof of Theorem 4.6. $\square$

Under Martin's axiom, we have the same situation as above (actually for the same reason); recall that, on the other hand, restricted versions of Martin's axiom such as the one for σ-centered forcings can be consistent with the dual[127] Borel Conjecture (see Theorem 3.29).

THEOREM 4.8. *Assume MA. Then both BC and dBC fail (and even weak BC and weak dBC fail).*

PROOF. Under MA, all cardinal characteristics of Cichoń's diagram are large, so the conclusion follows as in the proof of Theorem 4.6 (or the proof of Theorem 4.7). $\square$

Note that every model of MA has the same Cichoń's diagram as the amoeba model (see Figure 23): amoeba forcing is c.c.c., so $\text{add}(\mathcal{N})$ is large.

4.12.5. *The nine combinations of truth values for (weak) BC/dBC.* We briefly present (see Table 5) what we know about the nine combinations of the status of (weak) BC/dBC (for the definitions, see Section 2.7.6, and Section 3.7.7, respectively) in models of ZFC.

Actually, it seems to be open whether all nine combinations are consistent.

4.13. Stronger versions of BC+dBC (such as $\mathcal{E}$-BC). Recall the definitions in Section 1.5: for $\mathcal{I} \subseteq \mathcal{P}(2^\omega)$, a set $X \subseteq 2^\omega$ is $\mathcal{I}$-shiftable if $X + Z \neq 2^\omega$ for each $Z \in \mathcal{I}$; moreover, $\mathcal{I}$-BC ("the Borel Conjecture with respect to $\mathcal{I}$") is the assertion that there are *no uncountable* $\mathcal{I}$-shiftable sets.

[126] Again, we can alternatively take a finite support iteration of amoeba forcing, and obtain the same result (as well as the same Cichoń's diagram).

[127] ... but not with the Borel Conjecture: *MA(countable)*, i.e., Martin's axiom for Cohen forcing, makes $\text{cov}(\mathcal{M})$ large, i.e., is already sufficient to yield the failure of BC (see Theorem 2.2).

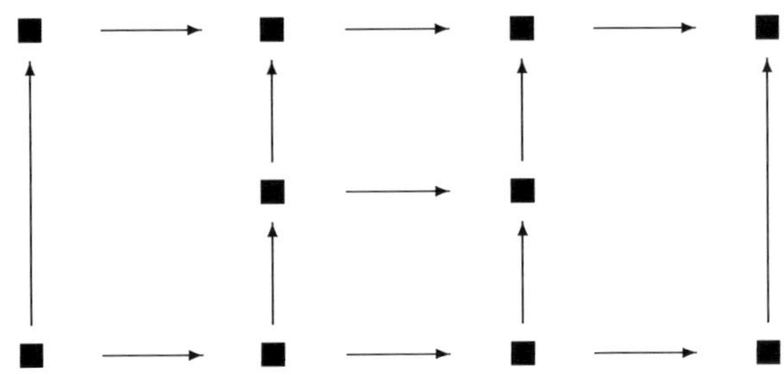

FIGURE 23. ($\square = \aleph_1$, $\blacksquare = \aleph_2$) Cichoń's diagram in the amoeba model (and under MA)

TABLE 5. Nine combinations of (weak) BC/dBC

BC	dBC	name of model	theorem	Cichoń	potential candidates
none	none	amoeba	Theorem 4.7	Figure 23	
weak	none	random	Corollary 2.37	Figure 10	
yes	none	Laver-random	Theorem 2.31	Figure 7	
none	weak	?			Miller?
weak	weak	Sacks	Theorem 2.36/3.32	Figure 21	
yes	weak	?			Laver?
none	yes	Cohen	Theorem 3.10	Figure 15	
weak	yes	?			Miller? dBC w/o Cohens?
yes	yes	BC+dBC	Theorem 4.1	Figure 19	

It is obvious from the definition that the notion of $\mathcal{I}$-shiftable gets weaker if $\mathcal{I}$ is replaced by any smaller ideal (or collection) $\mathcal{J}$:

PROPOSITION 4.9. *Let* $\mathcal{I}, \mathcal{J} \subseteq \mathcal{P}(2^\omega)$. *Then*

$$\mathcal{J} \subseteq \mathcal{I} \implies \mathcal{I}\text{-shiftable} \subseteq \mathcal{J}\text{-shiftable}.$$

This, in turn, means, that there is a direct connection between ideals (or arbitrary collections of sets of reals) and Borel Conjectures: the smaller the ideal (with respect to inclusion), the stronger the respective Borel Conjecture:

PROPOSITION 4.10. *Let* $\mathcal{I}, \mathcal{J} \subseteq \mathcal{P}(2^\omega)$. *Then*

$$\mathcal{J} \subseteq \mathcal{I} \implies \mathcal{J}\text{-BC implies } \mathcal{I}\text{-BC}.$$

It follows that for any collection $\mathcal{J}$ which is contained in both the ideal $\mathcal{M}$ of meager sets and the ideal $\mathcal{N}$ of measure zero sets, both $\mathcal{M}$-BC (i.e., BC) and $\mathcal{N}$-BC (i.e., dBC) is implied by $\mathcal{J}$-BC.

Since the *intersection*[128] *ideal* $\mathcal{M} \cap \mathcal{N}$ is the largest such collection, it is natural to consider the $(\mathcal{M} \cap \mathcal{N})$-BC as the weakest such Borel Conjecture which is (potentially strictly) stronger than BC+dBC; however, it is unknown whether it is consistent:

QUESTION 4.11. Is the Borel Conjecture with respect to $\mathcal{M} \cap \mathcal{N}$ consistent? (More explicitly: Is there a model of ZFC such that for each uncountable set X there exists a set Z which is both meager and measure zero such that $X + Z = 2^\omega$?)

If yes, is BC+dBC equivalent to $(\mathcal{M} \cap \mathcal{N})$-BC (or strictly weaker)?

Yet another, even smaller such ideal is the σ-ideal $\mathcal{E}$ generated by the closed measure zero sets (i.e., the collection of subsets of F_σ measure zero sets). Indeed, $\mathcal{E} \subseteq \mathcal{M} \cap \mathcal{N}$.

Therefore – by Proposition 4.10 – $\mathcal{E}$-BC implies BC+dBC. Again, we can ask:

QUESTION 4.12. Is $\mathcal{E}$-BC consistent? If yes, is $\mathcal{E}$-BC strictly stronger than $(\mathcal{M} \cap \mathcal{N})$-BC (or at least strictly stronger than BC+dBC)?

Note that the following is conceivable: it could be the case that a potential proof of the consistency of $(\mathcal{M} \cap \mathcal{N})$-BC or $\mathcal{E}$-BC yields a much easier proof of the joint consistency of BC and dBC than the one in [**GKSW14**] (see Theorem 4.1).

We also do not know whether there is any (small translation-invariant σ-ideal) $\mathcal{J} \subseteq \mathcal{M} \cap \mathcal{N}$ such that the Borel Conjecture with respect to $\mathcal{J}$ is inconsistent for a non-trivial reason:

PROBLEM 4.13. Find an interesting $\mathcal{J} \subseteq \mathcal{P}(2^\omega)$ such that ZFC proves (in a non-trivial way) the existence of an uncountable $\mathcal{J}$-shiftable set!

5. Very meager sets and a strong form of the dual Borel Conjecture

In this section, we are going to discuss and motivate the notion of very meager (a weaker variant of strongly meager; see Definition 5.4), as well as the respective stronger variant of the dual Borel Conjecture, which holds both in the Cohen model (see Theorem 5.11) and in our model for BC+dBC presented in Section 4 (see Theorem 5.12; the proof of this theorem is the main content of [**SW16**]).

5.1. Strongly meager sets do not (always) form an ideal. Recall that a set $X \subseteq 2^\omega$ is strongly meager ($X \in \mathcal{SM}$) if $X + H \neq 2^\omega$ for each measure zero set H; note (see the discussion after Definition 1.6) that this is the case if and only if for each measure zero set H there is a real $t \in 2^\omega$ such that $X \subseteq t + (2^\omega \setminus H)$.

Consistently, the collection of strongly meager sets does not form an ideal (let alone a σ-ideal):

THEOREM 5.1. *Assume CH.*[129] *Then there are two strongly meager sets whose union is not strongly meager.*

PROOF. For the involved proof, see [**BS01**]. □

On the other hand, it is consistent that the collection of strongly meager sets forms a σ-ideal: this follows immediately from the consistency of the dual Borel Conjecture (see Theorem 3.10).

Moreover, this is also possible without dBC:

THEOREM 5.2. *Let $\lambda \geq \aleph_1$ be a regular cardinal. Then it is consistent that a set is strongly meager if and only if it is of size strictly less than λ.*

PROOF. For the proof, see [**BS02**, Theorem 2.3]. □

The following, however, might be open:

QUESTION 5.3. Is it consistent that the strongly meager sets are an ideal without being a σ-ideal?

[128] The intersection ideal $\mathcal{M} \cap \mathcal{N}$ and its interaction with translations has been extensively studied by Tomasz Weiss in, e.g., [**Wei17**] and [**Wei18**].

[129] In fact, the conclusion is also consistent with the negation of CH.

5.2. Very meager sets. The following weaker variant of the notion of strongly meager can be found, for instance, in Kysiak's master thesis (see [**Kys00**, Definicja 5.4]; in Polish):

DEFINITION 5.4. A set $X \subseteq 2^\omega$ is *very meager* ($X \in \mathcal{VM}$) if for each measure zero set $H \subseteq 2^\omega$ there is a countable set $T \subseteq 2^\omega$ such that $X \subseteq T + (2^\omega \setminus H)$.

Note that demanding a *singleton* $\{t\}$ instead of the *countable set* T in the above definition exactly yields the definition of strongly meager; therefore, $\mathcal{SM} \subseteq \mathcal{VM}$, i.e., every strongly meager set is very meager.

In contrast to the strongly meager sets, the very meager sets are closed under countable unions (which is very easy to show):

PROPOSITION 5.5. *The collection $\mathcal{VM}$ of very meager sets is a σ-ideal.*

In particular, the σ-ideal $\sigma\langle\mathcal{SM}\rangle$ generated by the strongly meager sets is contained in $\mathcal{VM}$.

However, it seems to be unknown whether the above two sets are consistently different (see also Question 5.10 and Theorem 5.11):

QUESTION 5.6. Do the σ-ideal generated by the strongly meager sets and the collection of very meager sets coincide in general?

The following fact will give an alternative formulation of being very meager:

LEMMA 5.7. *Let $X \subseteq 2^\omega$, and let $H \subseteq 2^\omega$ be a measure zero set. Then the following are equivalent:*

(1) *There exists a countable set $T \subseteq 2^\omega$ such that $X \subseteq T + (2^\omega \setminus H)$,*
(2) *the set X can be partitioned into countably many sets $(X_n)_{n\in\omega}$ such that $X_n + H \neq 2^\omega$ for each $n \in \omega$.*

PROOF. An easy computation shows the equivalence of the two assertions (even for arbitrary $H \subseteq 2^\omega$). □

The above lemma yields an equivalent representation of the collection $\mathcal{VM}$ of very meager sets (see below); together with the following explicit description of the collection $\mathcal{SM}$ of strongly meager sets and the σ-ideal $\sigma\langle\mathcal{SM}\rangle$ generated by the strongly meager sets, it allows us to give the following neat comparison of $\mathcal{SM}$, $\sigma\langle\mathcal{SM}\rangle$, and $\mathcal{VM}$ (in which "$\exists \bigcup_n X_n = X$" is supposed to abbreviate "the set X can be partitioned into countably many sets $(X_n)_{n\in\omega}$"):

$$\begin{array}{llllll}
X \in \mathcal{SM} & \Leftrightarrow & & \forall H \in \mathcal{N} & & X + H \neq 2^\omega \\
X \in \sigma\langle\mathcal{SM}\rangle & \Leftrightarrow & \exists \bigcup_n X_n = X & \forall H \in \mathcal{N} & \forall n \in \omega & X_n + H \neq 2^\omega \\
X \in \mathcal{VM} & \Leftrightarrow & \forall H \in \mathcal{N} & \exists \bigcup_n X_n = X & \forall n \in \omega & X_n + H \neq 2^\omega
\end{array}$$

From this, it is immediate that $\mathcal{SM} \subseteq \sigma\langle\mathcal{SM}\rangle \subseteq \mathcal{VM}$; in particular, $\mathcal{VM}$ is obtained from $\sigma\langle\mathcal{SM}\rangle$ by switching quantifiers: it is a (potentially) stronger requirement to belong to $\sigma\langle\mathcal{SM}\rangle$ than to $\mathcal{VM}$, since in case of $\sigma\langle\mathcal{SM}\rangle$ there has to be a "uniform" partition of X, in case of $\mathcal{VM}$ the partition may depend on the set $H \in \mathcal{N}$ (recall, however, that it is open whether $\sigma\langle\mathcal{SM}\rangle \subsetneq \mathcal{VM}$ is consistent; see Question 5.6).

REMARK 5.8. In fact, both the notion of strongly meager and the notion of very meager (see Definition 5.4) can be considered to be a "category analogue" of the notion of strong measure zero. This is because the two respective "versions" of strong measure zero coincide: according to the Galvin-Mycielski-Solovay characterization (see Theorem 1.4), X is strong measure zero if and only if for each meager set $M \subseteq 2^\omega$ there is a real $t \in 2^\omega$ such that $X \subseteq t + (2^\omega \setminus M)$; paralleling the definition of very meager, one could define a set X to be *very null* if for each meager set $M \subseteq 2^\omega$ there is a countable set $T \subseteq 2^\omega$ such that $X \subseteq T + (2^\omega \setminus M)$, but it is not too hard to prove that these two properties are equivalent (see [**Kys00**, Stwierdzenie 5.6]; in Polish), i.e., the notions of strong measure zero and very null coincide.

This can actually be derived from a special property of the ideal $\mathcal{M}$ of meager sets called *ω-translatability*[130] (which again yields the fact that the collection of strong measure zero sets forms a σ-ideal – of course, the latter fact is already an immediate consequence of its "elementary definition", i.e., the original Definition 1.1).

The ideal $\mathcal{N}$ of measure zero sets, on the other hand, lacks[131] this property (it is even not *2-translatable*; see [**Bar00**], as well as [**Kys03**] and [**Bal11**]); this makes the (stronger) fact that the collection of strongly meager sets does (consistently) not form an ideal (see Theorem 5.1) possible in the first place.

The notion of very meager does not appear often in the literature; published references for the definition we are aware of include [**KW04**, Definition 2.4], as well as [**BS03**, Definition 4]. A quite similar notion was considered by[132] Scheepers (see [**Sch93**]):

DEFINITION 5.9. A set $X \subseteq 2^\omega$ is *piecewise strongly meager* ($X \in \mathcal{PSM}$) if for each countable sequence $C_0, C_1, C_2, \ldots$ of closed nowhere dense subsets of 2^ω with $\lim_{n\to\infty} \mu(2^\omega \setminus C_n) = 0$, there is a countable sequence $t_0, t_1, t_2, \ldots$ of elements of 2^ω such that $X \subseteq \bigcup_{n\in\omega}(t_n + C_n)$.

It is straightforward to check that the collection of piecewise strongly meager sets forms a σ-ideal containing all strongly meager sets (recall that the same holds for the collection of very meager sets); moreover, it is not hard to show that every piecewise strongly meager set is very meager.

Altogether, we have the following inclusions:

$$\mathcal{SM} \subseteq \sigma\langle\mathcal{SM}\rangle \subseteq \mathcal{PSM} \subseteq \mathcal{VM}.$$

---

[130] We say that a translation-invariant ideal $\mathcal{I}$ is *2-translatable* if for each $Z \in \mathcal{I}$ there exists a $Z' \in \mathcal{I}$ such that

$$\forall x_1, x_2 \in 2^\omega\ \exists x \in 2^\omega \quad (Z + x_1) \cup (Z + x_2) \subseteq Z' + x; \tag{5.1}$$

moreover, $\mathcal{I}$ is *ω-translatable* if the above property holds for ω many reals (instead of the two in (5.1)), i.e., if for each $Z \in \mathcal{I}$ there exists a $Z' \in \mathcal{I}$ such that

$$\forall (x_i)_{i\in\omega} \subseteq 2^\omega\ \exists x \in 2^\omega \quad \bigcup_{i\in\omega}(Z + x_i) \subseteq Z' + x.$$

It is easy to check that the collection of $\mathcal{I}$-shiftable sets (see Definition 1.8) forms an ideal (a σ-ideal, respectively) provided that $\mathcal{I}$ is 2-translatable (ω-translatable, respectively).

Furthermore, if $\mathcal{I}$ is ω-translatable, then the notions of $\mathcal{I}$-shiftable and its respective variant when "countable set" is used in place of "singleton" (as in defining very meager instead of strongly meager; see Definition 5.4 and the discussion afterwards) coincide.

[131] In contrast to the fact that the strongly meager sets do not form an ideal (which only holds under additional assumptions such as CH), the fact that $\mathcal{N}$ is not 2-translatable is provable in ZFC.

[132] In [**Sch93**], the definition is given for subsets of the real line $\mathbb{R}$; we give the analogous definition for 2^ω.

5.3. Strong form of the dual Borel Conjecture. Note that – even if the above Question 5.6 has a negative answer – it might be the case that the dual Borel Conjecture implies that all very meager sets are countable:

QUESTION 5.10. Is the existence of an uncountable very meager set consistent with the dual Borel Conjecture?

In fact, it seems that all known models of the dual Borel Conjecture indeed satisfy the respective strong form of the dual Borel Conjecture; in particular, this holds true for the Cohen model.

5.4. No uncountable very meager sets in the Cohen model. The following theorem strengthens Carlson's result [**Car93**] that the dual Borel Conjecture holds in the Cohen model:

THEOREM 5.11. *Let $\mathbb{P}_{\omega_2}$ be the finite support iteration*[133] *of Cohen forcing of length ω_2.*

Then in a generic extension by $\mathbb{P}_{\omega_2}$, there are no uncountable very meager sets.

PROOF. This is a straightforward modification of the proof that dBC holds in the Cohen model (see Theorem 3.10, and Sections 3.3–3.6 devoted to its proof).

(For more details, have a look at [**Kys00**, Twierdzenie 5.17].) □

This shows that $\mathcal{SM} = \mathcal{VM}$ is independent from ZFC: it holds in the Cohen model, but fails under CH (recall Theorem 5.1 and Proposition 5.5).

5.5. No uncountable very meager sets in the model for BC+dBC. It has been shown by Shelah and the author of this paper (see [**SW16**]) that the strong form of the dual Borel Conjecture also holds[134] in the model for BC+dBC from [**GKSW14**] (compare with Theorem 4.1 and the overview of its proof given in Section 4):

THEOREM 5.12. *In the model for BC+dBC from* [**GKSW14**]*, there are no uncountable very meager sets.*

Let us point out that the result looks like an "asymmetric strengthening" of the joint consistency of BC and dBC, only being concerned with strengthening dBC. However, the respective "stronger version of BC" is equivalent to BC (and therefore holds in the model from [**GKSW14**] anyway), due to the fact that the two respective "versions" of strong measure zero coincide: being strong measure zero is actually the same as being "very null" (see Remark 5.8).

PROOF OF THEOREM 5.12. The proof is a slight variant of the proof of dBC in [**GKSW14**] and can be found in [**SW16**]; the main content of [**SW16**] is how to adapt the proof of dBC in [**GKSW14**] to show the stronger version of dBC.

In fact, [**SW16**, Section 2.A] is concerned with strengthening the fact that "Janus forcing kills all strongly meager sets" (see [**GKSW14**, Lemma 2.9]) to

[133] As for the fact that dBC holds in the Cohen model (see Theorem 3.10), this stronger version of dBC also holds for arbitrarily long finite support iterations (compare with Theorem 3.26), or, equivalently, in any model obtained by adding many Cohen reals "simultaneously".

[134] The author of this paper thanks Marcin Kysiak for asking him this question (at the Winterschool 2011 in Hejnice, Czech Republic). Kysiak wondered whether the model from [**GKSW14**] could be the "first model" in which the σ-ideal $\sigma\langle\mathcal{SM}\rangle$ generated by the strongly meager sets is strictly smaller than the collection $\mathcal{VM}$ of very meager sets. Theorem 5.12 shows that it is not.

"Janus forcing kills all very meager sets" (see [**SW16**, Lemma 2.3]); for this, the formulation of being very meager as in Lemma 5.7(2) is used.

Finally, [**SW16**, Section 2.B] is concerned with presenting the necessary modifications in [**GKSW14**, Section 5]: not only dBC holds (see the original [**GKSW14**, Lemma 5.2]), but there are no uncountable very meager sets in the final model (see [**SW16**, Lemma 2.5]); for this, the formulation of being very meager as in Lemma 5.7(1) is used. □

6. Borel Conjecture (or dual Borel Conjecture) together with a projective well-order of the reals

In this section, we present unpublished joint work with Sy D. Friedman (the material is essentially taken from the author's PhD thesis; see [**Woh13**, Chapter 4]): we show that the existence of a *projective well-order* of the real numbers is consistent with the Borel Conjecture (and the dual Borel Conjecture, respectively); actually, the respective well-orders are Δ^1_3 *definable*.

To prove our results, we describe how to apply the techniques in [**FF10**] and [**FFZ11**]; the presentation is by far not self-contained, but heavily relies on these two papers.

In Section 6.2, we show that the existence of a Δ^1_3 definable well-order of the reals is *consistent with BC*, using the machinery of [**FF10**].

In Section 6.3, we show that the existence of a Δ^1_3 definable well-order of the reals is *consistent with dBC*, using the machinery of [**FFZ11**].

6.1. Historical context. Quoting from the introduction of [**FF10**] (which is – according to the authors – the first work on projective well-orders and cardinal characteristics of the continuum), we give some historical information:

> If $V = L$ then there exists a Σ^1_2 well-ordering of the reals. Furthermore, by Mansfield's Theorem (see [**Jec03**, Theorem 25.39]) the existence of a Σ^1_2 well-ordering of the reals implies that every real is constructible. Using a finite support iteration of ccc posets, L. Harrington showed that the existence of a $\boldsymbol{\Delta^1_3}$ wellordering of the reals is consistent with the continuum being arbitrarily large (see [**Har77**, Theorem A]). S. D. Friedman showed that Martin's Axiom (and not CH) is consistent with the existence of a Δ^1_3 definable wellordering of the reals (see [**Fri00**] and see [**Har77**] for the corresponding boldface result).

Martin's Axiom makes all cardinal characteristics in Cichoń's diagram large (see Figure 23), and is (therefore) incompatible with BC as well as dBC (see Theorem 4.8). So the results mentioned are clearly not sufficient to get a model of "BC (or dBC) + a projective well-order of the reals".

6.2. BC in the presence of a projective well-order of the reals. In this section, we describe how to combine Laver's proof of Con(BC) with the methods from the paper "Cardinal characteristics and projective wellorders" by Vera Fischer and Sy D. Friedman (see [**FF10**]) in order to obtain a model of ZFC satisfying "BC + there exists a projective well-order of the reals":

THEOREM 6.1. *The existence of a Δ^1_3 definable well-order of the reals is consistent with the Borel Conjecture (and $2^{\aleph_0} = \aleph_2$).*

By the Judah-Shelah-Woodin theorem (see Theorem 2.34), there is a model of the Borel Conjecture with large continuum (i.e., $2^{\aleph_0} > \aleph_2$). The proof demonstrates that BC remains valid when *many random reals are added (side-by-side)* to Laver's model of BC.

Since the known methods for getting projective well-orders of the reals seem to be based on *iteration* (and the method of adding random reals side-by-side is not an iteration), it is unclear to me whether it is possible to obtain such a model with a projective well-order:

QUESTION 6.2. Is the existence of a Δ^1_3 definable[135] well-order of the reals consistent with the Borel Conjecture and $2^{\aleph_0} \geq \aleph_3$?

PROOF OF THEOREM 6.1. For the key points of Laver's proof of Con(BC), see Theorem 2.6 and the subsequent discussion.

To obtain a model of BC, it is sufficient to proceed as follows:

(1) start with a *model of CH*,
(2) perform a *countable support iteration* (of length ω_2),
(3) make sure that (at least) **cofinally** many of the iterands "*kill old (uncountable) strong measure zero sets*" (e.g., Laver forcing does),
(4) *avoid resurrection* of "sets that have been killed"; this is guaranteed provided the tail of the forcing iteration has the *Laver property*, which can be enforced as follows:
    (a) make sure that **all** the iterands have the *Laver property*,
    (b) use some type of forcing iteration that *preserves the Laver property* (e.g., a countable support iteration of proper forcings),
(5) make sure that ω_1 *is preserved*,
(6) make sure that ω_2 *is preserved.*

So let us check that the above can be arranged within the framework of [**FF10**] (which leads to a Δ^1_3 definable well-order of the reals).

The (template for the) forcing iteration used in [**FF10**] is defined as follows: according to [**FF10**, Section 5], $\mathbb{P}_{\omega_2}$ is a countable support iteration, with iterands $\dot{\mathbb{Q}}_\alpha = \dot{\mathbb{Q}}^0_\alpha * \dot{\mathbb{Q}}^1_\alpha$, where

I. $\dot{\mathbb{Q}}^0_\alpha$ is a $\mathbb{P}_\alpha$-name for an **arbitrary** proper forcing notion (of cardinality at most $\aleph_1$),
II. $\dot{\mathbb{Q}}^1_\alpha$ is either (a name for) the trivial poset, or $\dot{\mathbb{Q}}^1_\alpha = \dot{\mathbb{K}}^0_\alpha * \dot{\mathbb{K}}^1_\alpha * \dot{\mathbb{K}}^2_\alpha$, where
    i. $\dot{\mathbb{K}}^0_\alpha$ is composed of "club shooting" forcings of the form $\mathcal{Q}(S)$ (with $S \subseteq \omega_1$ stationary, co-stationary),
    ii. $\dot{\mathbb{K}}^1_\alpha$ is (a name for) a "localization" forcing $\mathcal{L}(\phi_\alpha)$",
    iii. $\dot{\mathbb{K}}^2_\alpha$ is (a name for) a "coding" forcing $\mathcal{C}(Y_\alpha)$ (whose conditions are perfect trees, i.e., $\mathcal{C}(Y_\alpha)$ is similar to Sacks forcing).

We now perform an iteration according to this template, and let $\dot{\mathbb{Q}}^0_\alpha$ to be (a name for) Laver forcing for all α (note that this is allowed by the template since Laver forcing is proper; see (I)).

In this way, we obtain a model with a Δ^1_3 definable well-order of the reals. In order to confirm that BC holds in this final model, we have to check that items (1)–(6) above are satisfied:

(1) our ground model is L, hence CH is satisfied;

[135]...or any other complexity in the projective hierarchy

(2) by definition, $\mathbb{P}_{\omega_2}$ is a countable support iteration;
(3) *cofinally many iterands kill* old (uncountable) strong measure zero sets, since we have chosen all our $\dot{\mathbb{Q}}^0_\alpha$'s to be Laver forcing;
(4) the tails of the iteration have the *Laver property*:
    (a) all involved forcings have the *Laver property* (and are proper, or at least *S-proper*, for a fixed stationary set $S \subseteq \omega_1$ that belongs to the ground model):
        I. $\dot{\mathbb{Q}}^0_\alpha$ is always Laver forcing (which has the *Laver property*, and is *proper*);
        II. $\dot{\mathbb{Q}}^1_\alpha$ is either the trivial poset (hence has the *Laver property*, and is *proper*), or we have $\dot{\mathbb{Q}}^1_\alpha = \dot{\mathbb{K}}^0_\alpha * \dot{\mathbb{K}}^1_\alpha * \dot{\mathbb{K}}^2_\alpha$, where
            (i) $\dot{\mathbb{K}}^0_\alpha$ doesn't add new reals by [**FF10**, Lemma 9], hence vacuously has the *Laver property*;
            moreover, it is *S-proper* (see [**FF10**, Section 4]);
            (ii) $\dot{\mathbb{K}}^1_\alpha$ doesn't add new reals by [**FF10**, Lemma 4], hence again vacuously has the *Laver property*;
            moreover, it is *proper* by Lemma [**FF10**, Lemma 3];
            (iii) also $\dot{\mathbb{K}}^2_\alpha$ has the *Laver property*, but [**FF10**, Lemma 8] is not quite sufficient for that:
                it only shows that $\dot{\mathbb{K}}^2_\alpha$ is ω^ω-bounding; but actually, it is implicit in the proof of [**FF10**, Lemma 8] that $\dot{\mathbb{K}}^2_\alpha$ has the Laver property (and hence the "Sacks property"); just directly use the finite sets d_k instead of their maxima; they have size 2^k, so the limit of the "fusion sequence" indeed forces that $\dot{f}$ is not just bounded but contained in a 2^k-slalom of the ground model (yielding the Laver property);
            moreover, $\dot{\mathbb{K}}^2_\alpha$ is *proper* by Lemma [**FF10**, Lemma 7];
    (b) the *Laver property is preserved* under countable support iterations of *(S-)proper* forcings (analogous to [**FF10**, Lemma 18]);
(5) ω_1 is preserved since the iteration is S-proper (by [**FF10**, Lemma 18]);
(6) ω_2 is preserved since the iteration is $\aleph_2$-cc. □

6.3. dBC in the presence of a projective well-order of the reals. In this section, we describe how to combine Carlson's proof of Con(dBC) with the methods from the paper "Projective wellorders and mad families with large continuum" by Vera Fischer, Sy D. Friedman and Lyubomyr Zdomskyy (see [**FFZ11**]) in order to obtain a model of ZFC satisfying "dBC + there exists a projective well-order of the reals":

THEOREM 6.3. *The existence of a Δ^1_3 definable well-order of the reals is consistent with the dual Borel Conjecture (and both $2^{\aleph_0} = \aleph_2$ and $2^{\aleph_0} = \aleph_3$).*

Let us remark that it is possible to get such a model with even larger[136] continuum: it just makes the preparatory part of the forcing machinery (see the below proof of Theorem 6.3) more complicated (apart from that, the proof is essentially the same).

[136]In general, it is no problem to get models of dBC with large continuum: we can actually use a finite support iteration (of, e.g., Cohen forcing) longer than ω_2 (just recall Theorem 3.26 and note that δ can be arbitrarily large there).

PROOF OF THEOREM 6.3. We describe the version with $2^{\aleph_0} = \aleph_3$ (closely following the framework in [**FFZ11**], which is concerned with large continuum as well); the case $2^{\aleph_0} = \aleph_2$ is supposed to be similar (alternatively, one can easily derive it from the version with $2^{\aleph_0} = \aleph_3$, as demonstrated in Remark 6.4).

For the key points of Carlson's proof of Con(dBC), see Theorem 3.10 and the subsequent discussion; note that it is no problem to perform a finite support iteration of length more than ω_2 in Carlson's argument.

To obtain a model of dBC, it is sufficient to proceed as follows:

(1) start with a *model of CH*,
(2) perform a *finite support iteration* (of length $\geq \omega_2$),
(3) make sure that (at least) **cofinally** many of the iterands "*kill old (uncountable) strongly meager sets*" (e.g., Cohen forcing does),
(4) *avoid resurrection* of "sets that have been killed"; this is guaranteed provided the tail of the forcing iteration is *precaliber* $\aleph_1$, which can be enforced as follows:
  (a) make sure that **all** the iterands have *precaliber* $\aleph_1$,
  (b) use some type of forcing iteration that *preserves precaliber* $\aleph_1$ (e.g., finite support iteration).

So let us check that the above can be arranged within the framework of [**FFZ11**] (which leads to a Δ^1_3 definable well-order of the reals).

The (involved) forcing machinery used in [**FFZ11**] consists of two parts (see [**FFZ11**, Section 2, Step 3]):

I. (a preparatory part) the poset $\mathbb{P}_0 := \mathbb{P}^0 * \mathbb{P}^1 * \mathbb{P}^2$ (it is ω-distributive according to [**FFZ11**, Lemma 1]),
II. the finite support iteration $\langle \mathbb{P}_\alpha, \dot{\mathbb{Q}}_\gamma \colon \alpha \leq \omega_3, \gamma < \omega_3 \rangle$, starting with the above $\mathbb{P}_0$; and, for each $\alpha < \omega_3$, $\dot{\mathbb{Q}}_\alpha$ is a (name for a) σ-centered poset.

In this way, we obtain a model with a Δ^1_3 definable well-order of the reals, as shown in [**FFZ11**].

We claim that it is easy to arrange that dBC holds in this final model (alternatively, we can even argue that dBC holds true "automatically"); so let us go through items (1)–(4) above to make sure that dBC holds:

(1) the *model of CH* that is our "ground model for Carlson's proof" is not the actual ground model $V = L$, but rather the forcing extension by the forcing $\mathbb{P}_0$ from (I): note that it is still a model of CH since the forcing $\mathbb{P}_0$ is ω-distributive (hence adds no reals);
(2) after $\mathbb{P}_0$, the iteration $\langle \mathbb{P}_\alpha, \dot{\mathbb{Q}}_\gamma \colon \alpha \leq \omega_3, \gamma < \omega_3 \rangle$ is a finite support iteration of length ω_3 (see (II));
(3) it is no problem to explicitly add Cohen reals in between (along the way up to ω_3); then *cofinally many* of the iterands perform the task of *killing* old (uncountable) strongly meager sets;

    alternatively, one can argue as follows: as described in [**FFZ11**, Section 2, Step 3, Case 2], cofinally many of the $\dot{\mathbb{Q}}_\alpha$'s will be forcings similar to Hechler forcing, which add Cohen reals anyway (recall that the "Hechler real modulo 2" is a Cohen real); so we actually do not have to change the machinery in [**FFZ11**] at all to obtain a model of dBC;
(4) the tails of the iteration are *precaliber* $\aleph_1$:

(a) all the iterands $\dot{\mathbb{Q}}_\alpha$ are σ-centered (see (II)), hence in particular *precaliber* $\aleph_1$;

(b) *precaliber* $\aleph_1$ *is preserved* under finite support iterations. □

REMARK 6.4. As mentioned above, the respective model for which $2^{\aleph_0} = \aleph_2$ can be obtained in a similar way.

However, we can also derive it directly from the $2^{\aleph_0} = \aleph_3$ case: just collapse the continuum (which is $\aleph_3$) to $\aleph_2$ by a $<\aleph_2$-closed forcing; neither new reals, nor new sets of reals of size $\aleph_1$ are added by the collapse; consequently, dBC remains true, and the "definition of the well-order" is not influenced by the collapse either.

6.4. BC+dBC together with a projective well-order of the reals? With Theorem 4.1 in mind, it is natural to ask whether the respective techniques can be combined:

QUESTION 6.5. Is the existence of a projective well-order of the reals consistent with BC+dBC?[137]

7. Marczewski Borel Conjecture (MBC)

This section is concerned with s_0-shiftable sets and the Marczewski Borel Conjecture. First, we recall the Marczewski ideal s_0 (associated to Sacks forcing) and discuss some of its basic properties, as well as s_0-shiftable sets and MBC (see Section 7.1). Subsequently, most of the section is dedicated to a sketch of the proof of the ZFC theorem (from joint work [**BW**] with Jörg Brendle) that there are no s_0-shiftable sets of reals of size continuum (see Section 7.2). Finally, we discuss the status of MBC in several models of ZFC (see Section 7.3): it holds true under CH (which immediately follows from the above ZFC theorem) as well as in the Cohen model, but fails, e.g., in the Sacks model.

7.1. The Marczewski ideal s_0 and MBC. We will discuss the well-known σ-ideal s_0 of Marczewski null sets, which is associated to Sacks forcing (see Section 7.1.1), and sketch the proofs of some basic properties of s_0 (see Section 7.1.2). Moreover, we will talk about s_0-shiftable sets and the Marczewski Borel Conjecture (see Section 7.1.3).

7.1.1. *The Marczewski ideal s_0 and Sacks forcing.* A (non-empty) set $P \subseteq 2^\omega$ is *perfect* if it is closed and has no isolated points.

The Marczewski ideal s_0 is the following classical ideal on the real numbers:

DEFINITION 7.1. A set $Y \subseteq 2^\omega$ is *Marczewski null* ($Y \in s_0$) if for every perfect set $P \subseteq 2^\omega$ there is a perfect set $Q \subseteq P$ with $Q \cap Y = \emptyset$.

Recall the notion of *Sacks forcing* $(\mathbb{S}, \leq)$, the set of all perfect subtrees of $2^{<\omega}$ with the partial order of inclusion (i.e., $q \leq p$ if $q \subseteq p$). Two elements $p_0, p_1 \in \mathbb{S}$ are *incompatible* if there is no $r \in \mathbb{S}$ with $r \leq p_0$ and $r \leq p_1$ (otherwise they are *compatible*). An *antichain* is a collection of pairwise incompatible elements. Note that an antichain is *maximal* if and only if every element of $\mathbb{S}$ is compatible with some element of the antichain. For $p \in \mathbb{S}$, let

$$[p] := \{x \in 2^\omega : x \restriction n \in p \text{ for each } n \in \omega\}$$

[137]...or, thinking of Theorem 5.12, even with BC and the strengthening of dBC discussed in Section 5?

denote the *body* of p, i.e., the set of branches through the tree p. Note that the set of perfect sets in 2^ω is exactly the set of bodies of perfect trees in $2^{<\omega}$ (i.e., the set $\{[p] : p \in \mathbb{S}\}$), and that $q \le p$ if and only if $[q] \subseteq [p]$; moreover, two conditions $p_0, p_1 \in \mathbb{S}$ are incompatible if and only if $|[p_0] \cap [p_1]| \le \aleph_0$, i.e., if the intersection of the corresponding perfect sets is small (on the other hand, $|[p_0] \cap [p_1]| = \mathfrak{c}$ for compatible conditions p_0 and p_1). Also, for $p \in \mathbb{S}$ and $t \in 2^\omega$, let

$$p + t := \{\sigma + t \restriction |\sigma| : \sigma \in p\}$$

denote the tree p translated by the real t, and note that $[p + t] = [p] + t$.

Observe that

$$Y \in s_0 \iff \forall p \in \mathbb{S} \quad \exists q \le p \quad [q] \cap Y = \emptyset;$$

we will use this "forcing notation" from now on.

7.1.2. *Properties of the Marczewski ideal* s_0. It is well-known and quite easy to prove that s_0 is a translation-invariant σ-ideal: the σ-closure can be shown by constructing a fusion sequence of perfect sets (as in the proof that Sacks forcing satisfies Axiom A, the Sacks property, etc.).

Clearly, no perfect set (and hence no uncountable Borel set) is in s_0. On the other hand, each set of size less than the continuum belongs to s_0 (i.e., $\mathrm{non}(s_0) = \mathfrak{c}$):

PROPOSITION 7.2. *Let* $Y \subseteq 2^\omega$ *with* $|Y| < \mathfrak{c}$. *Then* $Y \in s_0$.

PROOF. Split the given perfect set into "perfectly many" (hence continuum many) disjoint perfect sets: one of them is going to miss all elements of Y. □

Note that the following is a straightforward consequence of Proposition 7.2:

$$Y \in s_0 \iff \forall p \in \mathbb{S} \quad \exists q \le p \quad |[q] \cap Y| < \mathfrak{c}; \tag{7.1}$$

in other words, replacing "disjoint" by "small intersection" does not change the notion of Marczewski null.

Since there are no (uncountable) Borel sets in s_0 (in contrast to the ideals $\mathcal{M}$ and $\mathcal{N}$), it even takes some effort to show that s_0 contains sets of size continuum; the proof[138] of this fact is the blueprint for (the proofs of) Lemma 7.8 and Lemma 7.12 (as well as Theorem 7.4) below:

THEOREM 7.3. *There is a set* $Y \in s_0$ *with* $|Y| = \mathfrak{c}$.

PROOF. Fix a maximal antichain $\{q_\alpha : \alpha < \mathfrak{c}\} \subseteq \mathbb{S}$ of size continuum. By induction on $\alpha < \mathfrak{c}$, pick

$$y_\alpha \in [q_\alpha] \setminus \bigcup_{\beta<\alpha} [q_\beta]; \tag{7.2}$$

this is possible since $|[q_\alpha] \cap [q_\beta]| \le \aleph_0$ for any $\beta < \alpha < \mathfrak{c}$ (due to the fact that $\{q_\alpha : \alpha < \mathfrak{c}\}$ is an antichain). Define

$$Y := \{y_\alpha : \alpha < \mathfrak{c}\}.$$

By construction, $|Y| = \mathfrak{c}$; moreover, it is easy to argue that $Y \in s_0$: fix $p \in \mathbb{S}$; by (7.1), it is enough to find $q \le p$ with $|[q] \cap Y| < \mathfrak{c}$; by the maximality of the antichain $\{q_\alpha : \alpha < \mathfrak{c}\}$, we can fix $\alpha < \mathfrak{c}$ such that p is compatible with q_α, i.e., we can pick $q \le p$ with $q \le q_\alpha$; for each $\gamma > \alpha$, we have $y_\gamma \notin [q_\alpha]$ (see (7.2)), hence $y_\gamma \notin [q]$; therefore $[q] \cap Y \subseteq \{y_\gamma : \gamma \le \alpha\}$, hence $|[q] \cap Y| < \mathfrak{c}$, as desired. □

---

[138]The proof of Theorem 7.3 presented here is essentially due to Thilo Weinert (see also [**Woh13**, Lemma 6.3]); for a somewhat different proof, see [**Mil84**, Theorem 5.10].

Recall that both the ideal $\mathcal{M}$ of meager sets and the ideal $\mathcal{N}$ of measure zero sets have a basis consisting of Borel sets, hence $\mathrm{cof}(\mathcal{M}) \leq \mathfrak{c}$ and $\mathrm{cof}(\mathcal{N}) \leq \mathfrak{c}$; it is also consistent that these cofinalities are strictly smaller than $\mathfrak{c}$.

On the other hand, we have $\mathfrak{c} = \mathrm{non}(s_0) \leq \mathrm{cof}(s_0)$ in ZFC (see Proposition 7.2 as well as the discussion in Section 1.6.1 and Figure 1). Obviously, s_0 has no Borel basis, so there is no good reason why it should have a basis of size continuum (i.e., why $\mathrm{cof}(s_0)$ should be $\mathfrak{c}$); indeed, a slightly more complicated version of the proof of Theorem 7.3 yields the following:

THEOREM 7.4. *ZFC proves that* $\mathrm{cof}(s_0) > \mathfrak{c}$.

PROOF (SKETCH). The proof is analogous to the proof of Theorem 7.3, with the following modification: to show that $\mathrm{cof}(s_0) > \mathfrak{c}$, assume towards a contradiction that $\{X_\alpha : \alpha < \mathfrak{c}\} \subseteq s_0$ is a basis for s_0. Instead of (7.2), we pick

$$y_\alpha \in [q'_\alpha] \setminus \bigcup_{\beta<\alpha} [q_\beta],$$

where $q'_\alpha \leq q_\alpha$ is chosen such that $[q'_\alpha] \cap X_\alpha = \emptyset$ (which is possible since $X_\alpha \in s_0$). Finally, $Y \in s_0$, but for each $\alpha < \mathfrak{c}$, we have $Y \not\subseteq X_\alpha$, contradicting our assumption that $\{X_\alpha : \alpha < \mathfrak{c}\}$ is a basis. □

REMARK 7.5. According to [**JMS92**, Introduction, right before Theorem 1.3], $\mathrm{cof}(s_0) > \mathfrak{c}$ was noted by Fremlin; the slightly stronger result that $\mathrm{cf}(\mathrm{cof}(s_0)) > \mathfrak{c}$ holds true in ZFC is shown in [**JMS92**, Theorem 1.3].

Actually, there are many other *Marczewski-like ideals* such as the Laver ideal ℓ_0 (which is associated to Laver forcing as s_0 is associated to Sacks forcing) or the Miller ideal m_0 (associated to Miller forcing); see Brendle's paper [**Bre95**] for more on this.

For a quite general investigation of the cofinalities of these Marczewski-like ideals, see [**BKW17**]: there you can find alternative proofs for Theorem 7.4 (and $\mathrm{cf}(\mathrm{cof}(s_0)) > \mathfrak{c}$, respectively); moreover, $\mathrm{cf}(\mathrm{cof}(\ell_0)) > \mathfrak{c}$ and $\mathrm{cf}(\mathrm{cof}(m_0)) > \mathfrak{c}$ are established there in ZFC.

Also see [**Woh13**, Chapter 7] for some tentative investigations about Borel Conjectures with respect to these Marczewski-like ideals.

7.1.3. *s_0-shiftable sets and MBC.* We now consider the notion of $\mathcal{I}$-shiftable (see Definition 1.8) for the instance of $\mathcal{I}$ being the Marczewski ideal s_0:

A set $X \subseteq 2^\omega$ is s_0-shiftable if $X + Y \neq 2^\omega$ for each $Y \in s_0$.

Since s_0 is a (proper) translation-invariant σ-ideal, each countable set is s_0-shiftable (see also the discussion in Section 1.5). As always, the question arises whether there are any uncountable s_0-shiftable sets; we give a new name to the respective Borel Conjecture (see Definition 1.9):

DEFINITION 7.6. The *Marczewski Borel Conjecture (MBC)* is the s_0-BC, i.e., the assertion that there is *no uncountable* s_0-shiftable set.

Recall that both $\mathcal{M}$-BC (i.e., the Borel Conjecture) and $\mathcal{N}$-BC (i.e., the dual Borel Conjecture) fail under CH; the argument is, for instance, as follows (compare with Corollary 1.11): both the ideal $\mathcal{M}$ of meager sets and the ideal $\mathcal{N}$ of measure zero sets have a basis consisting of Borel sets, so both $\mathrm{cof}(\mathcal{M}) \leq \mathfrak{c}$ and $\mathrm{cof}(\mathcal{N}) \leq \mathfrak{c}$ hold true (and hence $\mathrm{cov}(\mathcal{M}) = \mathrm{cof}(\mathcal{M}) = \mathfrak{c}$ and $\mathrm{cov}(\mathcal{N}) = \mathrm{cof}(\mathcal{N}) = \mathfrak{c}$ under CH);

therefore Proposition 1.10 can be applied, yielding the existence of uncountable $\mathcal{M}$-shiftable (i.e., strong measure zero) and $\mathcal{N}$-shiftable (i.e., strongly meager) sets.

The situation for the Marczewski ideal is quite different: we have $\mathrm{cof}(s_0) > \mathfrak{c}$ in ZFC (see Theorem 7.4), so $\mathrm{cov}(s_0) < \mathrm{cof}(s_0)$ holds for sure, and we cannot apply Proposition 1.10, i.e., there is no obvious way to construct an uncountable s_0-shiftable set under CH.

Indeed, the opposite is true: ZFC proves (see Theorem 7.7 below) that there is no s_0-shiftable set of size continuum (in other words, *weak MBC* holds in ZFC; for information about weak BC and weak dBC, see Section 2.7.6 and Section 3.7.7, as well as Sections 4.12.4 and 4.12.5); in particular, MBC follows from CH (see Corollary 7.18), which in turn establishes the consistency of the Marczewski Borel Conjecture.[139]

7.2. No s_0-shiftable sets of size continuum. One of the main results of the paper [**BW**] is the ZFC theorem[140] that there are no s_0-shiftable sets of reals of size continuum (which trivially implies MBC under CH):

THEOREM 7.7 (Brendle-Wohofsky). *Let $X \subseteq 2^\omega$ with $|X| = \mathfrak{c}$. Then X is* not *s_0-shiftable (i.e., there is a $Y \in s_0$ such that $X + Y = 2^\omega$).*

In the following, we are going to give a rough sketch of the proof; for the details, we refer the reader to the original paper [**BW**]. (For the generalization of Theorem 7.7 to arbitrary Polish groups, see Section 7.2.7.)

Part of the author's PhD thesis (see [**Woh13**, Chapter 6]) deals with exploring s_0-shiftable sets (mainly under CH); several concepts and results presented here can already be found there; in particular, Lemma 7.8 is reminiscent of the notion of *Sacks dense ideal* (see [**Woh13**, Definition 6.9 and Lemma 6.10]).

7.2.1. *Reducing the problem to finding a certain dense set in Sacks forcing.* We first reduce the problem of proving "X is not s_0-shiftable" to finding a dense and translation-invariant set $D \subseteq \mathbb{S}$ with the property that (the bodies of) fewer than continuum many trees from D do not cover X:

LEMMA 7.8. *Let $X \subseteq 2^\omega$, and let $D \subseteq \mathbb{S}$ be a collection of Sacks trees with the following properties:*

(1) *for each $p \in \mathbb{S}$, there is a $q \leq p$ with $q \in D$ ("density"),*
(2) *for each $p \in D$ and $t \in 2^\omega$, we have $p + t \in D$ ("translation-invariance"),*
(3) *if $\{p_\alpha : \alpha < \mu\} \subseteq D$ with $\mu < \mathfrak{c}$, then $X \not\subseteq \bigcup_{\alpha<\mu}[p_\alpha]$.*

Then there is a $Y \in s_0$ such that $X + Y = 2^\omega$.

PROOF (SKETCH). The proof is an enhanced version of the proof of Theorem 7.3.

Let $2^\omega = \{z_\alpha : \alpha < \mathfrak{c}\}$ be an enumeration of the reals, and fix a maximal antichain $\{q_\alpha : \alpha < \mathfrak{c}\}$ *inside* D. By induction on $\alpha < \mathfrak{c}$, pick

$$x_\alpha \in X \setminus \bigcup_{\beta<\alpha} (z_\alpha + [q_\beta]);$$

this is possible due to (2) and (3). For each $\alpha < \mathfrak{c}$, let $y_\alpha := x_\alpha + z_\alpha$, and define

$$Y := \{y_\alpha : \alpha < \mathfrak{c}\}.$$

[139] I thank Thilo Weinert for suggesting to consider the Marczewski Borel Conjecture and the question whether it is consistent.

[140] The proof of the theorem has been completed in 2015.

It is straightforward to check that $X+Y=2^\omega$, and (using (7.1), as well as (1) and the maximality of $\{q_\alpha : \alpha<\mathfrak{c}\}$ within D) that $Y\in s_0$.

For more details, see [**BW**, Lemma 8]. □

7.2.2. *Transitive versions of being Marczewski null.* Recall the characterization of being Marczewski null from (7.1):

$$Y\in s_0 \iff \forall p\in\mathbb{S}\quad \exists q\le p\quad |[q]\cap Y|<\mathfrak{c};$$

requiring that not only the body of the tree q itself, but also all its translates are "almost disjoint" from the set in question, yields "transitive" versions of the notion of Marczewski null:

Definition 7.9. A set $Y\subseteq 2^\omega$ is *$<\mu$-transitively Marczewski null* if

$$\forall p\in\mathbb{S}\quad \exists q\le p\quad \forall t\in 2^\omega\quad |([q]+t)\cap Y|<\mu.$$

In (7.1), the cardinal $\mathfrak{c}$ can be replaced by any smaller cardinal without changing the notion; in Definition 7.9, however, the situation is not so clear:

Question 7.10. To which extent does "being $<\mu$-transitively Marczewski null" depend on μ? For which[141] μ is it strictly stronger than Marczewski null?

In case the continuum is regular (i.e., $\mathrm{cf}(\mathfrak{c})=\mathfrak{c}$), the following key lemma (for a sketch of the proof, see Section 7.2.5 below) is sufficient to finish the proof:

Lemma 7.11. *Let $X\subseteq 2^\omega$ with $|X|=\mathfrak{c}$. Then there exists an $X'\subseteq X$ with $|X'|=\mathfrak{c}$ such that X' is $<\mathfrak{c}$-transitively Marczewski null.*

Proof of Theorem 7.7 (for regular $\mathfrak{c}$). Apply Lemma 7.11 to obtain a $<\mathfrak{c}$-transitively Marczewski null set $X'\subseteq X$ with $|X'|=\mathfrak{c}$. This gives rise to a set D asked for by Lemma 7.8: by definition, we can fix a family $(q_p : p\in\mathbb{S})$ such that for each $p\in\mathbb{S}$, we have $q_p\le p$, and

$$\forall t\in 2^\omega\quad |([q_p]+t)\cap X'|<\mathfrak{c}; \tag{7.3}$$

let $D:=\{q_p+t : p\in\mathbb{S},\ t\in 2^\omega\}\subseteq\mathbb{S}$; it is easy to check that D satisfies the three properties in Lemma 7.8 (for (3) with respect to X', regularity of $\mathfrak{c}$ is used). So Lemma 7.8 yields a set $Y\in s_0$ such that $X'+Y=2^\omega$, i.e., X' (and hence X) is not s_0-shiftable. □

Note that the conclusion of Lemma 7.11 is not strong enough to carry out the above proof in case $\mathfrak{c}$ is singular: a continuum sized set might be covered by fewer than continuum many sets each of size strictly below the continuum. In the singular case, we have to use a more complicated alternative lemma instead (see Lemma 7.15).

[141] Requiring disjointness would be definitely too much in the transitive case: the empty set is the only set which is <1-transitively Marczewski null. On the other hand, even finite instances might be worth exploring, i.e., $\le\mu$-transitively Marczewski null for $\mu\in\omega\setminus\{0\}$.

7.2.3. *No s_0-Luzin sets.* Recall that an uncountable set is *Luzin* if its intersection with any meager set is countable. More generally, let us say that a set (of size continuum) is *$\mathcal{I}$-Luzin* if its intersection with any set from $\mathcal{I}$ is strictly smaller than the continuum (e.g., $\mathcal{M}$-Luzin sets are generalized Luzin sets, and $\mathcal{N}$-Luzin sets are generalized Sierpiński sets).

Under certain[142] cardinal characteristics assumptions (in particular, under CH), it is straightforward to construct $\mathcal{M}$-Luzin sets and $\mathcal{N}$-Luzin sets. The following lemma says that there are no s_0-Luzin sets (in[143] ZFC):

LEMMA 7.12. *Let $X \subseteq 2^\omega$ with $|X| = \mathfrak{c}$. Then there exists an $X' \subseteq X$ with $|X'| = \mathfrak{c}$ such that $X' \in s_0$.*

PROOF (SKETCH). As the proof of Lemma 7.8, this is a generalized version of the proof of Theorem 7.3.

The idea is as follows: *in case* $X \in s_0$, let $X' := X$, and we are finished; *otherwise* (i.e., in case $X \notin s_0$), use (7.1) to fix $p \in \mathbb{S}$ satisfying

$$\forall q \leq p \quad |[q] \cap X| = \mathfrak{c}. \tag{7.4}$$

Now perform the usual proof (to construct a Marczewski null set of size continuum) *inside of* $[p] \cap X$, as follows: fix a maximal antichain $\{q_\alpha : \alpha < \mathfrak{c}\}$ *below* p, and let $X' := \{x_\alpha : \alpha < \mathfrak{c}\}$, where

$$x_\alpha \in X \cap [q_\alpha] \setminus \bigcup_{\beta < \alpha} [q_\beta]$$

for each $\alpha < \mathfrak{c}$ (this is possible due to (7.4) and the fact that $\{q_\alpha : \alpha < \mathfrak{c}\}$ is an antichain below p); as usual, it is quite straightforward to check that $X' \in s_0$.

For the details, see [**BW**, Lemma 11]. □

7.2.4. *Skew trees: incompatible to their translates.* The following notion was defined in [**Bla81**]:

DEFINITION 7.13. A Sacks tree $p \in \mathbb{S}$ is *skew* if for every $n \in \omega$, there is at most one splitting node of level n in p.

It is quite easy to check that being skew is both *dense* (i.e., for every $p \in \mathbb{S}$, there is a $q \leq p$ such that q is skew) and *translation-invariant* (i.e., for every skew $q \in \mathbb{S}$ and $t \in 2^\omega$, also $q + t$ is skew).

Moreover, it is straightforward to verify that a skew tree is incompatible to each of its translates:

LEMMA 7.14. *Let $p \in \mathbb{S}$ be skew, and let $t \in 2^\omega$ with $t \neq 0$. Then*

$$|[p] \cap ([p] + t)| \leq 2$$

(in particular, p and $p + t$ are incompatible).

PROOF. See [**BW**, Lemma 14]. □

[142] In fact, $\operatorname{cov}(\mathcal{M}) = \mathfrak{c}$ is sufficient to construct an $\mathcal{M}$-Luzin set, whereas $\operatorname{cov}(\mathcal{N}) = \mathfrak{c}$ is sufficient to construct an $\mathcal{N}$-Luzin set. (This is due to the fact that these ideals have bases consisting of Borel sets, and so, e.g., the assumption $\operatorname{cov}(\mathcal{M}) = \mathfrak{c}$ implies $\operatorname{cov}(\mathcal{M}) = \operatorname{cof}(\mathcal{M}) = \mathfrak{c}$, which guides the inductive construction.)

[143] This might be viewed as connected to the fact that ZFC proves $\operatorname{cov}(s_0) \leq \mathfrak{c} < \operatorname{cof}(s_0)$ (see Theorem 7.4).

7.2.5. *The regular case: the proof of the main lemma.* Recall Lemma 7.11 (which was enough to finish the proof of Theorem 7.7 for regular $\mathfrak{c}$): any set X of size $\mathfrak{c}$ has a subset $X' \subseteq X$ of size $\mathfrak{c}$ which is $<\mathfrak{c}$-transitively Marczewski null. We are going to provide the main ideas of its proof:

PROOF OF LEMMA 7.11 (SKETCH). We can assume without loss of generality that $X \in s_0$ (see Lemma 7.12). We distinguish two cases.

1st Case: For each skew $p \in \mathbb{S}$, we have $|[p] \cap X| < \mathfrak{c}$. Then it is easy to conclude that X is $<\mathfrak{c}$-transitively Marczewski null, due to the fact that being skew is dense and translation-invariant.

2nd Case: Fix a skew tree $p \in \mathbb{S}$ such that $|[p] \cap X| = \mathfrak{c}$. Define

$$X' := [p] \cap X.$$

Since p is skew, Lemma 7.14 implies that $\{p + t : t \in 2^\omega\} \subseteq \mathbb{S}$ is an antichain. To argue that X' is $<\mathfrak{c}$-transitively Marczewski null, fix $q \in \mathbb{S}$; we will find an $r \leq q$ such that[144]

$$\forall t \in 2^\omega \quad |([r] + t) \cap X'| \leq \aleph_0.$$

In case q is incompatible with $p + t$ for each $t \in 2^\omega$, it is easy to check that $r := q$ does the job. *Otherwise*, fix $t_0 \in 2^\omega$ and $q' \leq q$ with $q' \leq p + t_0$; by our assumption, $X \in s_0$ (hence also $X' + t_0 \in s_0$), i.e., we can fix $r \leq q'$ such that $[r] \cap (X' + t_0) = \emptyset$; since $\{p + t : t \in 2^\omega\}$ is an antichain, it is easy to check that r is as desired.

For more details, see [**BW**, Lemma 15]. □

7.2.6. *The singular case.* The following lemma allows to finish the proof of Theorem 7.7 in case $\mathfrak{c}$ is singular (compare with the last paragraph of Section 7.2.2).

The proof of the lemma is more complicated than the one of Lemma 7.11 (which was sufficient for the regular case). Note that, even though the conclusion of Lemma 7.15 is stronger, it is not a strengthening of Lemma 7.11 because there is the additional assumption that $\mathfrak{c}$ is singular.

LEMMA 7.15. *Assume $\mathfrak{c}$ is singular. Let $X \subseteq 2^\omega$ with $|X| = \mathfrak{c}$. Then there exists an $X' \subseteq X$ with $|X'| = \mathfrak{c}$ and $\mu < \mathfrak{c}$ such that X' is $<\mu$-transitively Marczewski null.*

PROOF. See [**BW**, Lemma 16]. □

In other words, for singular $\mathfrak{c}$, the sizes of the respective intersections in Definition 7.9 are bounded below $\mathfrak{c}$ (by Lemma 7.11, this is also the case for $\mathfrak{c}$ being a successor cardinal: the sizes are bounded by the predecessor of $\mathfrak{c}$).

We do not know, however, whether the same holds true in case $\mathfrak{c}$ is a regular limit[145] cardinal:

QUESTION 7.16. Does Lemma 7.15 hold true if $\mathfrak{c}$ is weakly inaccessible?

---

[144]In fact, X' is even shown to be $<\aleph_1$-transitively Marczewski null in this case.

[145]For the proof of Theorem 7.7 we do not care since it works due to regularity of $\mathfrak{c}$, no matter whether the sizes are bounded in $\mathfrak{c}$ or not.

7.2.7. *Generalization to Polish groups.* In the following, we briefly discuss why Theorem 7.7 generalizes to arbitrary Polish groups $(G, +)$ (in particular, to the classical real line).

Since $(G, +)$ carries a topology, there is the notion of perfect set: $P \subseteq G$ is *perfect* if it is closed and has no isolated points; so it makes sense to (re)define the Marczewski ideal s_0: a set $Y \subseteq G$ is *Marczewski null* if for each perfect set $P \subseteq G$ there is a perfect set $Q \subseteq P$ with $Q \cap Y = \emptyset$. Moreover, the group structure[146] of $(G, +)$ allows to (re)define the notion of s_0*-shiftable* and *Marczewski Borel Conjecture* (MBC).

Theorem 7.7 turns out to hold true in any Polish group $(G, +)$: there exists no $X \subseteq G$ of size continuum which is s_0-shiftable (and hence MBC holds under CH). The proof is completely analogous to the one for $(2^\omega, +)$; the only essential modification concerns the material presented in Section 7.2.4 involving the notion of skewness of a tree (whose combinatorial definition is very specific to 2^ω); so a generalized definition of skewness (replacing Definition 7.13) has to be adopted:

DEFINITION 7.17. A set $Z \subseteq G$ is *skew* if for all $x, y, v, w \in Z$ we have

$$x \neq y \ \wedge \ v \neq w \ \wedge \ \{x, y\} \neq \{v, w\} \implies x - y \neq v - w.$$

In fact, it is quite easy to see that being skew is translation-invariant, and that the analogue of Lemma 7.14 follows from the definition. The main (new) part of the proof concerns the *density* of skewness: for each perfect set $P \subseteq G$, there is a perfect set $Q \subseteq P$ which is skew (see [**BW**, Lemma 22]).

7.3. The status of the Marczewski Borel Conjecture. Finally, we discuss several models of the Marczewski Borel Conjecture, as well as models in which the Marczewski Borel Conjecture fails.

7.3.1. *MBC follows from CH.* The consistency of the Marczewski Borel Conjecture immediately follows from Theorem 7.7 (which says – in ZFC – that there is no s_0-shiftable set of size continuum):

COROLLARY 7.18. *Assume CH; then MBC holds (i.e., a set is s_0-shiftable if and only if it is countable). In particular, MBC is consistent.*

Note that it is enough to establish Theorem 7.7 for (the easier case of) regular $\mathfrak{c}$ in order to derive the above corollary (and, in turn, the consistency of MBC).

7.3.2. *MBC without CH.* In fact, the Marczewski Borel Conjecture is also consistent with continuum larger than $\aleph_1$ (indeed, arbitrarily large continuum):

THEOREM 7.19. *MBC holds in the Cohen model (more precisely: after adding λ many Cohen reals to a model of GCH, where $\lambda \geq \omega_2$ with $\mathrm{cf}(\lambda) \geq \omega_1$).*

PROOF. See [**BW**, Theorem 4 and Corollary 31]. □

7.3.3. *Failure of MBC.* On the other hand, the failure of the Marczewski Borel Conjecture is consistent as well:

---

[146]Note that there are two versions of s_0-shiftable (and MBC) in case of a non-abelian group. However, the results hold true for both versions, i.e., there is no s_0-shiftable set of size $\mathfrak{c}$ of either type; we do not claim though that the two notions necessarily coincide.

THEOREM 7.20. *MBC fails in the Sacks model.*[147] *Moreover, the failure of MBC is consistent with arbitrarily large continuum.*

PROOF. In the Sacks model, $\mathrm{cov}(s_0) = \mathfrak{c} = \aleph_2$ holds true[148] (see [**JMS92**, Theorem 1.2]), i.e., fewer than continuum many Marczewski null sets do not cover all the reals; it easily follows (see Proposition 1.12) that each set of size less than $\mathfrak{c}$ (in particular, each set of size $\aleph_1$) is s_0-shiftable (i.e., MBC fails).

Moreover, $\mathrm{cov}(s_0) = \mathfrak{c}$ is also known to be consistent with $\mathfrak{c}$ arbitrarily large (see [**Vel91**, Theorem 1 and Proposition 3]), which proves the second part of the theorem. □

Together with Theorem 7.7, the proof of the above theorem shows that it is consistent with arbitrarily large continuum that the collection of s_0-shiftable sets is exactly the ideal of sets of reals of size $< \mathfrak{c}$.

7.3.4. *Questions.* Note that in every model considered above, being s_0-shiftable is only a matter of size; we do not know whether this is always the case:

QUESTION 7.21. Is it consistent that there exists a cardinal λ such that there are both s_0-shiftable sets of size λ and sets of size λ which are not s_0-shiftable?

If the above question has a positive answer, we could ask more:

QUESTION 7.22. Is it consistent that the collection of s_0-shiftable sets does not form an ideal?

Finally, let us mention that we have not investigated s_0-shiftable sets in models other than the ones discussed above:

PROBLEM 7.23. Explore the status of the Marczewski Borel Conjecture (or, more generally, the collection of s_0-shiftable sets) in various models of ZFC.

7.3.5. *Last, but not least:*

Happy Birthday, Borel Conjecture!

Acknowledgments

The author wishes to thank Martin Goldstern, Marlene Koelbing, Jonathan Schilhan, David Schrittesser, and Thilo Weinert for many fruitful and inspiring conversations and helpful advice.

Moreover, he is grateful to Tomek Bartoszyński, Sy D. Friedman, and Otmar Spinas for reading parts of the paper and giving valuable suggestions, as well as to the anonymous referee for the careful reading.

[147] I.e., after a countable support iteration of Sacks forcing of length ω_2, starting with a model of CH; also compare with Theorem 4.4 (where the Sacks model is discussed in the context of BC/dBC): note that all cardinal characteristics of Cichoń's diagram are small, whereas $\mathrm{cov}(s_0)$ is large (which is used here to show the failure of the Marczewski Borel Conjecture).

[148] Intuitively speaking, this is because Sacks reals tend to avoid sets in s_0 since being disjoint from such sets is a dense property in Sacks forcing (by definition of s_0); since those "dense sets" are not really in the respective ground model, one has to refine this idea by a reflection argument etc.

References

[Bal11] R. Balka, *Duality between measure and category in uncountable locally compact abelian Polish groups*, Real Anal. Exchange **36** (2010/11), no. 2, 245–256. MR3016714

[Bar00] T. Bartoszyński, *A note on duality between measure and category*, Proc. Amer. Math. Soc. **128** (2000), no. 9, 2745–2748, DOI 10.1090/S0002-9939-00-05762-2. MR1766605

[Bar03] T. Bartoszyński, *Remarks on small sets of reals*, Proc. Amer. Math. Soc. **131** (2003), no. 2, 625–630, DOI 10.1090/S0002-9939-02-06567-X. MR1933355

[Bau83] J. E. Baumgartner, *Iterated forcing*, Surveys in set theory, London Math. Soc. Lecture Note Ser., vol. 87, Cambridge Univ. Press, Cambridge, 1983, pp. 1–59, DOI 10.1017/CBO9780511758867.002. MR823775

[BJ95] T. Bartoszyński and H. Judah, *Set theory*, A K Peters, Ltd., Wellesley, MA, 1995. On the structure of the real line. MR1350295

[BKW17] J. Brendle, Y. Khomskii, and W. Wohofsky, *Cofinalities of Marczewski-like ideals*, Colloq. Math. **150** (2017), no. 2, 269–279, DOI 10.4064/cm7113-5-2017. MR3719461

[Bla81] A. Blass, *A partition theorem for perfect sets*, Proc. Amer. Math. Soc. **82** (1981), no. 2, 271–277, DOI 10.2307/2043323. MR609665

[Bla10] A. Blass, *Combinatorial cardinal characteristics of the continuum*, Handbook of set theory. Vols. 1, 2, 3, Springer, Dordrecht, 2010, pp. 395–489, DOI 10.1007/978-1-4020-5764-9_7. MR2768685

[Bor19] E. Borel, *Sur la classification des ensembles de mesure nulle* (French), Bull. Soc. Math. France **47** (1919), 97–125. MR1504785

[Bre95] J. Brendle, *Strolling through paradise*, Fund. Math. **148** (1995), no. 1, 1–25, DOI 10.4064/fm-148-1-1-25. MR1354935

[BRS96] T. Bartoszyński, A. Rosłanowski, and S. Shelah, *Adding one random real*, J. Symbolic Logic **61** (1996), no. 1, 80–90, DOI 10.2307/2275599. MR1380678

[BS01] T. Bartoszyński and S. Shelah, *Strongly meager sets do not form an ideal*, J. Math. Log. **1** (2001), no. 1, 1–34, DOI 10.1142/S0219061301000028. MR1838340

[BS02] T. Bartoszyński and S. Shelah, *Strongly meager and strong measure zero sets*, Arch. Math. Logic **41** (2002), no. 3, 245–250, DOI 10.1007/s001530000068. MR1901186

[BS03] T. Bartoszyński and S. Shelah, *Strongly meager sets of size continuum*, Arch. Math. Logic **42** (2003), no. 8, 769–779, DOI 10.1007/s00153-003-0184-0. MR2020043

[BS10] T. Bartoszyński and S. Shelah, *Dual Borel conjecture and Cohen reals*, J. Symbolic Logic **75** (2010), no. 4, 1293–1310, DOI 10.2178/jsl/1286198147. MR2767969

[BW] J. Brendle and W. Wohofsky, *Borel Conjecture for the Marczewski ideal*, to appear in the Proceedings of the American Mathematical Society.

[Car93] T. J. Carlson, *Strong measure zero and strongly meager sets*, Proc. Amer. Math. Soc. **118** (1993), no. 2, 577–586, DOI 10.2307/2160341. MR1139474

[Cor89] P. Corazza, *The generalized Borel conjecture and strongly proper orders*, Trans. Amer. Math. Soc. **316** (1989), no. 1, 115–140, DOI 10.2307/2001276. MR982239

[CP04] K. Ciesielski and J. Pawlikowski, *The covering property axiom, CPA*, Cambridge Tracts in Mathematics, vol. 164, Cambridge University Press, Cambridge, 2004. A combinatorial core of the iterated perfect set model. MR2176267

[FF10] V. Fischer and S. D. Friedman, *Cardinal characteristics and projective wellorders*, Ann. Pure Appl. Logic **161** (2010), no. 7, 916–922, DOI 10.1016/j.apal.2009.11.003. MR2601019

[FFZ11] V. Fischer, S. D. Friedman, and L. Zdomskyy, *Projective wellorders and mad families with large continuum*, Ann. Pure Appl. Logic **162** (2011), no. 11, 853–862, DOI 10.1016/j.apal.2011.04.001. MR2817561

[Fri00] S. D. Friedman, *Fine structure and class forcing*, De Gruyter Series in Logic and its Applications, vol. 3, Walter de Gruyter & Co., Berlin, 2000. MR1780138

[GJS93] M. Goldstern, H. Judah, and S. Shelah, *Strong measure zero sets without Cohen reals*, J. Symbolic Logic **58** (1993), no. 4, 1323–1341, DOI 10.2307/2275146. MR1253925

[GKMS] Martin Goldstern, Jakob Kellner, Diego Mejía, and Saharon Shelah, *Cichoń's maximum without large cardinals*, to appear in J. Eur. Math. Soc. (JEMS).

[GKS19] M. Goldstern, J. Kellner, and S. Shelah, *Cichoń's maximum*, Ann. of Math. (2) **190** (2019), no. 1, 113–143, DOI 10.4007/annals.2019.190.1.2. MR3990602

[GKSW] Martin Goldstern, Jakob Kellner, Saharon Shelah, and Wolfgang Wohofsky, *An overview of the proof in "Borel Conjecture and Dual Borel Conjecture"*, unpublished notes, arxiv:1112.4424.

[GKSW14] M. Goldstern, J. Kellner, S. Shelah, and W. Wohofsky, *Borel conjecture and dual Borel conjecture*, Trans. Amer. Math. Soc. **366** (2014), no. 1, 245–307, DOI 10.1090/S0002-9947-2013-05783-2. MR3118397

[GMS79] F. Galvin, J. Mycielski, and R. M. Solovay, *Strong measure zero sets*, Abstract 79T-E25, Notices of the American Mathematical Society **26** (1979), A–280.

[GMS17] F. Galvin, J. Mycielski, and R. M. Solovay, *Strong measure zero and infinite games*, Arch. Math. Logic **56** (2017), no. 7-8, 725–732, DOI 10.1007/s00153-017-0541-z. MR3696064

[Gol] M. Goldstern, *Cohen reals and strong measure zero sets*, MathOverflow, 2011, https://mathoverflow.net/questions/63497.

[Gol93] M. Goldstern, *Tools for your forcing construction*, Set theory of the reals (Ramat Gan, 1991), Israel Math. Conf. Proc., vol. 6, Bar-Ilan Univ., Ramat Gan, 1993, pp. 305–360. MR1234283

[Gol98] M. Goldstern, *A taste of proper forcing*, Set theory (Curaçao, 1995; Barcelona, 1996), Kluwer Acad. Publ., Dordrecht, 1998, pp. 71–82. MR1601976 (99f:03069)

[GS13] F. Galvin and M. Scheepers, *Borel's conjecture in topological groups*, J. Symbolic Logic **78** (2013), no. 1, 168–184, DOI 10.2178/jsl.7801110. MR3087068

[Hal96] A. Halko, *Negligible subsets of the generalized Baire space* ${\omega_1}^{\omega_1}$, ProQuest LLC, Ann Arbor, MI, 1996. Thesis (Ph.D.)–Helsingin Yliopisto (Finland). MR2715037

[Hal17] L. J. Halbeisen, *Combinatorial set theory*, Springer Monographs in Mathematics, Springer, Cham, 2017. With a gentle introduction to forcing; Second edition of [MR3025440]. MR3751612

[Har77] L. Harrington, *Long projective wellorderings*, Ann. Math. Logic **12** (1977), no. 1, 1–24, DOI 10.1016/0003-4843(77)90004-3. MR465866

[HS01] A. Halko and S. Shelah, *On strong measure zero subsets of* ${}^{\kappa}2$, Fund. Math. **170** (2001), no. 3, 219–229, DOI 10.4064/fm170-3-1. MR1880900

[HS20] P. Hein and O. Spinas, *Antichains of perfect and splitting trees*, Arch. Math. Logic **59** (2020), pp. 367–388.

[HWZ16] M. Hrušák, W. Wohofsky, and O. Zindulka, *Strong measure zero in separable metric spaces and Polish groups*, Arch. Math. Logic **55** (2016), no. 1-2, 105–131, DOI 10.1007/s00153-015-0459-2. MR3453581

[HZ16] M. Hrušák and J. Zapletal, *Strong measure zero sets in Polish groups*, Illinois J. Math. **60** (2016), no. 3-4, 751–760. MR3707641

[HZ20] M. Hrušák and O. Zindulka, *Strong measure zero in Polish groups*, to appear in Centenary of the Borel Conjecture, Contemporary Mathematics, vol. 755, Amer. Math. Soc., Providence, RI, 2020.

[Jec03] T. Jech, *Set theory*, Springer Monographs in Mathematics, Springer-Verlag, Berlin, 2003. The third millennium edition, revised and expanded. MR1940513

[JMS92] H. Judah, A. W. Miller, and S. Shelah, *Sacks forcing, Laver forcing, and Martin's axiom*, Arch. Math. Logic **31** (1992), no. 3, 145–161, DOI 10.1007/BF01269943. MR1147737

[JS89] H. Judah and S. Shelah, MA(σ*-centered): Cohen reals, strong measure zero sets and strongly meager sets*, Israel J. Math. **68** (1989), no. 1, 1–17, DOI 10.1007/BF02764965. MR1035877

[JS90] H. Judah and S. Shelah, *The Kunen-Miller chart (Lebesgue measure, the Baire property, Laver reals and preservation theorems for forcing)*, J. Symbolic Logic **55** (1990), no. 3, 909–927, DOI 10.2307/2274464. MR1071305

[JSW90] H. Judah, S. Shelah, and W. H. Woodin, *The Borel conjecture*, Ann. Pure Appl. Logic **50** (1990), no. 3, 255–269, DOI 10.1016/0168-0072(90)90058-A. MR1086456

[KKLW] Y. Khomskii, M. Koelbing, G. Laguzzi, and W. Wohofsky, *Laver trees in the generalized Baire space*, in preparation.

[Kor18] M. Korch, *Measure and convergence: special subsets of the real line and their generalizations*, Ph.D. thesis (2018).

[Kun83] K. Kunen, *Set theory*, Studies in Logic and the Foundations of Mathematics, vol. 102, North-Holland Publishing Co., Amsterdam, 1983. An introduction to independence proofs; Reprint of the 1980 original. MR756630

[Kun11] K. Kunen, *Set theory*, Studies in Logic (London), vol. 34, College Publications, London, 2011. MR2905394

[KW04] M. Kysiak and T. Weiss, *Small subsets of the reals and tree forcing notions*, Proc. Amer. Math. Soc. **132** (2004), no. 1, 251–259, DOI 10.1090/S0002-9939-03-07026-6. MR2021269

[Kys00] M. Kysiak, *On Erdős-Sierpiński duality between Lebesgue measure and Baire category (in Polish)*, Master's thesis (2000).

[Kys03] M. Kysiak, *Another note on duality between measure and category*, Bull. Polish Acad. Sci. Math. **51** (2003), no. 3, 269–281. MR2017003

[Lav76] R. Laver, *On the consistency of Borel's conjecture*, Acta Math. **137** (1976), no. 3-4, 151–169, DOI 10.1007/BF02392416. MR422027

[LSV93] A. Louveau, S. Shelah, and B. Veličković, *Borel partitions of infinite subtrees of a perfect tree*, Ann. Pure Appl. Logic **63** (1993), no. 3, 271–281, DOI 10.1016/0168-0072(93)90151-3. MR1237234

[Luz14] N. Luzin, *Sur un problème de M. Baire*, Hebdomadaires Seances Acad. Sci. Paris **158** (1914), 1258–1261.

[Mah13] P. Mahlo, *Über Teilmengen des Kontinuums von dessen Mächtigkeit*, Sitzungsber. Sächs. Akad. Wiss. Leipzig Math.-Natu. Kl. **65** (1913), 283–315.

[Mil83] A. W. Miller, *Mapping a set of reals onto the reals*, J. Symbolic Logic **48** (1983), no. 3, 575–584, DOI 10.2307/2273449. MR716618

[Mil84] A. W. Miller, *Special subsets of the real line*, Handbook of set-theoretic topology, North-Holland, Amsterdam, 1984, pp. 201–233. MR776624

[Paw96a] J. Pawlikowski, *A characterization of strong measure zero sets*, Israel J. Math. **93** (1996), 171–183, DOI 10.1007/BF02761100. MR1380640

[Paw96b] J. Pawlikowski, *Every Sierpiński set is strongly meager*, Arch. Math. Logic **35** (1996), no. 5-6, 281–285, DOI 10.1007/s001530050045. MR1420258

[Paw96c] J. Pawlikowski, *Laver's forcing and outer measure*, Set theory (Boise, ID, 1992), Contemp. Math., vol. 192, Amer. Math. Soc., Providence, RI, 1996, pp. 71–76, DOI 10.1090/conm/192/02350. MR1367136

[PR95] J. Pawlikowski and I. Recław, *Parametrized Cichoń's diagram and small sets*, Fund. Math. **147** (1995), no. 2, 135–155. MR1341727

[Sch93] M. Scheepers, *Piecewise strongly meager sets*, Unpublished notes (1993).

[Sch19] M. Scheepers, *On a hypothesis for $\aleph_0$-bounded groups*, Topology Appl. **258** (2019), 229–238, DOI 10.1016/j.topol.2019.02.058. MR3924515

[She95] S. Shelah, *Every null-additive set is meager-additive*, Israel J. Math. **89** (1995), no. 1-3, 357–376, DOI 10.1007/BF02808209. MR1324470

[She98] S. Shelah, *Proper and improper forcing*, 2nd ed., Perspectives in Mathematical Logic, Springer-Verlag, Berlin, 1998. MR1623206

[She06] S. Shelah, *Non-Cohen oracle C.C.C*, J. Appl. Anal. **12** (2006), no. 1, 1–17, DOI 10.1515/JAA.2006.1. MR2243849

[Sie28] W. Sierpiński, *Sur un ensemble non dénombrable, dont toute image continue est de mesure nulle*, Fund. Math. **11** (1928), 302–304.

[SW] O. Spinas and W. Wohofsky, *A Sacks amoeba preserving distributivity of $\mathcal{P}(\omega)/fin$*, to appear in Fundamenta Mathematicae.

[SW16] S. Shelah and W. Wohofsky, *There are no very meager sets in the model in which both the Borel conjecture and the dual Borel conjecture are true*, MLQ Math. Log. Q. **62** (2016), no. 4-5, 434–438, DOI 10.1002/malq.201600002. MR3549562

[Vel91] B. Veličković, *CCC posets of perfect trees*, Compositio Math. **79** (1991), no. 3, 279–294. MR1121140

[Wei17] T. Weiss, *Properties of the intersection ideal $\mathcal{M} \cap \mathcal{N}$ revisited*, Bull. Pol. Acad. Sci. Math. **65** (2017), no. 2, 107–111, DOI 10.4064/ba8098-8-2017. MR3731016

[Wei18] T. Weiss, *More remarks on the intersection ideal $\mathcal{M} \cap \mathcal{N}$*, Comment. Math. Univ. Carolin. **59** (2018), no. 3, 311–316, DOI 10.14712/1213-7243.2015.252. MR3861554

[Woh] W. Wohofsky, *Does the Galvin-Mycielski-Solovay theorem hold for uncountable cardinals?*, in preparation.

[Woh13] W. Wohofsky, *Special sets of real numbers and variants of the Borel Conjecture*, Ph.D. thesis, Technische Universität Wien, 2013.

Kurt Gödel Research Center for Mathematical Logic, Universität Wien, Währinger Strasse 25, 1090 Wien, Austria

Email address: `wolfgang.wohofsky@gmx.at`

Contemporary Mathematics
Volume **755**, 2020
https://doi.org/10.1090/conm/755/15176

Selection principles in the Laver, Miller, and Sacks models

Lyubomyr Zdomskyy

ABSTRACT. This article is devoted to the interplay between forcing with fusion and combinatorial covering properties. We discuss known instances of this interplay as well as present a new one, namely that in the Laver model for the consistency of the Borel's conjecture, the Hurewicz property is preserved by finite products of metrizable spaces.

1. Introduction

Combinatorial covering properties (or selection principles) arguably arose from the study of special sets of reals. These resolved many classical questions in general topology and measure theory. As a result, information about special sets of reals is included in standard topology textbooks, such as Kuratowski's Topology [**28**]. The most influential survey on special sets of reals is, probably, Miller's chapter [**35**] in the Handbook of Set-Theoretic Topology. The most recent monograph on this topic is written by Bukovsky, see [**14**]. It complements nicely the classical book [**7**] of Bartoszynski and Judah. This theory still finds interesting applications in general topology, see, e.g., [**23**] for the interplay between λ-sets and homogeneity.

A typical example of the evolution of special sets of reals into selection principles comes from *strong measure zero* sets (SMZ sets in what follows): $X \subset \mathbb{R}$ is SMZ if for every sequence $\langle \epsilon_n : n \in \omega \rangle$ of positive reals there exists a sequence $\langle a_n : n \in \omega \rangle$ of "centers" such that $X \subset \bigcup_{n\in\omega} B(a_n, \epsilon_n)$, where $B(a, \epsilon) = \{x \in \mathbb{R} : |x - a| < \epsilon\}$. SMZ sets were introduced by Borel ca. a century ago who conjectured [**13**] that only countable sets have such property, i.e., that there are only trivial examples of SMZ sets. This conjecture has been consistently refuted by Sierpiński [**51**] in 1928, but the question whether Borel's conjecture (BC in what follows) is consistent was answered only in 1976 by Laver in his seminal paper [**29**]. This outstanding result was the first[1] instance when a forcing, adding a real, was iterated with countable supports without collapsing cardinals. This work of Laver can be thought of as a "birth" of fusion, the latter being a kind of a gentle weakening of the countable completeness which allows to add new reals and nonetheless "treat" countably many

2010 *Mathematics Subject Classification.* Primary 03E35, 54D20; Secondary 54C50, 03E05.
Key words and phrases. Menger space, Hurewicz space, Laver forcing, Miller forcing, Sacks forcing, fusion.

The author would like to thank the Austrian Science Fund FWF (Grants I 2374-N35 I 3709-N35) for generous support of this research.

[1]According to our colleagues who worked in set theory already in the 70s.

dense sets with a single condition, and thus it is also one of the motivations behind Baumgartner's axiom A and Shelah's theory of proper forcing.

In 1938 Rothberger worked on the question whether having SMZ is a topological property, see [**43**], and introduced the following topological counterpart of SMZ sets which is now known as Rothberger's property and is one of the most intensively studied covering properties: For every sequence $\langle \mathcal{U}_n : n \in \omega \rangle$ of open covers of X there exists a sequence $\langle U_n : n \in \omega \rangle$ such that $U_n \in \mathcal{U}_n$ for all n and the collection $\{U_n : n \in \omega\}$ is a cover of X. Three years later he has shown in [**44**] that $\mathfrak{b} = \omega_1$ implies the existence of a SMZ set which is not Rothberger and hence that having SMZ is not a topological property. Recall that $\mathfrak{b}$ is the minimal cardinality of a subset of ω^ω which cannot be covered by a σ-compact one. In the same paper he shows that the BC is equivalent to all Rothberger metrizable spaces being countable. The Rothberger property is an example of strong combinatorial covering properties in the sense that one selects a *single* element of each cover, aiming at a "diagonalizing" cover of certain kind. These proved to be important in the study of local properties of spaces of continuous functions, see [**2**] and references therein.

Another source of selection principles was the dimension theory, where the following property takes its origin: A topological space X has the *Menger* property (or, alternatively, is a Menger space) if for every sequence $\langle \mathcal{U}_n : n \in \omega \rangle$ of open covers of X there exists a sequence $\langle \mathcal{V}_n : n \in \omega \rangle$ such that each $\mathcal{V}_n$ is a finite subfamily of $\mathcal{U}_n$ and the collection $\{\cup \mathcal{V}_n : n \in \omega\}$ is a cover of X. This property was introduced by Hurewicz, and the current name (the Menger property) is used because Hurewicz proved in [**24**] that for metrizable spaces his property is equivalent to one property of a base considered by Menger in [**31**]. If in the definition above we additionally require that $\{\cup \mathcal{V}_n : n \in \omega\}$ is a γ-*cover* of X (this means that the set $\{n \in \omega : x \notin \cup \mathcal{V}_n\}$ is finite for each $x \in X$), then we obtain the definition of the *Hurewicz property* introduced in [**25**]. Each σ-compact space is obviously a Hurewicz space, and Hurewicz spaces have the Menger property. Contrary to a conjecture of Hurewicz the class of metrizable spaces having the Hurewicz property appeared to be much wider than the class of σ-compact spaces [**26**, Theorem 5.1]. Moreover, there always exists a subspace of ω^ω of cardinality $\mathfrak{b}$ all of whose finite powers are Hurewicz, see [**8, 9**]. Thus we can say that there are always *non-trivial* examples of metrizable Hurewicz spaces, as contrasted with the behaviour of Rothberger spaces in the Laver model. The properties of Menger and Hurewicz are classical examples of *weaker* combinatorial covering properties, i.e., properties obtained by selecting finitely many elements from each cover in a given sequence.

The Rothberger property and SMZ sets in the measure theory as well as properties of Menger and Hurewicz in the dimension theory, respectively, are not among the most intensively investigated ones in these areas of mathematics. The reason is that their study required methods developed decades after their introduction. Namely, one of the most fascinating features of these properties as well as many other selection principles is their behavior in various models of set theory, which makes forcing and cardinal characteristics one of the main tools in the area. This became apparent from at least the publication of [**29**], in [**6, 16, 17, 36, 50**] one can find more recent works along these lines. In what follows we discuss the affect of three classical forcing notions, namely Laver, Miller, and Sacks ones, on selection principles. The unifying theme of all of them is *fusion* based on the fact that the conditions are certain trees on the natural numbers.

The next after [**29**] application of the ***Laver forcing*** to combinatorial covering properties seems to be [**18**] where it is proved that in the Laver model, the properties α_1 and α_2 introduced by Arkhangel'skiĭ in [**1**] coincide. Recall that a space X satisfies α_2 (resp. α_1) if for every countable family $\mathcal{S}$ of non-trivial sequences converging to a point $x \in X$ there exists a sequence T converging to x such that $|T \cap S| = \omega$ (resp. $S \subset^* T$) for all $S \in \mathcal{S}$. How is this related to covering properties? It has been proved in [**58**] that for a metrizable space X, property α_1 of $C_p(X)$, the space of all continuous functions $f : X \to \mathbb{R}$ with the topology inherited from the Tychonoff product $\mathbb{R}^X$, is equivalent to X having the Hurewicz property with respect to all countable Borel covers. Due to the fact that (unlike the open sets) Borel sets are closed under countable intersections, this property is equivalent to $S_1(\mathcal{B}_\Gamma, \mathcal{B}_\Gamma)$ asserting that for every sequence $\langle \mathcal{U}_n : n \in \omega \rangle$ of Borel γ-covers of X there exists $\langle U_n \in \mathcal{U}_n : n \in \omega \rangle$ which is a γ-cover of X. Replacing here "Borel" with "clopen" one gets the property $S_1(\mathcal{C}_\Gamma, \mathcal{C}_\Gamma)$ equivalent to α_2 of $C_p(X)$, see [**46**]. One of the most intriguing conjectures along these lines is due to Scheepers [**48**] and says that for subsets of $\mathbb{R}$ the properties $S_1(\mathcal{C}_\Gamma, \mathcal{C}_\Gamma)$ and $S_1(\Gamma, \Gamma)$ coincide, the definition of the latter obtained from that of $S_1(\mathcal{C}_\Gamma, \mathcal{C}_\Gamma)$ be replacing "clopen" with "open". It follows from the above that for metrizable spaces we have $S_1(\mathcal{B}_\Gamma, \mathcal{B}_\Gamma) = S_1(\mathcal{C}_\Gamma, \mathcal{C}_\Gamma)$ in the Laver model, and by

$$S_1(\mathcal{B}_\Gamma, \mathcal{B}_\Gamma) \to S_1(\Gamma, \Gamma) \to S_1(\mathcal{C}_\Gamma, \mathcal{C}_\Gamma)$$

the Scheepers' Conjecture is true in the Laver model. However, it is totally unclear what happens under CH. Moreover, if there is $X \subset \mathbb{R}$ satisfying $S_1(\mathcal{C}_\Gamma, \mathcal{C}_\Gamma)$ but not $S_1(\Gamma, \Gamma)$, then this is also obviously true in any forcing extension with the same reals, and hence we can get such a set in a model of CH. Thus if the negative answer to Scheepers' question is consistent, it is also consistent with CH.

Using the key lemma of [**29**] allowing to analyze names for reals in the Laver model, Miller and Tsaban proved [**36**] that all $X \subset \mathbb{R}$ satisfying $S_1(\mathcal{B}_\Gamma, \mathcal{B}_\Gamma)$ must have size $\leq \omega_1 < \mathfrak{b} = \mathfrak{c}$ in this model, and hence this is also true for $S_1(\mathcal{C}_\Gamma, \mathcal{C}_\Gamma)$. This answered the question whether in ZFC there exists a set of reals satisfying $S_1(\mathcal{C}_\Gamma, \mathcal{C}_\Gamma)$ of size $\mathfrak{b}$ in the negative. Their application of Laver's analysis of names of reals found further application to covering properties of products.

One of the basic questions about a topological property is whether it is preserved by finite products. In case of combinatorial covering properties we know that the strongest possible negative result is consistent: Under CH there exist $X, Y \subset \mathbb{R}$ which have the γ-space property with respect to countable Borel covers[2], whose product $X \times Y$ is not Menger, see [**37**, Theorem 3.2]. Thus the product of spaces with the strongest combinatorial covering property considered thus far might fail to have even the weakest one. This implies that no positive results for combinatorial covering properties can be obtained outright in ZFC. Unlike the vast majority of topological and combinatorial consequences under CH, the latter one does not follow from any equality among cardinal characteristics of the continuum, see [**6**] and the discussion on [**37**, p. 2882]. However, there are many other negative results stating that under certain equality among cardinal characteristics (e.g., $cov(\mathcal{N}) = cof(\mathcal{N})$,

[2]This property is stronger than $S_1(\mathcal{B}_\Gamma, \mathcal{B}_\Gamma)$ and we refer the reader to [**37**] for its definition.

$\mathfrak{b} = \mathfrak{d}$, etc.[3]) there are spaces $X, Y \subset \mathbb{R}$ with some combinatorial covering property such that $X \times Y$ is not Menger, see, e.g., [**4**, **39**, **54**].

Regarding the positive results, until recently the most unclear situation was with the Hurewicz property and the weaker ones. There are two reasons why a product of Hurewicz spaces X, Y can fail to be Hurewicz. In the first place, $X \times Y$ may simply fail to be a Lindelöf space, i.e., it might have an open cover $\mathcal{U}$ without countable subcover. This may indeed happen: in ZFC there are two normal spaces X, Y with a covering property much stronger than the Hurewicz one such that $X \times Y$ does not have the Lindelöf property, see [**55**, Section 3]. However, the above situation becomes impossible if we restrict our attention to metrizable spaces. This second case turned out to be sensitive to the ambient set-theoretic universe: under CH there exists a Hurewicz space whose square is not Menger, see [**26**, Theorem 2.12]. The above result has been achieved by a transfinite construction of length ω_1, using the combinatorics of the ideal of measure zero subsets of reals. This combinatorics turned out [**54**, Theorem 43] to require much weaker set-theoretic assumptions than CH. In particular, under the Martin Axiom there are Hurewicz subspaces of the irrationals whose product is not Menger.

The following theorem, which is the main result which we are going to prove in this article, shows that an additional assumption in the results from [**26**, **54**] mentioned above is really needed.

THEOREM 1.1. *In the Laver model for the consistency of the Borel's conjecture, the Hurewicz property is preserved by finite products of metrizable spaces. Consequently, the product of any two Hurewicz spaces is Hurewicz if it is Lindelöf.*

Let us recall that for metrizable spaces, the separability is equivalent to being Lindelöf, see, e.g., [**19**, Theorem 4.1.15].

This theorem is a further improvement of the main result of [**40**] which states that product of two Hurewicz subspaces of $\mathbb{R}$ is Menger in the Laver model. The conclusion of Theorem 1.1 does not follow from Borel's conjecture: If we add ω_2 many random reals over the Laver model then Borel's conjecture still holds by [**7**, Section 8.3.B] and we have $cov(\mathcal{N}) = cof(\mathcal{N})$, and hence in this model there exists a Hurewicz set of reals whose square is not Menger, see [**54**]. Thus Borel's conjecture is consistent with the existence of a Hurewicz set of reals with even non-Menger square.

Theorem 1.1 seems to be an instance of a more general phenomena, namely that proper posets with fusion affect the behavior of combinatorial covering properties. This happens because sets of reals with certain combinatorial covering properties are forced to have a rather clear structure, which suffices to prove positive preservation results. For instance, the core of the proof of Theorem 1.1 is that Hurewicz subspaces of the real line are concentrated in a sense around their "simpler" subspaces.

Later we shall discuss similar behavior of combinatorial covering properties in the Miller and Sacks models. Now, however, we would like to present a "side effect" of the proof of Theorem 1.1 for maximal almost disjoint families of subsets of ω. Recall that an infinite $\mathcal{A} \subset [\omega]^\omega$ is called a *mad family*, if $|A_0 \cap A_1| < \omega$ for any distinct $A_0, A_1 \in \mathcal{A}$, and for every $B \in [\omega]$ there exists $A \in \mathcal{A}$ such that $|B \cap A| = \omega$.

[3]We refer the reader to [**10**] for the definitions and basic properties of cardinal characteristics of the continuum which are mentioned but are not used in the proofs in this article.

In [**15**, Theorem 2.1] Brendle constructed under CH a mad family $\mathcal{A}$ on ω such that the Mathias forcing[4] $M_{\mathcal{F}(\mathcal{A})}$ associated to the filter

$$\mathcal{F}(\mathcal{A}) = \{F \subset \omega : \exists \mathcal{A}' \in [\mathcal{A}]^{<\omega}(\omega \setminus \cup\mathcal{A}' \subset^* F)\}$$

adds a dominating real. In the same paper Brendle asked whether such a mad family can be constructed outright in ZFC. This question has been answered in the affirmative in [**22**] and later independently also in [**17**] using different methods. However, the mad families constructed there had topological copies of the Cantor set inside, and hence by a simple absoluteness argument are destroyed (i.e., are not maximal any more) by any forcing adding new reals. One can refine Brendle's question in the following way: Suppose that a mad family $\mathcal{A}$ cannot be destroyed by some very "mild" forcing $\mathbb{P}$, i.e., it remains maximal in $V^{\mathbb{P}}$, must then $M_{\mathcal{F}(\mathcal{A})}$ add dominating reals? This refinement makes sense in cases when one is interested in destroying mad families without adding dominating reals and takes some $\mathbb{P}$ which does not add them. Indeed, if $\mathcal{A}$ is already destroyed by $\mathbb{P}$, there is no need to use its Mathias forcing for its destruction in a hypothetic construction of a model where, e.g., $\mathfrak{b}$ should stay small. In what follows we concentrate on the case $\mathbb{P}$ being the Cohen forcing $\mathbb{C}$. Mad families $\mathcal{A}$ which remain maximal in $V^{\mathbb{C}}$ will be called Cohen-indestructible. The following theorem has been proved in [**3**]

THEOREM 1.2. *$\mathfrak{p} = cov(\mathcal{N}) = \mathfrak{c}$ implies the existence of a Cohen-indestructible mad family $\mathcal{A}$ such that $M_{\mathcal{F}(\mathcal{A})}$ adds a dominating real.*

Recall from [**27**] that a mad family $\mathcal{A}$ is called *ω-mad* if for every sequence $\langle X_n : n \in \omega\rangle$ of elements of $\mathcal{F}(\mathcal{A})^+$ there exists $A \in \mathcal{A}$ such that $|A \cap X_n| = \omega$ for all n. Cohen-indestructible mad families are closely related to ω-mad ones, see [**30**] or [**27**, Theorem 4]: Every ω-mad family is Cohen-indestructible, and if $\mathcal{A}$ is Cohen-indestructible, then for every $X \in \mathcal{F}(\mathcal{A})^+$ there exists $Y \subset X$, $Y \in \mathcal{F}(\mathcal{A})^+$, such that $\mathcal{A} \restriction Y = \{A \cap Y : A \in \mathcal{A}, A \cap Y \text{ is infinite}\}$ is ω-mad as a mad family on Y.

The proof of Theorem 1.2 actually gives an ω-mad family. The next theorem shows that $\mathfrak{b} = \mathfrak{c}$ would not suffice in Theorem 1.2 in a strong sense.

THEOREM 1.3. *In the Laver model for the consistency of the Borel conjecture, for every ω-mad family $\mathcal{A}$ the poset $M_{\mathcal{F}(\mathcal{A})}$ preserves all ground model unbounded sets. In particular, if $\mathcal{A}$ is Cohen-indestructible, then there exists $X \in \mathcal{F}(\mathcal{A})^+$ such that $M_{\mathcal{F}(\mathcal{A})\restriction X}$ preserves all ground model unbounded sets, where $\mathcal{F}(\mathcal{A}) \restriction X$ denotes the filter on ω generated by the centered family $\{F \cap X : F \in \mathcal{F}(\mathcal{A})\}$.*

Theorem 1.3 improves our earlier result proved in [**3**] asserting, under the same premises, that $M_{\mathcal{F}(\mathcal{A})\restriction X}$ keeps the ground model unbounded. In our proof of Theorem 1.3 we shall not work with the Mathias forcing directly, but rather use the following characterization obtained in [**17**]: For a filter $\mathcal{F}$ on ω the poset $M_{\mathcal{F}}$ adds no dominating reals (resp. preserves all ground model unbounded sets) iff $\mathcal{F}$ has the Menger (resp. Hurewicz) covering property when considered with the topology inherited from $\mathcal{P}(\omega)$, which is identified with the Cantor space 2^ω via characteristic functions. Thus Theorem 1.3 says that if A is an ω-mad family, then $\mathcal{F}(\mathcal{A})$ is Hurewicz, whereas its predecessor from [**3**] gave only the Menger property. Let us note that there are ZFC examples of Menger non-Hurewicz filters on ω, see [**42**].

[4]Since we shall not analyze this poset directly but rather use certain topological characterizations, we refer the reader to, e.g., [**15**] for its definition.

Theorem 1.3 also intuitively explains why the open question whether there exists an ω-mad family outright in ZFC (it is known and rather easy that they exist under $\mathfrak{b} = \mathfrak{c}$, e.g., in the Laver model) is so difficult: Any such ZFC example should "often" give a mad family whose dual filter is Hurewicz, and it is even unknown how to get one with the Menger filter in ZFC, see [**22**]. This motivates the following

QUESTION 1.4. Is there a ZFC example of a mad family A such that $\mathcal{F}(\mathcal{A})$ is Hurewicz?

If the answer to this problem is affirmative, then it should probably require some new ideas since with the existing methods it is unclear even how to get a Menger one. On the other hand, a negative answer might be surprisingly easy.

Superficially the conditions in the ***Miller forcing*** introduced in [**34**] are very similar to those in the Laver one, the only difference being that the splitting does not have to occur so often. However, these posets have completely different affects on the behavior of selection principles discussed above. In this context the Miller model has been first investigated in [**59**], where it has been proved that for sets of reals, the Rothberger property implies the Hurewicz one in this model. This is of course also true in the Laver model because the Borel's conjecture holds there. On the contrary, in the Miller model there are uncountable Rothberger spaces (see [**38**] for a much more stronger result and references therein), so here this implication holds for different, qualitative (and not cardinality) reasons. Let us note that, e.g., under CH there are Rothberger subspaces of $\mathbb{R}$ which are not Hurewicz, e.g., Luzin sets, see [**47**].

Another result proved in [**59**] is that every Menger space in the Miller model is *Scheepers*, where the latter property is defined in the same way as the Menger one, with the following additional requirement: The selected sequence $\{\cup\mathcal{V}_n : n \in \omega\}$ should be an ω-cover of X, which by the definition means that for every finite $F \subset X$ there exists n such that $F \subset \cup\mathcal{V}_n$. It is rather straightforward to check that if all finite powers of X have the Menger property, then X is Scheepers. Thus the Miller model was a natural place to address the question whether the Menger property is preserved by finite products of metrizable spaces. The following theorem is the main result of [**60**].

THEOREM 1.5. *In the Miller model, the product of any two Menger spaces is Menger provided that it is Lindelöf. In particular, in this model the product of any two Menger metrizable spaces is Menger.*

Of course Theorem 1.5 is not true in ZFC as there are equalities between cardinal characteristics which yield even Hurewicz sets of reals with non-Menger product, see our discussion of products of Hurewicz spaces above. Surprisingly, there are also inequalities between cardinal characteristics which imply that the Menger property is not productive even for sets of reals, see [**52**]. Let us also note that in the Laver model (more generally, under $\mathfrak{b} = \mathfrak{d}$) the Menger property is not preserved by products of metrizable spaces, see [**39**] and [**53**] for more recent results along these lines.

As we can see in [**60**], a big part of the proof of Theorem 1.1 (as well as all the results from [**59**] mentioned above) requires only the inequality $\mathfrak{u} < \mathfrak{g}$ which holds in the Miller model. However, we do not know the answer to the following

QUESTION 1.6. Is the Menger property preserved by finite products of metrizable spaces under $\mathfrak{u} < \mathfrak{g}$? If yes, can $\mathfrak{u} < \mathfrak{g}$ be weakened to the Filter Dichotomy or NCF?

We refer the reader to [**10**, § 9] for corresponding definitions. Let us note that it is not enough to assume $\mathfrak{u} < \mathfrak{d}$ in Theorem 1.5 as witnessed by the model constructed in [**12**], see [**53**]. Natural places to look for a possible negative answer to Question 1.6 are models constructed in [**11, 32, 33**]. Better understanding of these models and possible simplification of the methods used there would be at least as important as a solution of Question 1.6.

What about the products of Hurewicz spaces in the Miller model? The product of finitely many Hurewicz subspaces of 2^ω is Menger by Theorem 1.5, and thus the Miller model seemed for a while to be the best candidate for a model where the Hurewicz property is preserved by finite products of metrizable spaces. The next theorem proved in [**41**] refutes this expectation.

THEOREM 1.7. *In the Miller model there are two γ-subspaces X, Y of 2^ω such that $X \times Y$ is not Hurewicz. In particular, in this model the Hurewicz property is not preserved by finite products of metrizable spaces.*

γ-Spaces were introduced in [**21**] where it was proved that a Tychonoff space X is a γ-space if and only if $C_p(X)$ has the Frechet-Urysohn property, i.e., for every $f \in C_p(X)$ and $A \subset C_p(X)$ with $f \in \bar{A}$ there exists $\langle f_n : n \in \omega\rangle \in A^\omega$ converging to f. It is well-known that γ-spaces have the Hurewicz property in all finite powers, see, e.g., [**26**, Th. 3.6 and Fig. 2] and references therein.

The proof of Theorem 1.7 is based on the fact that if $X \subset 2^\omega$, $X \in V$, and X is a γ-space in V, then X remains a γ-space in the forcing extension by an iteration of the Miller forcing with countable supports. Previously this was only known for Cohen and random forcing, see [**37**] and [**49**], respectively. Let us note that Cohen forcing fails to preserve Hurewicz spaces, see the discussion in [**37**] after Problem 4.1 therein. This motivates the following

QUESTION 1.8. Does Miller forcing (resp. countable support iterations thereof) preserve the Hurewicz property of ground model spaces which do not contain topological copies of 2^ω? What about (Sierpiński) subspaces of 2^ω, provided that they exist in the ground model[5]?

Let us note that $(2^\omega)^V$ is not Hurewicz in the Miller model. This can be shown in the same way as that $(2^\omega)^V$ is not Menger in the Sacks model. The latter fact is proved in [**6**] and is attributed there to A. Miller.

Unlike the Laver and Miller models, in the ***Sacks model*** introduced in [**45**] we have the best possible non-preservation by products results: Countable support iterations of the Sacks forcing preserve γ-subspaces of $\mathbb{R}$ [**41**], while being non-Menger is obviously preserved by any forcing which does not add unbounded reals. Thus in the Sacks model there exist γ-spaces $X, Y \subset 2^\omega$ with non-Menger product.

The new feature of the Sacks model is that there are rather few Menger sets of reals in this model by the following result obtained in [**20**].

THEOREM 1.9. *There are $\mathfrak{c} = \omega_2$ many Menger subspaces of 2^ω in the Sacks model.*

[5]Recall that an uncountable $S \subset 2^\omega$ is called a *Sierpiński subspace* if $|S \cap Z| \le \omega$ for every Z of Lebesgue measure 0. It is known [**26**, Theorem 2.10] that Sierpiński subspaces are Hurewicz.

Let us note that by the methods developed in [**57**] there are at least $\mathfrak{c}^{\mathfrak{d}}$ (resp. $\mathfrak{c}^{\mathfrak{b}}$) many Menger (resp. Hurewicz) subspaces of 2^ω, and hence Theorem 1.9 fails in any model of $\mathfrak{c} = \mathfrak{d}$ including the Laver and Miller one. However, its analogue for Hurewicz spaces might still be true in the Miller model as $\mathfrak{c}^{\mathfrak{b}} = \omega_2^{\omega_1} = \omega_2$ there. This motivates the following

QUESTION 1.10. How many Hurewicz spaces $X \subset 2^\omega$ are there in the Miller model? How many concentrated subsets of 2^ω are there in the Miller model? In particular, is there a concentrated subset X of 2^ω of size ω_2 in the Miller model? If not, must every Hurewicz subspace of 2^ω of size ω_2 contain a topological copy of 2^ω?

A subset $X \subset 2^\omega$ is called *concentrated* on a countable $Q \subset 2^\omega$ if $|X \setminus U| \le \omega$ for any open $U \supset Q$. It is an easy exercise that if X is concentrated on $Q \subset X$ then X is Rothberger, and hence in the Miller model concentrated sets are Hurewicz [**59**], which naturally relates the different parts of Question 1.10.

2. Proofs of Theorems 1.1 and 1.3.

We shall first introduce several notions crucial for the proof of Theorem 1.1. The first two items of the following definition are due to A. Medini. The third one is an ad hoc notion suitable to our needs.

DEFINITION 2.1.
- $X \subset 2^\omega$ is called *countably controlled* if for every $Q \in [X]^\omega$ there exists a G_δ-subset $R \supset Q$ of 2^ω such that $R \subset X$;
- $X \subset 2^\omega$ is called *σ-compactly controlled* if for every σ-compact $Q \subset X$ there exists a G_δ-subset $R \supset Q$ of 2^ω such that $R \subset X$;
- $X \subset 2^\omega$ satisfies property (†) if for every function R assigning to each countable subset Q of X a G_δ-subset $R(Q) \supset Q$ of 2^ω, there exists a family $\mathsf{Q} \subset [X]^\omega$ of size $|\mathsf{Q}| = \omega_1$ and a mapping $K : \mathsf{Q} \to \mathcal{P}(2^\omega)$ assigning to every $Q \in \mathsf{Q}$ a σ-compact subset $K(Q)$ of $R(Q)$, such that $X \subset \bigcup_{Q \in \mathsf{Q}} K(Q)$.

Let us note that under CH any $X \subset 2^\omega$ satisfies (†).

As suggested by Definition 2.1 in what follows we shall work a lot with G_δ subsets of 2^ω containing certain countable sets, and hence we need to introduce some auxiliary notation. Given a countable sequence $\langle Q_n : n \in \omega\rangle$ of finite non-empty subsets of 2^ω and $u \in \omega^{\uparrow\omega}$, we set

$$R(\langle Q_n : n \in \omega\rangle, u) := \{x \in 2^\omega : \exists^\infty n \in \omega \exists q \in Q_n\ (|x - q| < 1/u(n))\},$$

where $\exists^\infty$ stands for "exists infinitely many" and we identify 2^ω with the standard Cantor subset of $[0,1]$ whenever we use expressions of the form $|a - b|$. As usually, for $x, y \in \omega^\omega$ notation $x \le^* y$ means $\{n : x(n) > y(n)\}$ is finite.

OBSERVATION 2.2. *Let $Q = \bigcup_{n\in\omega} Q_n \subset 2^\omega$ be a countable union of finite non-empty sets and $R \supset Q$ be a G_δ-set. Then there exists $u \in \omega^{\uparrow\omega}$ such that $R(\langle Q_n : n \in \omega\rangle, u) \subset R$*

PROOF. Let $R = \bigcap_{n\in\omega} O_m$, where each O_m is open. For every m we can find $u_m \in \omega^{\uparrow\omega}$ such that for every n and $x \in 2^\omega$, if there exists $q \in Q_n$ such that $|q - x| < 1/u_m(n)$, then $x \in O_m$. Now any $u \in \omega^{\uparrow\omega}$ such that $u_m \le^* u$ for all m, is as required. □

The following lemma is the key part of the proof of Theorem 1.1. Its proof is reminiscent of that of [**36**, Theorem 3.2]. We will use the notation from [**29**] with only differences being that smaller conditions in a forcing poset are supposed to carry more information about the generic filter, and the ground model is denoted by V. We shall work in $V[G_{\omega_2}]$, where G_{ω_2} is $\mathbb{P}_{\omega_2}$-generic and $\mathbb{P}_{\omega_2}$ is the iteration of length ω_2 with countable supports of the Laver forcing, see [**29**] for details. As usually, we shall denote $G_{\omega_2} \cap \mathbb{P}_\alpha$ by G_α, where $\alpha \leq \omega_2$. Following [**29**], for a condition $p \in \mathbb{P}_{\omega_2}$ we denote by $p(0)\langle 0\rangle$ the root of the condition $p(0)$, the latter being the initial coordinate of p.

A subset C of ω_2 is called an ω_1-*club* if it is unbounded and for every $\alpha \in \omega_2$ of cofinality ω_1, if $C \cap \alpha$ is cofinal in α then $\alpha \in C$.

LEMMA 2.3. *In the Laver model, if $X \subset 2^\omega$ has countably-controlled complement $2^\omega \setminus X$, then X satisfies* (†).

PROOF. Let R be such as in the definition of (†). By a standard closing-off argument there exists an ω_1-club $C \subset \omega_2$ such that for every $\alpha \in C$ the following conditions are satisfied:

- $X \cap V[G_\alpha] \in V[G_\alpha]$;
- For every $Q \in [X]^\omega \cap V[G_\alpha]$ the set $R(Q)$ is coded in $V[G_\alpha]$; and
- For every $Q \in [2^\omega]^\omega \cap V[G_\alpha]$ disjoint from X there exists a G_δ set $O \supset Q$ coded in $V[G_\alpha]$ such that $O \cap X = \emptyset$.

Let us fix $\alpha \in C$. We claim that $X \subset \bigcup_{Q\in\mathsf{Q}} W_Q$, where $\mathsf{Q} = [X]^\omega \cap V[G_\alpha]$ and $W_Q = \bigcup\{K : K$ is a compact subset of $R(Q)$ coded in $V[G_{\alpha+1}]\}$. This would complete our proof since it is easy to see (and well-known) that for every G_δ subset R of 2^ω and a family $\mathcal{W}$ of fewer than $\mathfrak{b}$ many compact subspaces of R, there exists a σ-compact subspace K of R containing $\bigcup\mathcal{W}$. Note that $\mathfrak{b} = \omega_2$ in $V[G_{\omega_2}]$ and there are ω_1 many compact sets coded in $V[G_{\alpha+1}]$.

Suppose that, contrary to our claim, there exists $p \in G_{\omega_2}$ and a $\mathbb{P}_{\omega_2}$-name $\dot{x}$ such that p forces $2^\omega \setminus \dot{X}$ to be countably-controlled and $\dot{x} \in \dot{X} \setminus \bigcup_{Q\in\mathsf{Q}} W_Q$. By [**29**, Lemma 11] there is no loss of generality in assuming that $\alpha = 0$. Applying [**29**, Lemma 14] to a sequence $\langle \dot{a}_i : i \in \omega\rangle$ such that $\dot{a}_i = \dot{x}$ for all $i \in \omega$, we get a condition $p' \leq p$ such that $p'(0) \leq^0 p(0)$, and a finite set U_s of reals for every $s \in p'(0)$ with $p'(0)\langle 0\rangle \leq s$, such that for each $\varepsilon > 0$, $s \in p'(0)$ with $p'(0)\langle 0\rangle \leq s$, and for all but finitely many immediate successors t of s in $p'(0)$ we have

$$p'(0)_t\,\hat{}\,p' \restriction [1, \omega_2) \Vdash \exists u \in U_s\ (|\dot{x} - u| < \varepsilon). \tag{1}$$

Without loss of generality we may assume that $U_s \setminus X \neq \emptyset$ for all $s \in p'(0)$, $p'(0)\langle 0\rangle \leq s$. Since $2^\omega \setminus X$ is countably-controlled, there exists an enumeration $\langle s_n : n \in \omega\rangle$ of $S' := \{s \in p'(0) : p'(0)\langle 0\rangle \leq s\}$ and $\phi \in \omega^{\uparrow\omega} \cap V$ such that $R' := R(\langle U_{s_n} \setminus X : n \in \omega\rangle, \phi)$ is disjoint from X. Fix a $\mathbb{P}_{\omega_2}$-name $\dot{N}$ for a natural number such that p' forces that $|\dot{x} - u| \geq 1/\phi(n)$ for all $n \geq \dot{N}$ and $u \in U_{s_n} \setminus \dot{X}$. Let $p'' \leq p'$ be such that $p'' \restriction [1, \omega_2) = p' \restriction [1, \omega_2)$ and $p''(0)$ is obtained from $p'(0)$ by throwing away for every $s_n \in S'$ finitely many of its successors (as well as all their extensions in $p'(0)$) so that

$$p''(0)_{s_n}\,\hat{}\,p'' \restriction [1, \omega_2) \Vdash \exists u \in U_{s_n}\ \big(|\dot{x} - u| < 1/(\phi(n) + 1)\big) \tag{2}$$

for all n such that $s_n \in S'' := \{s \in p''(0) : p''(0)\langle 0\rangle \leq s\}$. Replacing p'' with a yet stronger condition, if necessary, we may additionally assume that p'' decides $\dot{N}$ as some $N \in \omega$, and $s_n \in S''$ implies $n \geq N$. Thus we have that

$$(3) \qquad p'' \Vdash \forall u \in U_{s_n} \setminus X \left(|\dot{x} - u| \geq 1/\phi(n)\right)$$

for all n such that $s_n \in S''$. Combining Equations (1),(2), and (3) we get that for each $\varepsilon > 0$, $s \in S''$, and for all but finitely many immediate successors t of s in $p''(0)$ we have

$$(4) \qquad p''(0)_t{}^\frown p'' \restriction [1, \omega_2) \Vdash \exists u \in U_s \cap X\ (|\dot{x} - u| < \varepsilon).$$

Fix an enumeration $\langle t_n : n \in \omega\rangle \in V$ of S'' and find $\psi \in \omega^{\uparrow\omega} \cap V$ such that $R(\langle U_{t_n} \cap X : n \in \omega\rangle, \psi) \subset R(\bigcup_{n\in\omega} U_{t_n} \cap X)$. Replacing p'' with a stronger condition $p^{(3)}$ obtained by throwing away for every $t_n \in S''$ finitely many of its successors (as well as all their extensions in $p''(0)$), we can get

$$(5) \qquad p^{(3)}_{t_n}{}^\frown p^{(3)} \restriction [1, \omega_2) \Vdash \exists u \in U_{t_n} \cap X \left(|\dot{x} - u| < 1/2\psi(n)\right)$$

for all $t_n \geq p^{(3)}(0)\langle 0\rangle$, $t_n \in p^{(3)}(0)$ (the set of such t_n will be denoted by $S^{(3)}$). Let l be the first Laver generic, i.e., $l = \bigcap G_1 \subset \omega^{<\omega}$. Equation (5) implies that $p^{(3)}$ forces

$$\dot{x} \in K := \bigcap_{t_n \in l \cap S^{(3)}} \{z \in 2^\omega : \exists u \in U_{t_n}\ (|z - u| \leq 1/2\psi(n))\},$$

and K is obviously a compact subset of $R(\langle U_{t_n} \cap X : n \in \omega\rangle, \psi) \subset R(\bigcup_{n\in\omega} U_{t_n} \cap X)$ coded in $V[G_1]$. This contradicts our assumption on p and thus finishes our proof. □

The next lemma has been proved in [**40**]. We present its proof for the sake of completeness.

LEMMA 2.4. *Let $Y \subset 2^\omega$ be Hurewicz and $Q \subset 2^\omega$ countable. Then for every G_δ-subset O of $(2^\omega)^2$ containing $Q \times Y$ there exists a G_δ-subset $R \supset Q$ such that $R \times Y \subset O$.*

PROOF. Without loss of generality we shall assume that O is open. Let us write Q in the form $\{q_n : n \in \omega\}$ and set $O_n = \{z \in 2^\omega : \langle q_n, z\rangle \in O\} \supset Y$. For every n find a cover $\mathcal{U}_n$ of Y consisting of clopen subsets of 2^ω contained in O_n. Let $\langle \mathcal{U}'_k : k \in \omega\rangle$ be a sequence of open covers of Y such that each $\mathcal{U}_n$ appears in it infinitely often. Applying the Hurewicz property of Y we can find a sequence $\langle \mathcal{V}_k : k \in \omega\rangle$ such that $\mathcal{V}_k \in [\mathcal{U}'_k]^{<\omega}$ and $Y \subset \bigcup_{k\in\omega} Z_k$, where $Z_k = \bigcap_{m\geq k} \cup\mathcal{V}_m$. Note that each Z_k is compact and $Z_k \subset O_n$ for all $n \in \omega$ (because there exists $m \geq k$ such that $\mathcal{U}'_m = \mathcal{U}_n$, and then $Z_k \subset \cup\mathcal{V}_m \subset O_n$). Thus $Q \times Y \subset Q \times (\bigcup_{k\in\omega} Z_k) \subset O$. Since Z_k is compact, there exists for every k an open $R_k \supset Q$ such that $R_k \times Z_k \subset O$. Set $R = \bigcap_{k\in\omega} R_k$ and note that $R \supset Q$ and $R \times Y \subset R \times \bigcup_{k\in\omega} Z_k \subset O$. □

The next lemma demonstrates the relation between (†) and products with Hurewicz spaces.

LEMMA 2.5. *Suppose that $\mathfrak{b} > \omega_1$. Let $Y \subset 2^\omega$ be a Hurewicz space and $X \subset 2^\omega$ satisfy (†). Then $X \times Y$ is Hurewicz.*

PROOF. Fix a sequence $\langle \mathcal{U}_n : n \in \omega\rangle$ of covers of $X \times Y$ by clopen subsets of $(2^\omega)^2$. Thus $X \times Y \subset \bigcap_{n\in\omega} O_n$, where $O_n = \cup\mathcal{U}_n$. By Lemma 2.4 for every $Q \in [X]^\omega$ we can find a G_δ-subset $R_Q \supset Q$ such that $R_Q \times Y \subset \bigcap_{n\in\omega} O_n$. Since X satisfies (†), there exists $\mathsf{Q} \subset [X]^\omega$ of size $|\mathsf{Q}| = \omega_1$, and for every $Q \in \mathsf{Q}$ a σ-compact $K_Q \subset R_Q$, such that $K = \bigcup\{K_Q : Q \in \mathsf{Q}\}$ contains X as a subset. It follows that $X \times Y \subset K \times Y \subset \bigcap_{n\in\omega} O_n$. It is well-known and easy to see that a product of a compact space and a Hurewicz space is Hurewicz, as well as that the Hurewicz property is preserved by unions of fewer than $\mathfrak{b}$ many subsets of 2^ω, see [**56**] and references therein. Thus $K \times Y$ is Hurewicz and $\mathcal{U}_n$ is an open cover thereof, therefore for all $n \in \omega$ there exists $\mathcal{V}_n \in [\mathcal{U}_n]^{<\omega}$ such that $\{\cup\mathcal{V}_n : n \in \omega\}$ is a γ-cover of $K \times Y$, and hence it is also a γ-cover of $X \times Y$ as well. □

Finally, we can prove the characterization of Hurewicz subspaces of 2^ω which holds in the Laver model and implies Theorem 1.1.

PROPOSITION 2.6. *In the Laver model, for a subspace X of 2^ω the following conditions are equivalent:*

(1) *X is Hurewicz;*
(2) *X satisfies (†);*
(3) *$2^\omega \setminus X$ is countably controlled;*
(4) *$2^\omega \setminus X$ is σ-compactly controlled; and*
(5) *$X \times Y$ is Hurewicz for any Hurewicz subspace Y of 2^ω.*

PROOF. The equivalence between (1) and (4) is true in ZFC, see [**26**, Theorem 5.7] or [**5**, Theorem 1.2]. The implication (4) → (3) is obvious outright in ZFC, while (3) → (2) is established in Lemma 2.3. The implication (2) → (5) is proved in Lemma 2.5 and thus requires only $\mathfrak{b} > \omega_1$. And finally, (5) → (1) is again obvious in ZFC, take, e.g., Y to be a singleton. □

We are now in a position to present the

Proof of Theorem 1.3. By Proposition 2.6 it is enough to show that $\mathcal{F}(\mathcal{A})^+$ is countably controlled: This would give that $\mathcal{P}(\omega) \setminus \mathcal{F}(A)^+$ is Hurewicz, and the map $\mathcal{P}(\omega) \ni Z \mapsto \omega \setminus Z$ is easily seen to be a homeomorphism between $\mathcal{F}(A)$ and $\mathcal{P}(\omega) \setminus \mathcal{F}(A)^+$. Let us fix a countable $\mathcal{X} \subset \mathcal{F}(\mathcal{A})^+$ and find countable infinite $\mathcal{A}_0 \subset \mathcal{A}$ such that $A \cap X$ is infinite for all $A \in \mathcal{A}_0$ and $X \in \mathcal{X}$. It follows that

$$\{Z \in \mathcal{P}(\omega) : \forall A \in \mathcal{A}_0(|A \cap Z| = \omega)\}$$

is a G_δ subset of $\mathcal{P}(\omega)$ which contains $\mathcal{X}$ and is contained in $\mathcal{F}(\mathcal{A})^+$. □

By nearly the same argument as at the end of [**40**] ne can prove that Theorem 1.1 follows from Proposition 2.6. Again, we present its proof for the sake of completeness. A family $\mathcal{F} \subset [\omega]^\omega$ is called a *semifilter* if for every $F \in \mathcal{F}$ and $X \subset \omega$, if $|F \setminus X| < \omega$ then $X \in \mathcal{F}$.

The proof of the first part of Theorem 1.1 uses the characterization of the Hurewicz property obtained in [**59**]. Let $u = \langle U_n : n \in \omega\rangle$ be a sequence of subsets of a set X. For every $x \in X$ let $I_s(x, u, X) = \{n \in \omega : x \in U_n\}$. If every $I_s(x, u, X)$ is infinite (the collection of all such sequences u will be denoted by $\Lambda_s(X)$), then we shall denote by $\mathcal{U}_s(u, X)$ the smallest semifilter on ω containing all $I_s(x, u, X)$. By [**59**, Theorem 3], a Lindelöf topological space X is Hurewicz if and only if for every $u \in \Lambda_s(X)$ consisting of open sets, the semifilter $\mathcal{U}_s(u, X)$ is Hurewicz. The proof given there also works if we consider only those $\langle U_n : n \in \omega\rangle \in \Lambda_s(X)$ such that all U_n's belong to a given base of X.

Proof of Theorem 1.1. Suppose that X, Y are Hurewicz spaces such that $X \times Y$ is Lindelöf and fix $w = \langle U_n \times V_n : n \in \omega\rangle \in \Lambda_s(X \times Y)$ consisting of open sets. Set $u = \langle U_n : n \in \omega\rangle$, $v = \langle V_n : n \in \omega\rangle$, and note that $u \in \Lambda_s(X)$ and $v \in \Lambda_s(Y)$. It is easy to see that

$$\mathcal{U}_s(w, X \times Y) = \{A \cap B : A \in \mathcal{U}_s(u, X), B \in \mathcal{U}_s(v, Y)\},$$

and hence $\mathcal{U}_s(w, X \times Y)$ is a continuous image of $\mathcal{U}_s(u, X) \times \mathcal{U}_s(v, Y)$. By [**59**, Theorem 3] both of latter ones are Hurewicz, considered as subspaces of 2^ω, and hence their product is a Hurewicz space by Proposition 2.6. Thus $\mathcal{U}_s(w, X \times Y)$ is Hurewicz, being a continuous image of a Hurewicz space. It now suffices to use [**59**, Theorem 3] again. □

References

[1] A. V. Arhangel′skiĭ, *The spectrum of frequencies of a topological space and the product operation* (Russian), Trudy Moskov. Mat. Obshch. **40** (1979), 171–206. MR550259

[2] A. V. Arkhangel′skiĭ, *Topological function spaces*, Mathematics and its Applications (Soviet Series), vol. 78, Kluwer Academic Publishers Group, Dordrecht, 1992. Translated from the Russian by R. A. M. Hoksbergen. MR1144519

[3] Aurichi, L.; Zdomskyy, L. *Covering properties of ω-mad families*, Archive for Mathematical Logic **59** (2020), 445–452.

[4] L. Babinkostova, *On some questions about selective separability*, MLQ Math. Log. Q. **55** (2009), no. 5, 539–541, DOI 10.1002/malq.200810010. MR2568764

[5] T. Banakh and L. Zdomskyy, *Separation properties between the σ-compactness and Hurewicz property*, Topology Appl. **156** (2008), no. 1, 10–15, DOI 10.1016/j.topol.2007.12.017. MR2463818

[6] D. Barman and A. Dow, *Proper forcing axiom and selective separability*, Topology Appl. **159** (2012), no. 3, 806–813, DOI 10.1016/j.topol.2011.11.048. MR2868880

[7] T. Bartoszyński and H. Judah, *Set theory*, A K Peters, Ltd., Wellesley, MA, 1995. On the structure of the real line. MR1350295

[8] T. Bartoszynski and S. Shelah, *Continuous images of sets of reals*, Topology Appl. **116** (2001), no. 2, 243–253, DOI 10.1016/S0166-8641(00)00079-1. MR1855966

[9] T. Bartoszyński and B. Tsaban, *Hereditary topological diagonalizations and the Menger-Hurewicz conjectures*, Proc. Amer. Math. Soc. **134** (2006), no. 2, 605–615, DOI 10.1090/S0002-9939-05-07997-9. MR2176030

[10] A. Blass, *Combinatorial cardinal characteristics of the continuum*, Handbook of set theory. Vols. 1, 2, 3, Springer, Dordrecht, 2010, pp. 395–489, DOI 10.1007/978-1-4020-5764-9_7. MR2768685

[11] A. Blass and S. Shelah, *There may be simple $P_{\aleph_1}$- and $P_{\aleph_2}$-points and the Rudin-Keisler ordering may be downward directed*, Ann. Pure Appl. Logic **33** (1987), no. 3, 213–243, DOI 10.1016/0168-0072(87)90082-0. MR879489

[12] A. Blass and S. Shelah, *Ultrafilters with small generating sets*, Israel J. Math. **65** (1989), no. 3, 259–271, DOI 10.1007/BF02764864. MR1005010

[13] E. Borel, *Sur la classification des ensembles de mesure nulle* (French), Bull. Soc. Math. France **47** (1919), 97–125. MR1504785

[14] L. Bukovský, *The structure of the real line*, Instytut Matematyczny Polskiej Akademii Nauk. Monografie Matematyczne (New Series) [Mathematics Institute of the Polish Academy of Sciences. Mathematical Monographs (New Series)], vol. 71, Birkhäuser/Springer Basel AG, Basel, 2011. MR2778559

[15] J. Brendle, *Mob families and mad families*, Arch. Math. Logic **37** (1997), no. 3, 183–197, DOI 10.1007/s001530050091. MR1619601

[16] D. Chodounský, O. Guzmán González, and M. Hrušák, *Mathias-Prikry and Laver type forcing; summable ideals, coideals, and +-selective filters*, Arch. Math. Logic **55** (2016), no. 3-4, 493–504, DOI 10.1007/s00153-016-0476-9. MR3490916

[17] D. Chodounský, D. Repovš, and L. Zdomskyy, *Mathias forcing and combinatorial covering properties of filters*, J. Symb. Log. **80** (2015), no. 4, 1398–1410, DOI 10.1017/jsl.2014.73. MR3436375

[18] A. Dow, *Two classes of Fréchet-Urysohn spaces*, Proc. Amer. Math. Soc. **108** (1990), no. 1, 241–247, DOI 10.2307/2047719. MR975638

[19] R. Engelking, *General topology*, 2nd ed., Sigma Series in Pure Mathematics, vol. 6, Heldermann Verlag, Berlin, 1989. Translated from the Polish by the author. MR1039321

[20] Gartside, P.; Medini, A.; Zdomskyy, L., *The Vietoris hyperspace $\mathcal{K}(X)$ is hereditarily Baire if and only of X is co-Menger*, work in progress.

[21] J. Gerlits and Zs. Nagy, *Some properties of $C(X)$. I*, Topology Appl. **14** (1982), no. 2, 151–161, DOI 10.1016/0166-8641(82)90065-7. MR667661

[22] O. Guzmán, M. Hrušák, and A. Martínez-Celis, *Canjar filters*, Notre Dame J. Form. Log. **58** (2017), no. 1, 79–95, DOI 10.1215/00294527-3496040. MR3595342

[23] M. Hrušák and J. van Mill, *The existence of a connected meager in itself* CDH *space is independent of* ZFC, Proc. Amer. Math. Soc. **146** (2018), no. 6, 2689–2695, DOI 10.1090/proc/13434. MR3778168

[24] W. Hurewicz, *Über eine Verallgemeinerung des Borelschen Theorems* (German), Math. Z. **24** (1926), no. 1, 401–421, DOI 10.1007/BF01216792. MR1544773

[25] Hurewicz, W., *Über Folgen stetiger Funktionen,* Fund. Math. **9** (1927), 193–204.

[26] W. Just, A. W. Miller, M. Scheepers, and P. J. Szeptycki, *The combinatorics of open covers. II*, Topology Appl. **73** (1996), no. 3, 241–266, DOI 10.1016/S0166-8641(96)00075-2. MR1419798

[27] M. S. Kurilić, *Cohen-stable families of subsets of integers*, J. Symbolic Logic **66** (2001), no. 1, 257–270, DOI 10.2307/2694920. MR1825183

[28] C. Kuratowski, *Topologie. I et II* (French), Éditions Jacques Gabay, Sceaux, 1992. Part I with an appendix by A. Mostowski and R. Sikorski; Reprint of the fourth (Part I) and third (Part II) editions. MR1296876

[29] R. Laver, *On the consistency of Borel's conjecture*, Acta Math. **137** (1976), no. 3-4, 151–169, DOI 10.1007/BF02392416. MR422027

[30] V. I. Malykhin, *Topological properties of Cohen generic extensions* (Russian), Trudy Moskov. Mat. Obshch. **52** (1989), 3–33, 247; English transl., Trans. Moscow Math. Soc. (1990), 1–32 (1991). MR1056464

[31] Menger, K., *Einige Überdeckungssätze der Punktmengenlehre, Sitzungsberichte.* Abt. 2a, Mathematik, Astronomie, Physik, Meteorologie und Mechanik (Wiener Akademie) **133** (1924), 421–444.

[32] H. Mildenberger and S. Shelah, *The near coherence of filters principle does not imply the filter dichotomy principle*, Trans. Amer. Math. Soc. **361** (2009), no. 5, 2305–2317, DOI 10.1090/S0002-9947-08-04806-X. MR2471919

[33] Mildenberger, H., *Local Ramsey Spaces in Matet Forcing Extensions and Finitely Many Near-Coherence Classes*, submitted.

[34] J. E. Baumgartner, D. A. Martin, and S. Shelah (eds.), *Axiomatic set theory*, Contemporary Mathematics, vol. 31, American Mathematical Society, Providence, RI, 1984. MR763887

[35] A. W. Miller, *Special subsets of the real line*, Handbook of set-theoretic topology, North-Holland, Amsterdam, 1984, pp. 201–233. MR776624

[36] A. W. Miller and B. Tsaban, *Point-cofinite covers in the Laver model*, Proc. Amer. Math. Soc. **138** (2010), no. 9, 3313–3321, DOI 10.1090/S0002-9939-10-10407-9. MR2653961

[37] A. W. Miller, B. Tsaban, and L. Zdomskyy, *Selective covering properties of product spaces, II: γ spaces*, Trans. Amer. Math. Soc. **368** (2016), no. 4, 2865–2889, DOI 10.1090/tran/6581. MR3449260

[38] T. Orenshtein and B. Tsaban, *Linear σ-additivity and some applications*, Trans. Amer. Math. Soc. **363** (2011), no. 7, 3621–3637, DOI 10.1090/S0002-9947-2011-05228-1. MR2775821

[39] D. Repovš and L. Zdomskyy, *On M-separability of countable spaces and function spaces*, Topology Appl. **157** (2010), no. 16, 2538–2541, DOI 10.1016/j.topol.2010.07.036. MR2719396

[40] D. Repovš and L. Zdomskyy, *Products of Hurewicz spaces in the Laver model*, Bull. Symb. Log. **23** (2017), no. 3, 324–333, DOI 10.1017/bsl.2017.24. MR3732200

[41] D. Repovš and L. Zdomskyy, *Preservation of γ-spaces and covering properties of products*, Proc. Amer. Math. Soc. **147** (2019), no. 11, 4979–4985, DOI 10.1090/proc/14593. MR4011529

[42] D. Repovš, L. Zdomskyy, and S. Zhang, *Countable dense homogeneous filters and the Menger covering property*, Fund. Math. **224** (2014), no. 3, 233–240, DOI 10.4064/fm224-3-3. MR3194416

[43] Rothberger, F., *Eine Verschärfung der Eigenschaft C*, Fund. Math. **30** (1938), 50–55.

[44] F. Rothberger, *Sur les familles indénombrables de suites de nombres naturels et les problèmes concernant la propriété C* (French), Proc. Cambridge Philos. Soc. **37** (1941), 109–126. MR4281

[45] G. E. Sacks, *Forcing with perfect closed sets*, Axiomatic Set Theory (Proc. Sympos. Pure Math., Vol. XIII, Part I, Univ. California, Los Angeles, Calif., 1967), Amer. Math. Soc., Providence, R.I., 1971, pp. 331–355. MR0276079

[46] M. Sakai, *The sequence selection properties of $C_p(X)$*, Topology Appl. **154** (2007), no. 3, 552–560, DOI 10.1016/j.topol.2006.07.008. MR2280899

[47] M. Scheepers, *Combinatorics of open covers. I. Ramsey theory*, Topology Appl. **69** (1996), no. 1, 31–62, DOI 10.1016/0166-8641(95)00067-4. MR1378387

[48] M. Scheepers, *Sequential convergence in $C_p(X)$ and a covering property*, East-West J. Math. **1** (1999), no. 2, 207–214. MR1727383

[49] M. Scheepers, *Gerlits and function spaces*, Studia Sci. Math. Hungar. **47** (2010), no. 4, 529–557, DOI 10.1556/SScMath.47.2010.4.1161. MR2883130

[50] M. Scheepers and F. D. Tall, *Lindelöf indestructibility, topological games and selection principles*, Fund. Math. **210** (2010), no. 1, 1–46, DOI 10.4064/fm210-1-1. MR2720214

[51] Sierpiński, W., *Sur un ensemble non dénombrable, dont toute image continue est de mesure nulle*, Fund. Math. **11** (1928), 302–303.

[52] P. Szewczak and B. Tsaban, *Products of Menger spaces: a combinatorial approach*, Ann. Pure Appl. Logic **168** (2017), no. 1, 1–18, DOI 10.1016/j.apal.2016.08.002. MR3564373

[53] Szewczak, P.; Tsaban, B.; Zdomskyy, L., *Finite powers and products of Menger sets,* Fundamenta Mathematicae, to appear.

[54] M. Scheepers and B. Tsaban, *The combinatorics of Borel covers*, Topology Appl. **121** (2002), no. 3, 357–382, DOI 10.1016/S0166-8641(01)00078-5. MR1908999

[55] *Đuro Kurepa memorial volume*, Institut Mathématique, Belgrade, 1995. Publ. Inst. Math. (Beograd) (N.S.) **57(71)** (1995). MR1387348

[56] B. Tsaban, *Additivity numbers of covering properties*, Selection principles and covering properties in topology, Quad. Mat., vol. 18, Dept. Math., Seconda Univ. Napoli, Caserta, 2006, pp. 245–282. MR2395757

[57] B. Tsaban and L. Zdomskyy, *Scales, fields, and a problem of Hurewicz*, J. Eur. Math. Soc. (JEMS) **10** (2008), no. 3, 837–866, DOI 10.4171/JEMS/132. MR2421163

[58] B. Tsaban and L. Zdomskyy, *Hereditarily Hurewicz spaces and Arhangel'skiĭ sheaf amalgamations*, J. Eur. Math. Soc. (JEMS) **14** (2012), no. 2, 353–372, DOI 10.4171/JEMS/305. MR2881299

[59] L. Zdomsky, *A semifilter approach to selection principles*, Comment. Math. Univ. Carolin. **46** (2005), no. 3, 525–539. MR2174530

[60] L. Zdomskyy, *Products of Menger spaces in the Miller model*, Adv. Math. **335** (2018), 170–179, DOI 10.1016/j.aim.2018.06.016. MR3836661

Kurt Gödel Research Center for Mathematical Logic, University of Vienna, Währinger Strasse 25, A-1090 Wien, Austria.

Email address: `lzdomsky@gmail.com`

URL: `http://www.logic.univie.ac.at/~lzdomsky/`

Selected Published Titles in This Series

755 **Marion Scheepers and Ondřej Zindulka, Editors,** Centenary of the Borel Conjecture, 2020

754 **Susanne C. Brenner, Igor Shparlinski, Chi-Wang Shu, and Daniel B. Szyld, Editors,** 75 Years of Mathematics of Computation, 2020

753 **Matthew Krauel, Michael Tuite, and Gaywalee Yamskulna, Editors,** Vertex Operator Algebras, Number Theory and Related Topics, 2020

752 **Samuel Coskey and Grigor Sargsyan, Editors,** Trends in Set Theory, 2020

751 **Ashish K. Srivastava, André Leroy, Ivo Herzog, and Pedro A. Guil Asensio, Editors,** Categorical, Homological and Combinatorial Methods in Algebra, 2020

750 **A. Bourhim, J. Mashreghi, L. Oubbi, and Z. Abdelali, Editors,** Linear and Multilinear Algebra and Function Spaces, 2020

749 **Guillermo Cortiñas and Charles A. Weibel, Editors,** K-theory in Algebra, Analysis and Topology, 2020

748 **Donatella Danielli and Irina Mitrea, Editors,** Advances in Harmonic Analysis and Partial Differential Equations, 2020

747 **Paul Bruillard, Carlos Ortiz Marrero, and Julia Plavnik, Editors,** Topological Phases of Matter and Quantum Computation, 2020

746 **Erica Flapan and Helen Wong, Editors,** Topology and Geometry of Biopolymers, 2020

745 **Federico Binda, Marc Levine, Manh Toan Nguyen, and Oliver Röndigs, Editors,** Motivic Homotopy Theory and Refined Enumerative Geometry, 2020

744 **Pieter Moree, Anke Pohl, L'ubomír Snoha, and Tom Ward, Editors,** Dynamics: Topology and Numbers, 2020

743 **H. Garth Dales, Dmitry Khavinson, and Javad Mashreghi, Editors,** Complex Analysis and Spectral Theory, 2020

742 **Francisco-Jesús Castro-Jiménez, David Bradley Massey, Bernard Teissier, and Meral Tosun, Editors,** A Panorama of Singularities, 2020

741 **Houssam Abdul-Rahman, Robert Sims, and Amanda Young, Editors,** Analytic Trends in Mathematical Physics, 2020

740 **Alina Bucur and David Zureick-Brown, Editors,** Analytic Methods in Arithmetic Geometry, 2019

739 **Yaiza Canzani, Linan Chen, and Dmitry Jakobson, Editors,** Probabilistic Methods in Geometry, Topology and Spectral Theory, 2019

738 **Shrikrishna G. Dani, Surender K. Jain, Jugal K. Verma, and Meenakshi P. Wasadikar, Editors,** Contributions in Algebra and Algebraic Geometry, 2019

737 **Fernanda Botelho, Editor,** Recent Trends in Operator Theory and Applications, 2019

736 **Jane Hawkins, Rachel L. Rossetti, and Jim Wiseman, Editors,** Dynamical Systems and Random Processes, 2019

735 **Yanir A. Rubinstein and Bernard Shiffman, Editors,** Advances in Complex Geometry, 2019

734 **Peter Kuchment and Evgeny Semenov, Editors,** Differential Equations, Mathematical Physics, and Applications, 2019

733 **Peter Kuchment and Evgeny Semenov, Editors,** Functional Analysis and Geometry, 2019

732 **Samuele Anni, Jay Jorgenson, Lejla Smajlović, and Lynne Walling, Editors,** Automorphic Forms and Related Topics, 2019

731 **Robert G. Niemeyer, Erin P. J. Pearse, John A. Rock, and Tony Samuel, Editors,** Horizons of Fractal Geometry and Complex Dimensions, 2019

For a complete list of titles in this series, visit the
AMS Bookstore at **www.ams.org/bookstore/conmseries/**.